DYNAMICS
OF
AGRICULTURAL DEVELOPMENT

VOLUME III : POLICY PLANNING AND LIBERALISATION

Vol. I : Land Reforms, Growth and Equity
Vol. II : Technological Changes and Sustainable Development
Vol. III : Policy Planning and Liberalisation

About the Authors

Dr. K.S. Dhindsa is a a Professor of Economics at the Punjab School of Economics, Guru Nanak Dev University, Amritsar (India). Dr. Dhindsa has a brilliant career, who after graduating from Punjab University, completed the post-graduate diploma in Economic Development from Glasglow University. Thereafter, he obtained his Master Degree from Leeds University where he successfully completed his Ph.D. degree. He has published a fairly large number of research papers in National and International journals. For his post-doctorate he went to the School of Oriental and African Studies, University of London, during 1989-91. He was the Editor of *Indian Journal of Quantitative Economics*—A Journal of International Development, during 1994-96. He was the Dean of Faculty of Economics and Business during 1992-93 and was Chairman of Punjab School of Economics, during 1994-96.

Anju Sharma is a Lecturer in Economics in the Post-Graduate Department of Economics, Khalsa College, Amritsar (Punjab). She obtained her M.Phil and Ph.D. degreed from Punjab School of Economics, Guru Nank Dev University, Amristsar. She has carried out a variety of research in the field of Economics. She is the author of a book entitled. *Dynamics of Cropping Pattern : A Regional Study of Punjab.*
A dozen of her research papers have already been published in reputed Journals.

DYNAMICS OF AGRICULTURAL DEVELOPMENT

VOLUME-III

POLICY PLANNING AND LIBERALISATION

K.S. DHINDSA
ANJU SHARMA

CONCEPT PUBLISHING COMPANY, NEW DELHI-110059

ISBN : 81-7022-837-9 (Vol. III)
ISBN : 81-7022-834-4 (Set)

First Published 2001

Published and Printed by
Ashok Kumar Mittal
Concept Publishing Company
A/15-16, Commercial Block, Mohan Garden,
New Delhi-110059 (INDIA)

Phone : 5648039, 5649024
Fax : 091-11-5648053
Email : publishing@conceptpub.com

FOREWORD

Development of the agriculture sector in India has traversed a period of five decades since the country achieved Independence in 1947. During this period the very response system of governance has undergone a discernable change from governing an important colony to managing of an independent society that had an entity as a nation state, which was riddled with some formidable problems on economic, social, culture and political fronts. It was under this changed response scenario that agriculture sector assumed prime importance in the development process of the Indian economy so that (1) food and fiber needs of the society were met on a sustained basis, and (2) more than 80 per cent of the population living in rural areas who were wholly or partially dependent for their wherewithal on agriculture sector were provided with gainful employment opportunities. It is of course a different matter, the deriving inspiration from socialistic system of the then USSR, industrial sector was encouraged as a future in the development process and with the public sector achieving commanding heights, distributional social justice was aimed at for the populace on economic front. Yet, this approach and overt emphasis on industrial sector, specially the obssession with big industries, was not supposed to dilute the prominency of the agriculture sector; may be out of sheer compulsions of feeding the large and fast growing population of the country.

Thus, on one hand serious effort was made and are as yet afoot to tone up the agricultural sector for higher growth through strengthening the agricultural research system aimed at developing appropriate production technology, farm level extension of the new technology and agricultural education as well as institutional and policy support and on the other hand the agriculture sector was systematically squeezed to yield marketed surpluses, specially of foodgrain to feed the urban population through public distribution system. Thus, agriculture sector acted as a lead sector that yielded capital surpluses and wage good following a classic model of economic growth and development, though sluggish in pace, as postulated for agriculture sector dominated developing societies, this approach manifested through adoption of the process of investment planning and system of five year development plans.

The priority accorded to the agriculture sector was visible in the First Five Year Plan of 1951-52 through 1955-56. The explicit-emphasis, however, got diluted in the Second plan. Even the Third Five Year Plan did not absorb the reality of the situation and agriculture sector did not get adequate allocations. This happened because of good production response of the agriculture sector in the First Five Year Plan. By the time the mistake was realized it was too late and the country had virtually become a 'basket case' by the mid-sixties.

This impelled the policy makers to accord priority to the agriculture sector on technological improvement through expanding and strengthening the national agricultural research system, including the central and state level agricultural universities and research institutes. The research effort was complemented with reorganisation and toning up of extension network and also with reorganization of agricultural education. In order to boost sustained growth in agricultural production, supportive institutional and policy framework was put in position through a system of minimum support-prices and procurement of agricultural produce by the government. The creation of Food Corporation of India (FCI) and Public Distribution System (PDS) was a land mark in the history of Independent India.

The basic purpose of the system was three fold :

(i) Provision of price certainty to the agricultural producer at reasonable (remunerative) level to save him from vagaries of buyer's market in which stands helpless spectator in the post harvest period, when the buyers so often pooled to bring down the market prices to the disadvantage of the producer seller. With the introduction of the system of support/procure-ment prices, which by and by covered 22 commodities, the government stand obligated to lift whatever quantities the farmers offer for sale to the government at the announced prices. For this purpose Food Corporation of India acted and any loss was to be covered by the Central Government and profit was to be turned over to the government. The FCI acted direct as well as through the state agencies as their agents.

(ii) Supply of foodgrains, sugar, oil and raw material to the urban population at a reasonable prices through the Public distribution system fed through the procurement of foodgrains, specially wheat, rice and sugar appropriated through sugar levies on sugar mills. Raw materials like jute and cotton were procured from the market at support prices and also quite often at market prices through Jute Corporation of India (JCI) and Cotton Corporation of India (CCI) respectively for consumption of the public enterprises.

(iii) Creation of buffer stocks for food security because Indian sensetivities were deeply hurt in mid-sixties when it had become difficult for the government to buy foodgrain from the international market. In order to balance these three, often time conflicting, objectives, Agricultural Prices Commission (APC) was created, which was later renamed as Commission for Agricultural Costs and Prices (CACP), with expanded scope.

Initially the system worked well. In fact during first two years the FCI registered net profit. Yet, soon the ruling party's approach to appease the urban consumers resulted in subsidized issue prices at the PDS and foodgrain as well other commodities were supplied to all the consumers holding ration cards without targeting the distribution for the poor. As a consequence the budget provisions for food subsidies started escalating. Since the government opted not to increase issue prices for the consumers out of political reasons the brunt was borne by the farm sector. Government hesitated to increase support/procurement prices and turned the terms of trade against the farm sector. With a view to maximizing procurement of foodgrain at low prices (support/procurement prices) indirect approach was adopted which eliminated the private trader from the grain-markets. Drastically reducing the limit on food credit (selective credit), putting restrictions on holding of stocks, denying railways wagons to the traders and putting restriction on inter-state or inter-zone movement of grains the market prices were artificially brought down to the level of support/procurement prices. This process continued throughout during the past three decades with exception for only a few years intermitantly when farmers reaped bumper crops, stores/godowns of the government were full and consumer demand and liftings from PDS were low. These years too the farmers suffered on account of back of market clearance for their produce. Thus on one side the production was encouraged as a policy and on the other produce was procured for consumers on low prices to keep the urban voters happy. Thus a great deal of politics, rather cheap politics got mixed with the economics of the policy, which did not put the agriculture sector on an optimal growth path, put a drain on the budgetary resources of the government and created unwanted and unwarranted expectations in the minds of the urban consumers, which promoted pressure on the political system to keep consumption subsidized without any targeting for the poor.

As a consequence of these policy distortions, the agriculture sector remains net taxed or under negative protection according to some scholars. Although this view is widely accepted, yet there are other economists who believe that agriculture sector be brought under the income tax net. There is some logic in this argument too, because there are quite a large number of big farmer who earn huge incomes from farming, yet the sector as a whole is not capable of yielding any income tax, specially when collection costs are kept in view.

There are several other problems the agriculture sector faced as it progressed through last five decades and bore the impact of policy prescriptions at various stages of development. At this stage the farm sector is on crossroads, where production is not picking up as fast as is the need of the nation, weather as yet remains a very important factor and minor deficiencies in rain put the growth of the sector of the track, research and technology development has gone sluggish, policy has gone into a *rut*, farmers are coming under economic squeeze, consumers are becoming more demanding, population growth and improvements in money incomes are increasing the demand for food and fibers, government exchequer are coming under financial strain and above all due to progressive liberaliztion and globalisation of the agriculture market, the agricultural sector is going to face global competition. These are a few of the challenges that are staring at the face of the policy makers and market players in this sector.

It is against this background that the subject matter of this book assumes topical importance. The coverage of the subjects and issues like institution changes in historical perspective, reflections on input out put relationships and their impact on agricultural growth; impact of price support and market intervention, relationships between technology and growth as well as equity, implications and possibilities of agricultural taxation and changing terms of trade, impact of agricultural growth on labour absorption, specially with the adoption of improved technology, probable impact of liberalization and globalization on the agriculture sector as well as general review of macro and micro level agricultural policies followed through this period have a great relevance today if this sector is to be put on an optimum growth path in order to meet the fast growing demand for food and fibre in the national market.

This book is a good job done in this respect; which should prove to be a good material for students and researchers engaged in the field of agricultural economics and specially for those specializing in agricultural development policy.

S.S. JOHL
Prof. of Eminence in Economics (Retd.) and
former Vice Chancellor
Punjabi University, Patiala

PREFACE

K.S. DHINDSA

There exists at present a considerable body of literature on Indian agriculture in different research publications. Some of this material is accessible in a few edited volumes, but none of them covers all the major, constantly swelling aspects of this subject. So the necessity for editing a comprehensive book on Indian agricultural development was being felt since long. The present book "Dimensions of Agricultural Development" is a humble attempt towards this end. In this book an attempt has been made to include good quality articles of renowned economists published in various reputed Journals. These studies, although related to India's agricultural development can be read with keen interest by various economists and policy-makers who are interested in agricultural Development of other third world countries.

While selecting various articles for this volume, syllabi of agricultural economics for post-graduate classes, prepared recently by various universities of India was also kept in view. This book will be of immense use to the students, teachers, researchers and policy makers in India as well as abroad.

In order to make a detailed study of various dimensions of agricultural development in India, thirty three articles have been included in three different volumes. Articles included in these volumes cover both micro and macro aspects pertaining to Indian agriculture, examine the issues and problems empirically and also provide careful presentation of most of the relevant facts of India's agricultural development.

The first volume of the book deals with the issues relating to land reforms, growth and equity. The second volume comprises of various articles relating to technological development in agriculture and sustainability of agricultural development. Third volume deals with the policy and planning concerning agricultural development and the impact of recently concluded GATT round on Indian agriculture.

The main introduction of the book has been given in the first volume. Most of the articles included in these volumes are reprinted articles and some of them are specially invited papers. Details of the original source of each articles has been given in the notes at the end of the chapter.

In an edited book, the co-operation of all those people is very much required whose work we wish to include. We must express our thanks to the contributors for giving us the permission to reproduce their research papers in this book. We are also grateful to the publishers of the Journals who have permitted us to reproduce these papers from their publications.

Before finalising the contents of the book, we had discussions with professors Yoginder K. Alagh, G.S. Bhalla, G.K. Chadha, Krishna Bhardwaj, S.S. Johl, K.L. Krishna, V .M. Rao, Suresh S. Tendulkar and Autar Singh Dhesi. We gratefully acknowledge their valuable suggestions. Thanks are also due to the staff members of Rattan Tata Library, Delhi, School of Economics, Institute of Economic Growth Library (Delhi), National Council of Applied Economic Research (Delhi), Jawaharlal Nehru University Library (Delhi), and Punjab Agricultural University Library (Ludhiana) for providing us facilities to use the relevant research material.

Last but not the least, we are profoundly grateful to the publisher M/s. Concept Publishing Company, of their whole-hearted co-operation in bringing out this book in record time.

K.S. DHINDSA

ANJU SHARMA

CONTENTS

*Foreword by **Prof. S.S. Johl*** 5

Preface 11

List of Contributors 15

List of Tables 17

Introduction 21

—K.S. Dhindsa and Anju Sharma

SECTION I

POLICY AND PLANNING

1. Policies for Agriculture 35
   —Yoginder K. Alagh
2. State Policies and Regional Disparity in Indian Agriculture 48
   —K. Subbarao
3. Some Aspects of Wheat Price Policy in India 72
   —G.S. Raychaudhuri and *Raj Krishna*
4. Agricultural Price Policy : A Survey with Reference to Indian Foodgrain Economy 99
   —N. Krishnaji
5. Agricultural Holdings Tax : A Modified Scheme 134
   —Amaresh Bagchi
6. Edible Oil Crisis in India:Some Policy Options 182
   —S.S. Johl
7. Restructuring of Agricultural Planning and Policies for Alleviating Rural Poverty 204
   —N.S. Randhawa
8. A Simplified Approach to District Agriculture Plan Formulation 245
   —Autar S. Dhesi and *N.S. Bhullar*

SECTION II
LIBERALISATION

9. Agriculture Under GATT: What It Holds for India 277
—*Ashok Gulati* and *Anil Sharma*

10. Agricultural Pricing, Marketing and International Trade under New Economic Environment 297
—*M.S. Bhatia*

11. Farmers in Market Economy: Would Farmers Gain through Liberalisation? 319
—*V.M. Rao*

12. GATT: Implications for Agriculture with Special Reference to Punjab 334
—*Karan Singh, D.S. Sidhu, S.S. Grewal, Joginder Singh, D.R. Singh, O.P. Sahni* and *M.M. Verma*

Index 358

LIST OF CONTRIBUTORS

Yogider K. Alagh
Former Vice-Chancellor, Jawahar Lal Nehru University, Delhi

Amaresh Bagchi
Professor, National Institute of Public Finance Policy, New Delhi

M.S. Bhatia
Additional Economic Adviser, Ministry of Agriculture, Government of India, New Delhi

N.S. Bhullar
Lecturer, Department of Economics, Layalpur Khalsa College, Jalandhar.

Autar S. Dhesi
Visiting Professor (U.G.C.) Punjabi University, Patiala

K.S. Dhindsa
Professor, Punjab School of Economics, Guru Nanak Dev University, Amritsar

Ashok Gulati
Professor, National Council of Applied Economic Research, Parisila Bhawan, Indraprastha Estate, New Delhi

S.S. Grewal
Professor, Department of Economics and Sociology, Punjab Agricultural University, Ludhiana

S.S. Johl
Professor of Eminence (Retd.) Department of Economics and Sociology, Punjab Agricultural University, Ludhiana

Raj Krishna
Professor of Economics, Delhi School of Economics, Delhi University, Delhi

N. Krishnaji
Professor, Centre for Economics and Social Studies, Nizamiah Observator Campus, Begumpet, Hyderabad.

N.S. Randhawa
Former Head, Policy and Planning Unit, FAO, Consultant U.N. Organisations FAO, IFAD, World Bank and UNDP. Presently Senior Advisor CRRID , Madhya Marg, Chandigarh

G.S. Raychaudhuri
Professor (Retd.), Delhi School of Economics, Delhi University, Delhi

V. M. Rao
Former Member, Agricultural Costs and Prices Commission, New Delhi

Anju Sharma
Lecturer, Department of Economics, Khalsa College, Amritsar

Anil Sharma
c/o, Director, NCAER, Parisila Bhawan, Indraprastha Estate, New Delhi.

D.S. Sidhu
Professor (Retd.), Department of Economics and Sociology, Punjab Agricultural University, Ludhiana

D.R. Singh
c/o, Department of Economics and Sociology, Punjab Agricultural University, Ludhiana.

Joginder Singh
Professor, Department of Economics and Sociology, Punjab Agricultural University, Ludhiana

O.P. Sahni
c/o Department of Economics and Sociology, Punjab Agricultural University, Ludhiana.

Karan Singh
Professor, Department of Economics and Sociology, Punjab Agricultural University, Ludhiana

K. Subbarao
Senior Economist, Women in Development Division, Population and Human Resource Department, The World Bank, Washington, USA

M.M. Verma
c/o Department of Economics and Sociology, Punjab Agricultural University, Ludhiana.

LIST OF TABLES

2.1	Changes in the Relative Position of the Different States in Output and Procurement of Food Grains, 1967-81	51
2.2	Growth in Subsidies for Agricultural and Rural Development	54
2.3	Inter-State Differences between Andhra Pradesh and the Punjab in Average Costs of the Production of Paddy, 1978-79	56
2.4	Inter-State Distribution of Benefits from Support Price Policy, 1975-76	57
2.5	Relative Shares of the States in Input Subsidies	59
2.6	Actual and Potental Fertilizer Use in Different States	61
2.7	The Changing Position of Different States in Terms of Input Subsidies: A Summary	63
2.8	Summary Statement of the Changes in the Relative Position of the States in Institutional Finance of the Agricultural Sector, 1967-68 to 1977-78	64
2.9	Irrigation in Selected Backward and Advanced States	66
3.1	Mean Excess of the Procurement Price Fixed by the State Governments over the Price Recommended by the Agricultural Prices Commision	74
3.2	Procurement Price and Cost of Production of Wheat	76
3.3	Changes in Procurement Price and Cost of Production, Punjab and Uttar Pradesh	77
3.4	Farm Harvest Price and Cost of Production of Wheat	78
3.5	Procurement Price and Farm Harvest Price of Wheat	79
3.6	Procurement Price and Annual Average Wholsale Price of Wheat	79
3.7	Estimated Elasticity of the Wheat Procurement Price with Respect to the Average Wholesale Price in two Preceding Years in Different States (1965-66/1975-76)	81
3.8	Ratio of Wheat Procurement to Production, Indian States (1951-52/1976-77)	82
3.9	Wheat Zoning and Price Variation: India (1951-52/1974-75)	87
3.10	Weighted Average Realised Prices (WARP) and Wholesale Prices (PW} of Wheat: Indian States (1965-66/1975-76)	88
3.11	Wheat Acreage Elasticity Estimates	90
3.12	Wheat Supply Response Regressions (1975-58/1969-70)	93
3A.1	Proportion of All-India Wheat Output and Procurement in Different States (Average 1972-73 to 1974-75)	96

5.1	Receipts from Agricultural Income Tax and Land Revenue (Budget Estimates for 1978-79)	135
5.2	Current Receipts from Land Revenue and Estimated Yield of Modified AHT	149
5.3	Haryana Land Holding Tax—Rates of Tax	163
A-1	Average Output and Value of Principal Crops	171
A-2	Gross Cropped Area and Net Area Sown Statewise	176
A-3	Rateable Value of Land per acre Statewise	176
A-4	Estimate of Yield and Incidence of Modified Agricultural Holding Tax	177
6.1	Production and Availability of Edible Oils in India 1971-72 through 1986-87	183
6.2	Price behaviour of edible oils in India, 1976-77 through 1986-87	185
6.3	Per cent of Unsaturated fatty acids and essential fatty acids (linoleic) in different oils (on oil basis)	186
6.4	Oilseeds Area, Production and Yield All India	188
6.5	Comparative Economics of Production of Groundnut and Soya-bean in India and USA for 1981-82 and 1982-83	189
7.1	Poverty Situations in 1985	205
7.2	Rural and Urban Poverty in the 1980s	205
7.3	Some Characteristics of Small Farms, 1974-82	207
7.4	Changes in Poverty in the 1980s	208
7.5	Composite Index Numbers of Performance and Ten Selected Indicators	210
7.6	Composite Index and Human Development Index	213
8.1	Compound Growth Rate Per Cent Per Annum of Area, Yield and Production of Principal Crops of Jalandhar District (1969-70 through 1979-80)	248
8.2	Agricultural Indices of Julandhar District	251
8.3	Projected Land Utilisation Pattern (1990-91)	253
8.4	Alternative Yield Potentials of Selected Crops of Jalandhar in Kilograms	254
8.5	Average and Marginal Yields of Selected Crops (1969-70 to 1979-80)	255
8.6	Projected Yield per Hectare of Principal Crops in Kilograms	256
8.7	Projected Area of Principal Crops	258
8.8	Projection Gross Cropped Area under Principal Crops and Cropping Pattern (1990-91)/(1979-80)	259
8.9	Projected Production Scenarios (1990-91) and Growth of Production of Major Crops	261
8.10	Projected Agricultural Indices of Julandhar District	262
8.11	Projected Gross Value Added of Major Crops	263
8.12	Marginal Balances of Major Crops (1990-91)	264
8.13	Estimated Input Requirement of Jalandha District's Agricultural Plan	266

9.1	Product Specific and Non-Product Specific Aggregate Measure of Support (TE 1988-89 and TE 1992-93)	282
9.2	Extent of Support per Unit of Output Based on Support Prices	283
9.3	Extent of Support Per Unit of Output Based on Farm Harvest Prices	285
9.4	Prices of Various Agricultural Commodities in Relation to Wheat	290
10.1	Share of India in the World Production and Export of Principal Agricultural Commodities, 1992	299
10.2	Principal Agricultural Exports of India	301
10.3	Domestic and World Prices of Principal Agricultural Commodities in 1992	303
10.4	Compound Rates of Growth of Prices of Principal Agricultural Commodities During 1980-92	304
10.5	Non-Product Specific Aggregate Measure of Support (AMS) in Agriculture in India during 1986-88	308
10.6	Production Specific Market Price Support, 1986-88	309
10.7	Growth of World Export Trade	314
11.1	Growth Rates of Prices during April 1982 to September 1992	322
11.2	Month to Month changes in Prices During April 1982 to September 1992	323
11.3	Growth Rates of Area, Production and Yield	324
11.4	Wholesale Price Indices of Raw Cotton and Cotton Manufactures (September 1990 to March 1993)	239
12.1	Export Trade, 1991	348
12.2	Major Exporting and Importing Countries for Wheat	348
12.3	Major Exporting and Importing Countries, Rice	349
12.4	India's Production of Oil Seeds and Imports of Vegetable Oil	349
12.5	Cotton Exports from India	350
12.6	Punjab's Export Surplus	351
12.7	Comparative Position of Punjab and World Market Prices, Wheat, 1989 to 1994	352
12.8	Comparative Position of Punjab and World Market Prices, Rice, 1989-1993	353
12.9	Price of J-34 (Medium staple) Cotton	354
12.10	Price of MCU-8 (Long staple) Cotton	354

INTRODUCTION

K.S. DHINDSA & ANJU SHARMA

There is no exaggeration in saying that macro policies relating to agriculture followed by Indian Government have been successful in stimulating the use of modern inputs and thus have been very much helpful in enhancing the agricultural output especially of foodgrains. In various five years plans, the main objective on agricultural front was to further increase the production of principal crops. To achieve this objective, emphasis was laid on adoption of improved agricultural practices extension of area under irrigation and high yielding varieties of seeds, soil conservation, reclamation of waste land, consolidation of holdings and distribution of fertilisers and improved varieties of seeds etc. Besides this, the plans aimed diversification of agricultural economy and recognised the importance of livestock improvement. Emphasis was also laid on accelerating the pace of institutional reforms.

In various plans, incentive policies including the provisions of subsidies such as input subsidies, subsidy on institutional credit and price support were followed for stimulating the diffusion of modern technology. Fertilizer subsidy was given to the farmers to cover the difference between the domestic cost of fertiliser and the actual price at which it was supplied to the farmers. As regards irrigation a substantial part of the subsidies is being provided on the canal irrigation system in the form of providing canal water at lower rates. Electricity is also provided to the farmers at subsidised rates. With the expansion of commercial and regional rural banks, credit is also supplied to the rural people at cheap rates.

Public distribution system has also been introduced which aims at supplying some essential commodities like sugar, wheat, rice and edible oils etc. to the weaker section of the society at reasonable prices. This system is used as a instrument of poverty alleviation. Public distribution system supplies have a stabilising effect on open market prices by increasing the availability and checking the speculative tendencies. It is also argued that incentive policies have increased inter-regional disparities in Indian agriculture. The poorest states like Bihar, Uttar Pradesh, Orissa, Madhya Pradesh and Assam are relatively less covered by the public distribution system.

Another important measure of utmost importance is the announcement of procurement and support prices to ensure fair returns to the farmers so that even in years of glut the prices do not fall down and farmers do not suffer heavy losses. Government of India has spent a lot on the promotion of agricultural research and training to discover high yielding varieties of seeds, techniques for increasing the productivity of soil and counter the attacks of pests and insets. The triple function of agricultural research, education and extension is being implemented through various institutes, agricultural universities and project directorates etc. The ICAR which was set up in 1929 is the national body for coordinating research and extension education in the fields of agriculture, animal husbandry and fisheries sciences.

In 1991 Government of India has adopted new economic policy to improve the economic situation of the country. Some have argued that the economic reforms introduced since 1991 have not targeted the agricultural sector directly and hence not much may be expected of it under the new policy regime.

Excessive protection given to industry in the earlier policy has harmed in agriculture in several ways. It raised the prices of industrial products relative to agricultural products, thereby turned the terms of trade against agriculture. And this led to a progressive shift in investible resources away from agriculture. It also created an overvalued exchange rate which adversely affected exports of agricultural goods.

It is argued that instead of relying on expansion of subsidised supplies of various inputs, it would be much better to reduce the protectionist bias against agricultural sector by lowering protectionist barriers favouring industry and altering relative prices in favour of agriculture. This would create a potentially more profitable agriculture which would be able to bear the economic costs of technological modernisation and expansion. The reforms in the industrial and trade policy will reduce protection for the industry and hence investible resources may have a tendency to shift towards agriculture.

Under the reforms since 1991, the relative profitability of agriculture could only have increased, for industrial tarrifs have been lowered very substantially, while all other interventions favourable to agriculture remain intact.

For reducing rural poverty and unemployment, agricultural growth is very much important. In India, many schemes which would not only lead to the creation of productive assets in the economy but would also utilise more labour, especially of poor, were started during the plans. These programmes so far as rural areas are concerned included the Integrated Rural Development Programme, Drought Prone Area Programme, Crash Schemes for Rural Employment, Development of Villages and Small Scale Industries, National Rural Employment Programme, Rural Landless Employment Guarantee Scheme, Jawahar Rojgar Yojna, Employment Assurance Schemes introduced in 1993-94 and the District Rural Industries Project started in October 1994. All these programmes if honestly and efficiently implemented, are likely to help many people go above the poverty line.

The main determinant of the welfare of agricultural workers is the price of the foodgrains. Due to this reason, the prices of crops assumes a very important role in determining the welfare in an economy with high poverty.

For raising the welfare of poor farmers procurement prices have been raised almost on an annual basis. If the foodgrain stocks last and individuals possess the necessary purchasing power, the public distribution system can prevent starvation that might arise from price increases. It is not an instrument that can raise the consumption level of those currently with access to it. This can only be achieved through a rise in income for which task supply oriented rationing system is not equipped. The public distribution cannot be a safety net because it is maintained by a procurement policy which is based on raising offer prices to the surplus farmers i.e. a policy which is inherently inflationary.

Food subsidy given under the Public Distribution System is only very partially a consumption subsidy. The bulk of it goes to subsidise high-cost production and wastage and corruption of the Food Corporation of India. The F.C.I.'s costs are on procurement, storage, transportation, spoilage and pilferage. Storage facilities are inadequate, often leading to storage in the open and inevitable spoilage. Centralised procurement, storage and distribution add layers of cost that would be avoided in a decentralised system. Private traders should be allowed to carry foodgrain stocks without limit. The internal distribution of food can be left to private trade. The poor should be given food stamps distributed through local govt. bodies. The overall availability of foodgrain at any point of time can be ensured by individual state civil supplies departments. For improving productivity, agricultural policy must include the removal of pre-capitalist agrarian relations and encouragement of commercial farming, on the lines of the West Bengal reforms. Since much of the fertiliser subsidy goes to the fertiliser producers and not to the farmers, there is a need for creating a competitive environment among fertiliser producers so that they are under pressure to raise efficiency and curtail unit cost.

The public distribution cannot be a safety net because it is maintained by a procurement policy which is based on raising offer prices to the surplus farmers i.e. a policy which is inherently inflationary. It is also argued that in the wake of new economic policy the trade and price reforms alone would not be sufficient to improve agricultural growth. Without infrastructural development and institutional changes, the price incentives may not increase the aggregate supply response. For reducing rural poverty one should pay attention towards employment generation as well as labour productivity, together with the provision of non-farm sector employment. Reduction in regional disparities will also go a long way in improving human resources and reduction in overall poverty. So there is a need for higher investment for improving physical and human capital in poor states of India. There is also a need for involving panchayats, self help groups and community based organisations for strengthening government employment schemes and for helping small and marginal farmers and landless labourers.

Th present volume of the book 'Policy, Planning and Liberalisation' deals with the important issues relating to agricultural policy and planning adopted by the Indian government since independence. The first section of the volume is devoted to these particular issues. The second section deals with the articles relating to the impart of liberalisation and recently included GATT (WTO) agreement on Indian agriculture.

SECTION I

The articles included in Section-I discuss the macro-economic policy options appropriate for accelerating the development of Indian agriculture, the effect of macro policies on agricultural development, regional disparities in agricultural development and solution for solving edible oil crises in India. Besides this a couple of papers included in this section examine the issues related to agricultural price policies, procurement policies and supply response of foodgrains with respect to output/input prices and profitability.

It has been widely recognised that agricultural technology should be directed towards specific local conditions. Recently the planning commission has taken the initiative to identify homogeneous agro-climatic regions for crop specialisation. In his paper on Policies for Agriculture, Yoginder K. Alagh has also emphasised the need for Agro-climatic regional planning approach for agricultural growth. He stresses that, first the ground water and watershed development should be accorded high priority in order to insulate production from the vagaries of monsoon. Secondly, price rationalisation schemes for inputs like seeds, fertilisers pesticides etc. should be introduced. For instance, the price of pesticides should be reduced through regulated imports. Thirdly, the technologically efficient agro-processing and marketing should be made an important part of agricultural strategy, keeping in view the potential for employment generation, exports and higher income to the producers. Fourthly, since the poultry farming is emerging as an important activity for enhancing nutrition and providing employment, reduction in import duty of specified type is essential for improving the quality of poultry feed. In order to solve the problem of acute shortages of milk in lean summer months, particularly in cities attention should be paid on Operation-Flood having specified targets of the growth of Amul-type co-operatives in states like Bihar, Orissa, West Bengal and Assam. Since at present, one third of the estimated fish potential from marine and in land sources is being exploited, emphasis on this sector needs to be laid on (a) large scale fish production from the reservoir (b) liberalised leasing policy for Inland water area. Lastly, in order to provide adequate and uninterrupted credit to new and non-defaulting members, incentives and recognition should be given to local branch managers who process loan applications expeditiously.

While writing about 'State Policies and Regional Disparity in Indian Agriculture", K. Subbarao contends that regional differences in tennurial conditions and state policies with respect to irrigation, subsidies and prices have resulted in the emergence of small enclaves of growth and vast areas of low productivity and stagnation. According to the author, in incremental output of foodgrains and public procurement of wheat, the Eastern and North-Western states have contributed respectively smaller and larger proportion with respect to their gross cropped area. Therefore, the surplus states of North-west have mainly benefited from the price support programme either because their unit costs were higher and/or because their share in public procurement was very high. The disproportionately large share of benefits went to states with high and assured irrigation facilities.

The author argues that the pre-empting of fertiliser supplies by the states which are better off has led to low use of fertiliser in rainfed eastern part of India. Since a substantial part of the irrigation subsidies is provided on the canal irrigation system, benefits have gone to other than Eastern states having network of canal system. It is argued that a huge amount of funds allocated for subsidies have resulted into the reduction for public investment in critical irrigation works programmes. According to Subbarao, it can be said that the benefits emerging from a combination of the states incentive policy-comprising price support, input subsidies and subsidised credit-have accrued disproportionately to the prosperous states which already have access to water, power and fertilisers.

The paper by G.S.Raychaudhuri and Raj Krishna is related to the procurement and supply response of wheat to price and irrigation. The authors point out that the distribution of marketed surplus between government procurement and sale in the open market is expected to depend on the ratio of procurement prices to wholesale prices. If the procurement price remains substantially below the wholesale prices, the level of procurement will remain at low level. The elasticity of procurement with respect to output and to the ratio of the procurement price and the open market wholesale price has been found to be 2.32 and 2.72 respectively.

The authors have also estimated acreage and output elasticities of wheat crop with respect to output/input prices and irrigation variable. Output elasticity with respect to irrigation has been found to be higher than that of output/input prices. It has also been found that interstate variations of wholesale wheat prices have been higher in the zoning years as compared to free trade years. Thus the zoning policy followed by the government has increased the interstate price dispersions. The authors conclude that for satisfactory growth of wheat output, policy makers must maintain a favourable output/Input price regime, steady growth of irrigation capacity, input supplies and technology knowledge.

N.Krishnaji in his article writes about the agricultural price policy with reference to Indian foodgrain economy. The author points out that the aim of price policy in case of foodgrains should be the fulfilment of the overall food requirements of the country. While fixing support prices for foodgrains, the Commission for Agricultural Costs and Prices tries to determine the impact of it on cost of living, level of wages, industrial cost structure, producers and consumers. He emphasises that minimum support price must cover the total average cost and the imputed value of family land and family labour used in the process of production.

According to Krishnaji, relatively higher and rising agricultural prices have gone in favour of surplus producing big farmers and this indeed may be necessary to promote the agricultural growth at a satisfactory rate. Thus the subsidies and price support programmes initiated by the government have benefited mainly the surplus producing states. The operation of Public Distribution System has only benefited the urban poor and not the rural poor. The author also argues that since output response with respect to price is lower than that of output response with respect to increase in the levels of irrigation, a well balanced package of government instruments like promotion of technology, infrastructural investments including investment in irrigation, production subsidies is preferable to price/support programmes.

Amaresh Bagchi in his paper outlines the rationale and procedure levying agricultural holding tax. The author argues that super imposition of surcharges or cesses on the basic assessment would only increase the inequities which have resulted from various disparate practices followed in the case of settlement of land in different parts of the country over the years. Assessment of the productivity of land for each holding individually is neither practicable nor necessary. Bagchi contends that a tax based on an index, reflecting the productivity of land on an average, after making reasonable allowance for differences in soil and climate, is needed. The agricultural holding tax proposed by the Raj Committee was intended to provide model for such a tax. According to author, even with moderate rates, the yield from a modified Aggregate Holding Tax would be substantially higher than that from land revenue including cesses surcharges. Aggregate Holding Tax (AHT) would be a tax leviable on the holdings of farm families according to the aggregate Rateable Value (ARV) of each holding. The ARV of a farm holding would be determined on the basis of the rateable value per hectare of land comprised in the holding. Bagchi suggests that in order to simplify, the administrative part, the tax should be levied, on ownership holdings and not on the basis of operational holdings. Periodic reassessment gives incentives for higher productivity as raising the output above that level would not be subject to additional tax. At the same time those having productivity below average are penalised. Further, levying the tax on gross produce gives inducement to the farmers to reduce cost of cultivation and in turn improves the efficiency.

S.S. Johl in his paper on Edible Oil Crisis, gives an alternative solution for solving the problem of shortage of edible oils in India. The author points out that with the increasing population pressure, the demand for foodgrains is going to increase in future, so there is no greater scope for diverting area from cereal crops to oilseeds. Johl pleads for evolving the new varieties of oilseeds crops which are disease resistant and can be grown in dry land areas. Thus, with new varieties of oilseeds coupled with proper implementation of assured minimum support prices, it is possible to achieve a dramatic increase in oilseed production.

In the absence of a major breakthrough in the technology of oilseed crops, Johl suggests the development of dairy farming on scientific lines for solving the oil crisis. Following the pattern of western countries, butter oil instead of vanaspati ghee/edible oil can also be used for cooking purposes. In case of milk production, he stresses the need for shifting emphasis from buffaloes to cows of improved breeds, yielding milk upto 5000 litres per lactation of ten months. After extracting fat content from milk which can be used for making butter oil, milk powder etc., the standardised milk can be sold to the consumers at reasonably low prices. For rapid growth of dairy industry on scientific lines, proper handling of milk-its processing and marketing should be given the top most priority.

Increasing the productivity and purchasing power of a large number of marginal farmers and other rural poor is the urgent need of the hour. N.S. Randhawa, has addressed this issue in his paper entitled 'Restructuring of Agricultural Planning and Policies for Alleviating Rural Poverty ...". He points out that there is a severe dearth of appropriate technologies suited to small farmers operating under harsh and risky environment. Therefore, appropriate technologies, within the reach of small farmers are the main vehicles for their development. To achieve this objective, several changes will have to be made in the allocation of research funds and in research priorities. Biological and chemical technologies being land saving are more relevant for the small farmers. Also the support price policy need to be restructured according to the new weights assigned to small farmers in development strategy. The author concludes that since relative neglect of small farmers was inherent in the agricultural development strategy and the planning, the remedy is to correct the bias against small farmers. The restructuring of agricultural planning and policies is, therefore suggested as a remedy by according weights to the development of small farms.

Given the vast socio-economic diversity, the desirability of district planning has been widely acknowledged as an instrument of decentralised development in a large country like India. A.S.Dhesi and N.S.Bhullar in their article suggest a simplified operational methodology for formulating district agricultural plan with special reference to Jalandhar district. They mention that once the agricultural development potential is estimated, it becomes easy to plan the future development of agro-based industries in the district. Such type of planning, involving participation of local people can result in effective utilisation of local resources.

SECTION II

Articles included in this concluding section of the book deal with the challenges which the Indian agriculture has to face in the context of recently concluded GATT (WTO) agreement. The important issues discussed relate to the impact of gradual reduction/elimination of input subsidies, and patenting of seeds on India's agriculture and the future prospects for the exports of agricultural commodities in India.

The paper by Ashok Gulati and Anil Sharma analyse the likely impact on India's agriculture regarding India's commitments pertaining especially to Trade Related Intellectual Rights (TRIPS), market access, domestic support and export competition in the areas of agriculture. Under the market access commitments, all member countries of GATT are required to replace all types of non-tariff barriers with tariff barriers and the tariff levels are to be reduced by 38 and 24 per cent in case of developed and developing countries respectively. Maximum market access of 3 per cent of domestic consumption, rising to six per cent in the implementation period is to be given at reduced tariff rates. Export subsidies are also to be reduced by 36 and 24 per cent in case of developed and developing countries respectively. The authors point out that India gives only one indirect export subsidy under which profits from sale of exports under Income tax are exempted. Since developing countries are exempted from reducing export subsidies during the implementation period, so the commitments pertaining to reduction in export subsidies will not have adverse impact on agricultural export from India. To take advantage from the WTO agreement, according to Gulati and Sharma, India should remove export restrictions such as export quotas, canalisation, minimum export price etc. Consequent to the reduction in export subsidies, India would be in a position to export fine varieties of wheat, flowers fruits and vegetables and their processed items.

In his paper on Agriculture Pricing, Marketing and International

Trade under New Economic Environment, M.S. Bhatia points out that the international economic environment is expected to change significantly with the latest GATT agreement, which is now known as World Trade Organisation (WTO). The protectionist policies followed by various countries have to be changed with rational policies in which subsidy support is to be reduced and quantitative restriction are to be replaced by moderate tariff restrictions. According to Bhatia, if the prices in the Indian domestic market increase faster than the world prices, then the export of the commodities would become relatively less remunerative or unprofitable unless the local currency is devalued through readjustment of exchange rate proportionately. Thus, the most important variables that will decide the future of country's export performance are the relative increase in domestic to world prices and competitiveness of the country's exchange rate.

The author also pinpoints that as per the agreement, the country has the option to protect its own plant varieties. The Indian government is preparing a Sui-Generis through which it is proposed to protect the traditional and inalienable rights of the farmers to save, use, exchange, share and sell seeds on a non-commercial basis. The proposed legislation would allow researchers the right to free access to protect material or breed to develop new varieties without the authorisation of original breeder. The author further argues that GATT through reduction in tariff and increase in market access is likely to provide upward push to the rate of growth of world export trade. New trading environment would provide greater opportunities for diversification of exports which is very much essential to sustain growth of exports and markets.

V.M. Rao in his paper points out that there are several aspects of agriculture such as the Green Revolution areas or the newly emerging corporate activities in the field of horticulture, floriculture etc. for which liberalisation could be a boon. However, given the inequitable distribution of land and other agricultural resources, a situation which favours better and efficient use of resources concentrated in large farms, would not necessarily benefit the farming community consisting predominantly of small farmers. The author argues that the intervention by the government for providing infrastructure to organic research and extension and to regulate supplies of inputs critical for agricultural growth have been, on the whole, very helpful in promoting agricultural development. He further emphasises that the violent inter-seasonal fluctuations produce a disparity between the prices realised by the grower and those prevailing in the subsequent months which is often far in excess of the carrying cost of the commodity. These are the market imperfections which tend to heighten the spatial margins beyond that normally justified by freight and other incidentals. Both cotton and oil seeds crops holding considerable promise for backward areas, appear to have market structures in which the farmers bear the burden of fall in prices but do not receive their due when the prices are high. Such asymmetry is likely to be a rule rather than an exception when farmers deal with organised groups and the two parties differ markedly in bargaining power. The author argues that it is only through group action (co-operative marketing) that the farmers would be able to achieve a measure of parity with the manufacturer purchaser of his produce belonging to the organised industrial sector.

The paper by a panel of experts from Punjab Agricultural University examines the implications of GATT for Indian agriculture with special reference to Punjab. The authors point out that the main objective of GATT is to promote world trade by reducing barriers to it such as custom duties and quotas with special differential treatment to developing countries. Inter-zonal restrictions and other non-tariff measures which often generate disincentive signals for the farmers, are proposed to be removed and converted to tariffs. It is also pointed out that the total subsidy (product and non-product) given to Indian agriculture is much below the cut-off mark, so the Indian agricultural production will not be adversely affected on account of reduction in these subsidies. Moreover, marginal and small farmers which cultivate more than 30 per cent cultivable land, are exempted from reduced subsidies.

SECTION I

POLICY AND PLANNING

1

Policies for Agriculture

YOGINDER K. ALAGH

Indian agriculture has made significant strides during successive decades of planning. Foodgrains production has more than trebled; and, in spite of the rapid growth of population and recurring droughts the need to import food has virtually been eliminated. Net availability of foodgrains has increased from 150 kg. per capita on the eve of the planning era to 172 kg. per capita in 1987/88. Improvements in availability of sugar, eggs and milk have been even more impressive. Increased production of foodgrains has also enabled the building of buffer stocks. An expanding PDS aims at increasing the access of the poor to food at reasonable prices. Payment of part of the wages in foodgrains under the NREP and RLEGP also contributes to this objective.

The progress of agricultural growth has, however, been characterised by certain imbalances. The income levels of a large part of agricultural population, namely, marginal holders and agricultural labourers continue to be low and quite often lower than the poverty line income. Improvements in per capita availability of food items cannot still be considered adequate, given the unequal access to incomes and markets. Despite expansion in irrigation the agricultural economy is still far from being drought proof. Employment generation has been rather slow in recent years, particularly in agriculturally better developed areas.

The Eighth Plan will need to focus on these problems through a spatially and cropwise diversified strategy of accelerated development which emphasises specificities of different regions. The strategy will need to consist of the following components:

(a) a massive land development programme comprising soil

consolidation and land levelling, reclamation of *barani, usar* and other degraded lands and more optimal use plannings;

(b) a water management and operation strategy focussing on full utilisation of benefits from major and medium projects, a plan for improvement of water delivery systems, timely completion of major multi-purpose projects, phased programmes of drainage and flood control and a major priority to ground water development and conjunctive use;

(c) development of optimal landuse and cropping pattern plans for different agro-climatic regions and back-up with appropriate marketing, agro-processing, communication, technology extension and input supply infrastructure for the farmers;

(d) special plans for animal husbandry, fish and poultry development;

(e) greater use of modern technology such as computers for irrigation management, seed and pesticides practices including bio-tech processes, embryo transfer in animal husbandry and water saving devices including drips;

(f) economic and financial policy packages including rationalisation of taxes on equipment and inputs, streamlined credit arrangements and more effective delivery systems;

(g) development of systems management practices for the agricultural sector to make more effective use of available resources and to properly provide for output targets and input systems at the regional and local levels and timely adjustment in strategies in the light of changing circumstances;

(h) particular attention to the involvement of small and marginal farmers in growth and modernisation through better credit and input support;

(i) special emphasis on land reforms and records of rights in land; and

(j) a more widespread growth to maximise income and employment generation possibilities in the backward areas.

Policy initiatives are required to be taken in a number of areas to operationalise this strategy. Some of the more important policy requirements during are as follows:

Development and Management of Land and Water Resources

Groundwater and watershed development need to be accorded high priority in order to insulate production in the dryland areas from the vagaries of monsoon, to the extent possible. A number of experiments are presently underway. The cost of works in contour bunding, check dams, land levelling and digging of percolation tanks and soil conservation is often lower when undertaken by voluntary agencies than in Government projects. These groups lay considerable emphasise on raising contribution from the beneficiaries. Involvement of these agencies on a larger scale needs to be promoted. The irrigation of rural development schemes like NREP, RLEGP and IRDP with watershed development schemes would have to be actively pursued.

The technical aspects of soil conservation, land levelling, micro water harvesting techniques and appropriate organic cover for degraded land for nutrient replenishment, proper crop cycles or fodder/forestry cover, are by now well known. The central question is to deliver these technologies to the field. A modest beginning has been made in the Seventh Plan. These need to be very substantially accelerated. Also, such programmes are at present incorporated in a number of different schemes in the Central and the State Plans. Many of these schemes overlap. It is important that they are integrated with a major focus on delivery of end results, i.e., improved land management and land development with participation of the local community and with pachayati raj agencies.

Completion of the on-going major and medium irrigation projects needs to be given high priority. Mechanisms would have to be devised to provide adequate resources through larger Central sector involvement where necessary for proper planning and timely completion of large irrigation and flood control projects.

Delays in completion of projects needs to be minimised by simplifying the procedure for land acquisition, sorting out inter-State disputes in time, ensuring availability of key construction materials and expeditious clearance of these projects for investment by competent authorities.

The SPPP has already operationalised the concept that irrigation targets are not to be conceived in terms of "potential created", but in terms of the number of waterings to be delivered at the field in each crop. Identification on annual basis of the area to which water will now be delivered and working backwards to investment and action plans will have to become the general practice.

In respect of the Command Area Development (CAD) projects, it is important to identify a single agency responsible for delivering water from the headworks of the canal systems to the field. More emphasis has to be laid on the construction of field channels and of *warabandi* or alternate operation systems of canals. High priority needs to be given to modernisation of existing systems. Targets will have to be conceived in terms of number of waterings and irrigation efficiencies increased from the existing levels of around 35 to 40 per cent in many systems to around 60 per cent. Improved water regulatory systems, and aquifers modelled to allow for investment of conjunctive use for ground and surface water, are other avenues for improvement. Assessment of the achievement of the CAD programme should be done in terms of incremental agricultural production rather than benefited area as is being done currently.

The Ganga Flood Control Commission and the Brahmaputra Flood Control Board have submitted master plans and detailed reports for some projects. Both the plans emphasise the need for catchment area treatment. The concerned flood control agencies should prepare and execute detailed catchment area plans. Systematic studies on river morphology and river mechanics would need to be initiated to evolve long term measures for flood control. Apart from the Central and State schemes in this field, other land and water harvesting projects in the wasteland development project, forestry schemes, special employment schemes and area development schemes, have to be integrated with catchment area treatment.

The high priority given to groundwater development in recent years will have to be continued. A number of studies are now showing the working of groundwater markets in the country. The farmer has shown that he is willing to pay for reliable supplies of water and groundwater development policies which take into account the existing markets for such water and increase the prices to the farmers, should be a major objective of policies. In such areas provision of easier finance through Government, para statal and cooperative agencies should lead to larger supplies of ground water and more effective delivery to all segments of the agricultural population.

In areas where such markets have not developed and institutional system act as a constraint, the State will have to play a more direct role in meeting the requirements of small and marginal farmers.

The prices of pumping equipment vary in the different parts of the country and effort is required through standardisation and research in upgrading the energy efficiency of available pumpsets and bringing about uniformity in their prices. Instructions should be given to banking system that farmers taking loans/subsidies for pumpsets should be allowed to purchase equipment from any supplier, rather than agro-industries corporations/approved suppliers. There should be sample checks on diversion of loans from beneficiaries and exemplary punishment to culprits (Banking change).

Tank irrigation needs to be revived and strengthened in dryland areas. Studies have shown that existing minor irrigation tanks are not being properly maintained since the community labour earlier available for maintenance is no longer being supplied, in the context of increasing commercialisation of the rural economy and the breakdown of community maintenance practices. On the other hand, studies have also shown that the rate of return to desilting and improved operation of tanks is very high in terms of raising the groundwater level in the catchment area and the command of the tank. In the Eighth Plan, each State must conduct a quick survey of all available tanks in which information on principle characteristics like storage capacities, hydrology, size of ayacuts an ground water regime is stored at a central place. Given the high rate of return, these is a case for a percentage of the maintenance costs of the tanks being met through specially designed schemes, the balance to be made available by community effort. Such schemes should be continuously monitored with modern data processing techniques so that renovated tanks do not fall into disrepair.

Inputs-Seeds and Pesticides

The Task Force on the Special Foodgrains Production Programme recommended pricing strategies to meet the objective of cheaper inputs to farmers, rather than direct subsidies. In inputs like pesticides and seeds, subsidies can become a method supporting inefficient production units (high costs technical grade manufacturers or formulators) or high cost distributors in cooperative, State or private sectors and the benefits may not go to the farmers. Thus subsidies can support high cost pesticide supply or seed prices say higher than Rs. 450 per quintal. With a price rationalisation scheme, the farmer gets the benefit straight away. Apart from sectors like pesticides and seeds, other traded inputs to the agriculture sector, need a price rationalisation strategy. Plastics, water saving and energy saving devices, all need to be priced at international levels, rather than at higher prices, which the agricultural sector cannot afford. Thus the agricultural sector now needs to get the same kind of policy attention that the industrial sector was getting in the early years of the Seventh Plan in terms of policy and delivery system reforms.

The price of certified seeds of wheat and paddy should be reduced to efficiency levels. The Bureau of Industrial Costs and Prices (BICP) or the Chief Adviser (Costs) of the Ministry of Finance may be asked to estimate the efficiency cost of seed production. The National Seeds Corporation, the State Seeds Corporations and the Agricultural Universities may be asked to switch over to a system of efficiency pricing in a phased manner, say over a three year period. While the efficiency price is being worked out, costs of production higher than 50 per cent above the procurement price of grain may be written off as losses. However, if this proposal is not acceptable, a subsidy of around Rs. 200 per quintal may be required on HYV seeds for paddy and wheat given the existing levels of seed prices. During the transition period, the State Governments' and the Central Governments' budgets would have to meet the loss arising out of inefficient cost of production. The policies for (i) systematic supply plans and subsidies in short run, and (ii) medium term adjustment to efficiency pricing, need to be clearly set up.

In delivery systems, already policies have been developed for sectors, like seeds, of selective support and technological upgradation. The new seed policy providing for specialisation and selective import has been generally welcomed but there has been some ill-informed comments. First, the policy does not allow for import of seed for self-pollinated crops like wheat and paddy where our own strengths are great. Second, in crops like hybrid millets and oilseeds, the best available germplasm can be imported for development by Indian seed producers, whether in the public, the para statal or the private sector. Only in vegetables and flowers, where our own strengths are limited is relatively open import allowed, subject to very strict quarantine laws. Once import is allowed, tariffs have been rationalised to benefit the farmer through cheaper prices. He will also benefit through competition of alternative delivery systems. Finally, the ICAR and its agencies will experiment, with all seeds grown and the media will be used to inform the farmer of the quality characteristics of the seed.

The new seed policy should be supplemented with the announcement of a policy of strengthening our own capabilities on a selective basis. These selected centre should be given access to germplasm, equipment, access to international literature and other facilities to maintain our leadership. The ICAR, in consultation with the Planning Commission, should prepare for these centres a detailed plan with milestones.

Pesticides

Consumption of some of the pesticides need to be encouraged. The use of tariff policy to give measured protection to the domestic industry and to reduce prices to the farmer, if necessary through regulated imports, should be continued and supplemented with a policy on monitoring prices at the farm level and streamlining distribution system of formulated pesticides.

Agricultural Implements and Machinery

A mechanisation policy which makes small tools and implements available to the farmer, needs to be developed. Following measures are suggested to encourage the use of such agricultural implements and machinery:

(i) Farm mechanisation extension must become a part of the existing T&V system: this may require strengthening of training of T&V workers by agricultural universities/departments of mechanical engineering and State level cells strengthened;
(ii) Training for artisans producing implements at village level may be improved through TRYSEM scheme, i.e. welding technologies etc.;
(iii) A national fund may be designed to provide finance for productionising new products of farm implements designed by agricultural universities/institutes used by a majority of the peasants in the concenred areas. This fund may be run with the association of representative bodies of entrepreneurs and could provide for consultancy help to standardise new tools and implements and provide financial assistance for viable ideas;
(iv) Centres may be established or existing centres, such as the one at Ludhiana, strengthened to help to productionise such innovative ideas for production planning in the small scale sector;
(v) Fairs may be encouraged in which the latest trends in the equipment and technologies on a global plane with special

emphasis for small farmers, may be organised for those engaged in agricultural extension/equipment organisations of farmers;

(vi) Financial institutions may consider financial assistance to intending entrepreneurs, particularly trained unemployed persons, to set up hiring and leasing centres of taluka headquarters and other rural centres of equipment and technology packages; and

(vii) Components and materials used for farm equipment must be available to the producers at reasonable prices and tax and tariff policies used for this purpose.

Agro-processing

It is now widely recognised that policies which encourage more diversified and technologically efficient agro-processing and marketing, need to be an important part of the agricultural strategy, keeping in view their potential for employment generation, exports and higher incomes to the producers. A number of fiscal concessions for food processing and packaging industry have been announced by the Finance Minister in his Budget Speech. Excise on parts of refrigerated appliances and machinery as well as compressors used in refrigerated vans meant for transportation of food and dairy products, has been reduced from 40 to 15 per cent. The concessional rate of import duty of 40 per cent is to be extended to a few more specified items of food processing and packaging machinery such as transport refrigeration unit and machinery for eggs processing. Excise duty on skimmed milk powder and condensed milk, certain food preparations of fish, meat, tapioca and sago in metal containers, has been reduced from 15 to 10 per cent. Excise duty on *namkeens* and ready-to-eat mixes withdrawn.

However, this only represents a beginning. If food processing industry is to merge as a major foreign exchange earner, more effective steps are required for improving processing and packaging technology, for improving infrastructural facilities and for rationalising the taxation structure.

Established cooperatives with turnover of over Rs. 1 crore should be freely permitted to have technology and marketing tie-ups with foreign companies in the agro-processing area, without any licenses and only with registration with the Department of Food Processing. Large Indian cooperatives have the skill to bargain for and protect the interest of the small Indian farmer in transactions with foreign companies. In other cases, permission may be given on a case-by-case basis depending on export commitments and R&D plans in India.

Poultry Farming

Poultry farming is emerging as an important activity for enhancing nutrition and providing employment. The Finance Minister in his Budget Speech has announced a few incentives for the poultry farming. For instance, to encourage investment in poultry farming, tax exemption to the income from poultry farming at the rate of 33⅓ per cent of such income has been proposed. Further, to help improve the quality of poultry feed, import duty on the specified amino acids is to be reduced from the present level of 147.25 to 70 per cent.

However, a lot more needs to be done to enable poultry farming to realise its true potential. Shortages of various poultry feed ingredients like maize, soyabean extractions, rice, bran, fish meal, etc. with resultant high prices of feed ingredients, should be minimised. Critical components such as food casings, sodium caseinate and chicken flavour, poultry equipment and spares and components etc. should be made available at reasonable prices. Efficient research and development and training programmes with use of computer systems for breeding programmes and modern health practices in the private sector, should be integrated with university and ICAR Research programmes which are falling behind and tax exemptions provided for such expenditure.

Dairy Development

Production and availability of milk is beset with regional variations, besides acute shortages in lean summer months particularly in cities. Operation Flood should have specified targets of growth of Amul type cooperatives in States like Bihar, Orissa, West Bengal, Assam, North Eastern States and Himalayan regions of the Northern States. Methods must be found to develop dairying in these areas and plan funds committed for this purpose.

Some of the milk plants in cooperative/public sectors are incurring losses. Bureau of Industrial Costs and Prices (BICP) has been entrusted to conduct detailed studies on the losses of dairy federations. Most of the dairies are being run at low levels of capacity utilisation and with high costs. Efficiency pricing must be enforced in these dairies.

Fisheries

At present, only about one-third of the estimated fish potential from marine and inland sources is being exploited. In order to provide more nutritious food to the increasing population, to generate employment and additional income among the poor, and also to boost export of marine products, it is necessary that fish production is substantially stepped up.

Fishery development in the Exclusive Economic Zone and prawn culture in brackish water areas are the main thrust areas in the marine sector. Inland fishery, reservior fishery development, strengthening of the fishery cooperatives, building fishery estates will largely contribute to increase in fish production as well as improvement of the economic status of fishermen. Priority needs to be given to improving the access of fishermen to markets. Policies must be developed for larger investments in viable cold chains. Employment and rural development schemes should be used for maintaining and creating fish ponds and encouraging aquaculture. Emphasis in the sector needs to be on (i) exploitation of deep sea fishery resources by introducing adequate number of deep sea vessels; (ii) increasing the brackish water fish culture so as to take up semi-intensive methods of culturing prawns covering large brackish water areas; (iii) liberalised leasing policy for inland water areas; and (iv) large-scale fish production from the reservoirs. In many States, policies related to the use of land for aquaculture need review. Modern hatcheries which integrate seedlings with small ponds and fish farms must be developed.

The newly set-up Advisory Board on Fisheries must coordinate policies between different Ministries/State Governments in consultation with the Planning Commission. It must monitor Plan targets regarding fisheries. Improved beach landing crafts may be introduced on a larger scale for the benefit of small marine fishermen; raw materials for this must be made available at lower prices.

Plasticulture

New materials like plastics which optimise the use of scarce water and land and augment agricultural growth and employment will require to be introduced in the agricultural sector. An Action Plan is being implemented for fifteen districts for the use of plastics in canal lining, water conservation, agricultural packaging and modern water delivery systems and green houses. More optimal tariff, tax and subsidy regimes are required for accelerating the use of plastics in agriculture. Upgradation based on the best available techniques and building of long-term domestic capabilities, needs to be encouraged. Policies of graduated initial subsidies for such applications already in force must be implemented on a large scale particularly in water scarce areas.

Cooperative and Banking Support for Agriculture

Provision of adequate agricultural credit and its delivery is providing a major challenge for technological upgradation for agricultural and rural development. Cooperative credit went up by a meager 4.9 per cent in 1987/88. Share of priority advances is constant. The Crop Loan Manual must be examined to meet the genuine needs of the farmer in an automatic manner.

NABARD refinance for minor irrigation, land development, command area, farm mechanisation, horticulture, poultry, sheep breeding, piggery, fisheries, dairying and agricultural marketing only went up by 4.8 per cent in 1987-88. To ensure adequate and uninterrupted credit flow to the new and non-defaulting members of the PAC's in the area of ineligible CCBs, National Bank may implement non-overdue cover, minimum involvement of internal lendable resources, coverage of small farmers, etc. schemes. The non-overdue cover scheme must be implemented for each State. (In the States of Assam, West Bengal, Rajasthan, Karnataka, Haryana and Maharashtra, provision for this purpose in SPPP in 1988-89 has not been used).

Crop loan application by farmers or applications by voluntary organisations for land levelling, land development or soil conservation schemes and water development/harvesting/village tanks etc. Schemes should be expeditiously handled. Incentives and recognition should be given to local branch managers who are successful in processing such proposals expeditiously. Loan applicants should have easy access to designated higher levels for the resolution of their problems.

The Finance Minister in his Budget Speech has indicated reduction in the rate of interest charged on crop loans between Rs. 15,000 and Rs. 25,000 to 12 per cent from the existing maximum rate of 14 per cent. To ensure credit flow to the share-croppers in the long run, legal provision should be made and enforced for their identification and recording in the records for their identification and recording in the records of rights. They should also be given legal protection from eviction and legal provisions regarding sharing of cost of production and share of produce payable as lease rent be enacted and strictly followed. A group may be appointed representing different regions of the State to prepare forms, procedures, specific administrative and banking regulation steps, so that crop loan and medium term credit requirement of share-croppers can be sanctioned at the village level. These procedures should be notified early. All loans given to subsidised categories should be notified on panchayat ghars to allow transparency in systems.

Agricultural Research and Education

The G.V.K. Rao Committee has suggested a major reform of the agricultural research backup support for Indian agriculture. These suggestions, if judicially examined and implemented, would go a far way in releasing the tremendous potential energy available in the Indian Agricultural Research establishments for resolving the nations agricultural and rural development problems. G.V.K. Rao Committee has suggested that the ICAR should be devoid of political interference. As in the CSIR, the Governing Body of ICAR should truly be professional in nature and should primarily have a membership of scientists relating to agriculture and allied areas. The Governing Body of Council should act as a "Think Tank" providing leadership and guidance in all matters relating to Agricultural Research and Education.

More financial autonomy may be given to ICAR and its national level and other research institutes for streamlining financial procedures. ICAR should be allowed to permit its institutes' projects to retain their own earnings. Financial structure chosen for an institution should be such as to promote flexibility in operation and full autonomy consistent with the required degree of accountability to ensure a good return on the investments made on the institution. It is, therefore, necessary that ICAR should develop its own accounts service and financial rules and also simplify its working procedures on the lines of CSIR, Department of Atomic Energy. ICAR should be encouraged to charge for its services and keep such funds for future research.

Conclusion

In the next phase of Indian Agriculture, as we can see, there are many exciting tasks to be undertaken to develop policies and institutions to raise the income and employment levels of rural India. We can look forward to implementing these exciting vistas.

2

State Policies and Regional Disparity in Indian Agriculture

K. SUBBARAO

Introduction

It is generally agreed that Indian agricultural policy has been successful in stimulating technological change and the production of foodgrains to the point of eliminating the reliance on imports. This success on the production front has not resulted in increasing per capita consumption nor in reducing the percentage of the population surviving on an amount of food that is below the required nutritional intake. The poor may be expected to benefit from the enhanced production of foodgrains provided they get a share in the rising size of the basket of production, and/or food becomes relatively cheaper. In India, nearly 70 per cent of the nation's rural poor are located in the eastern states of Assam, Bihar, West Bengal and Orissa, as well as eastern Uttar Pradesh, and in the central states of Madhya Pradesh and Rajasthan. As such, an analysis of the regional pattern of agricultural growth and the implications of the states' incentive policies for regional equity may shed light on the extent to which poor farmers benefited from the developmental thrust. Furthermore, given the recent sharp rise in the wave of regionalism,[1] I believe an appraisal of the issues relating to regional equity in a large country such as India would be useful in itself. This article makes an attempt in this direction.

I begin by sketching the colonial history and the initial conditions in the next section, which is followed by a discussion of the recent trends in the regional pattern of agricultural output and surpluses. I then examine the regional distribution of gains from the states' major incentive policies for the agricultural sector. The impact of these policies on the poorer eastern states is discussed thereafter. Finally, the article concludes with a few remarks on the changing class character and role of the state in the context of the growing regional differentiation in Indian agriculture.

Initial Conditions: A Brief Historical Perspective

During the colonial period, government policy with respect to land settlement and the extent of commercial penetration into the agrarian economy differed from region to region depending on various considerations such as extraction of raw materials, mobilizing revenues, defence etc. The eastern states were the major contributors to the states' revenues. This extraction of revenues was facilitated by the creation of revenue collectors who had no interest in productive cultivation. These revenue collectors were later converted into absentee landlords, and a *zamindari* system of land tenure was ushered in throughout the eastern states. This led to sub-infeudation and growth of a parasitical class of landlords, merchants and money-lenders (Bhaduri 1983).

By contrast the western region, comprising the present Punjab, Haryana and western Uttar Pradesh as well as coastal Andhra Pradesh and Tamil Nadu, was characterized by a tenurial system based on peasant proprietorship which was more conducive to personal initiative and entrepreneurship. Furthermore, this region benefited from massive public investments in irrigation (Whitcombe 1983). These regional differences in tenurial conditions and the irrigation policy of the government resulted in the emergence of small enclaves of growth and vast areas of low productivity and stagnation.

The regional differences described above in colonial policy are also important from the viewpoint of the evolution of agrarian class formations in India. Much of the observed unevenness in the agrarian class configurations can be attributed to the underlying historical forces and colonial policy. In the eastern region there has been a broad polarization of classes into parasitical classes on the one hand and a vast mass of landless labourers and small and marginal farms on the other. The former classes had no interest in productive investments in agriculture, while the latter had no capacity (i.e. means) to invest. By contrast, in the north-western region and the coastal southern region the main agrarian class was comprised of the peasant proprietors and of occupancy tenants with heritable rights in land who had strong direct involvements in agriculture; intermediate parasitical classes were few. Thus the class basis of economic power in the two regions at the time of independence was different: in the eastern zone the class of rent-seeking landlords who often combined money-lending and trading wielded the power, while in the north-west the position of peasant proprietors was strong and that of occupancy tenants was strengthened by the tenancy laws. Both with respect to economic power of the agrarian classes and conditions of agricultural production, the post-independence era began with manifest regional inequity.

Recent Trends in Regional Inequity

Before examining the factual situation relating to inter-state disparities, a discussion on the selection of the state as a unit of analysis is in order. Since the states in India constitute heterogeneous economic regions, treatment of regional disparities at the state level may not reflect the pattern of regional distribution within each state. It would be desirable to conduct the analysis taking the district as the unit of observation. However, this is not possible because important disaggregated data are not available for certain crucial components (especially for the post-Green Revolution period) at the district level. Hence my analysis has had to be confined to the state level.

Despite this limitation, a study of regional disparities at the state level is of considerable importance for a large country like India, especially because the reduction in inter-state disparities has been one of the well recognized social objectives. The state being a political unit and agriculture being a state-managed concern under the Constitution, state governments play an important role in the allocation of resources for agriculture (Rao 1975:91), and farmers' lobbies from the different states actively influence the establishment of input and output prices. Moreover, it is possible to broadly classify the states into two types in terms of the evolution of agrarian structure discussed in the previous section: viz. the areas under the former *zamindari* (permanent) settlement in the eastern states, and the areas under the former *ryotwari* system (Punjab, Haryana and western Uttar Pradesh, Tamil Nadu etc.).

The factual situation pertaining to regional disparity after the introduction of new seed varieties is summarized in Table 2.1. The compound rates of growth of food grain production based on the

TABLE 2.1 : Changes in the Relative Position of the Different States in Output and Procurement of Foodgrains. 1967-81

States	*Per cent Growth Rates Based on Triennial Average of Production with Central Years*		*Relative Share of States in Average Production of Foodgrains for the Years*		*Shares in the Incremental Output of Foodgrains*	*Relative Shares of States in Public Procurement of Rice and Wheat*		*Average Procurement as a % of Average Production (Rice and Wheat)*	
	1968-69 to 1980-81	*1968-69 to 1975-76*	*1967-68 to 1969-70*	*1978-79 to 1980-81*		*1967-68 to 1969-70*	*1978-79 to 1980-81*	*1967-68 to 1969-70*	*1978-79 to 1980-81*
(1)	(2)	(3)	(4)	(5)	(6)	(7)	(8)	(9)	(10)
Andhra Pradesh	3.29	3.50	7.7	8.4	10.6	8.2	6.3	8.9	10.5
Assam	0.87	0.93	2.3	2.0	0.7	3.5	0.2	8.3	1.2
Bihar	0.90	0.83	8.9	7.6	3.0	1.0	0.6	0.9	0.9
Gujarat	2.80	0.23	3.2	3.6	5.0	0.5	Neg.	2.4	0
Haryana	3.43	0.49	4.1	4.9	7.6	7.4	15.6	18.5	39.7
Karnataka	2.24	2.04	5.5	5.9	7.2	1.3	0.7	3.1	3.1
Kerala	–0.14	0.54	1.4	1.1	0.1	1.7	Neg.	6.7	0.1
Madhya Pradesh	0.38	0.90	10.5	8.7	2.6	7.5	2.5	7.2	4.7
Maharashtra	6.11	3.54	7.5	8.3	11.4	3.7	0.3	10.1	1.1
Orissa	0.43	–1.11	5.3	4.3	0.9	4.9	0.6	5.8	1.8
Punjab	5.67	4.15	6.6	9.9	21.2	31.5	55.2	33.5	59.4
Rajasthan	1.49	2.16	5.5	5.4	5.3	0.6	1.2	2.3	4.9
Tamil Nadu	1.38	0.44	6.3	5.8	4.1	11.6	1.1	13.7	2.5
Uttar Pradesh	1.99	0.11	18.0	17.7	16.7	9.3	14.6	5.0	10.5
West Bengal	0.88	1.42	7.3	6.5	3.8	7.4	1.1	6.0	1.7
Coefficient of Variations(%)			57.0	57.0	88.0	112.0	208.0	91.0	173.0

Sources : Columns (2) and (3), Sawant 1983; other columns, calculated from the *Bulletin of Food Statistics* (various issues).

triennial average of production, taking the years from 1968-69 to 1980-81 as the central years, show that the states in the eastern and central regions (Assam, Bihar, Orissa, West Bengal and Madhya Pradesh) recorded much lower rates than most others, with the exception of Kerala which specializes in non-food grains (see columns [2] and [3]). In general, the relative contribution of the eastern and central states to India's total foodgrain production declined over the 1970s. Five states (Punjab, Haryana, Uttar Pradesh, Andhra Pradesh and Maharashtra), with a share of 40.5 per cent of India's total gross cropped areas, have accounted for 68 per cent of the incremental output of foodgrains since the Green Revolution. In contrast, the eastern and central states (excluding Uttar Pradesh), where India's rural poverty is concentrated, accounted for only 17 per cent of the *incremental* output, but their share in gross cropped area was as high as 41 per cent. There has also been a dramatic rise in the concentration of surpluses, as is revealed by the figures relating to public procurement of rice and wheat (see column [7]- [10]). Barely three states (Punjab, Haryana and Uttar Pradesh), with a share of 21.3 per cent of India's total gross cropped area, accounted for as much as 86 per cent of public procurement of rice and wheat, averaged for the years 1978-79 to 1980-81.

Since the inception of Green Revolution, there has been substantial growth in commercial agriculture in the advanced northwestern states and in Andhra Pradesh and Tamil Nadu in the South, where the incremental output and surpluses were concentrated, while peasant agriculture, which is pervasive in the eastern states and the semi-arid zone, participated little in this impressive national performance. It is interesting that while the coefficient of variation of the states' relative share in *output* has remained the same since the mid-1960s (57 per cent), that of public procurement (a proxy for *surpluses*) rose sharply from 112 to 208 per cent over this period.

An interesting recent development in the wheat economy of the north-western states needs comment. Recent trends in the physical yields of wheat suggest a decline in the response to technological inputs such as fertilizer. According to Grewal & Rangi (1983: 517-8), one kilogram of fertilizer nutrients yielded about eight kilograms of wheat during the period from 1967-68 to 1976-77, when the average dose of fertilizer used for wheat was about one-third of the recommended dose. During the succeeding period ending in 1981-82, this response was only 5.6 kilograms of wheat per kilogram of fertilizer application, when its use was two-thirds of the recommended dose. One can, therefore, expect that further use of fertilizer will add still less to wheat productivity in the Punjab. There may be many reasons for the decline in yield responsiveness, including the diversion into wheat of marginal lands not suitable for wheat cultivation.

As long as productivity per hectare rises, the commercial profitability per hectare may be expected to rise despite relative stability in the unit prices of inputs and outputs. However, when yield response to input use begins to decline, commercial profitability per hectare is threatened. The recent agitations for strengthening the incentive framework of the state may be at least in part due to this observed greying of the Green Revolution in wheat.

The state's incentive policies such as subsidies have a long history in India, and the agricultural sector today indeed abounds in a bewildering variety of incentives and special assistance programmes. The next section deals with the regional distribution of benefits from the state's major incentive policies.

Incentive Policies and Regional Equity: An Overview

It is indeed difficult to explicitly identify all the state-subsidized programmes of the Central and state governments for the agricultural sector. The more easily identifiable forms of state incentive policies are: (i) price supports, (ii) input subsidies, and (iii) subsidized institutional credit.

The rationale behind these subsidized programmes is obvious: the primary objective is of course to encourage the diffusion of new technology. By their very nature, incentives which involve heavy subsidies are expected to be short-term measures, and are not meant to be substitutes to long-term improvements in social and physical infrastructure, which alone ensures the viability of new technology (Hayami, Bennagen & Barker 1977).

Even in the short-term, a sharp rise in subsidies may cut into the resources available for productive investments and thus constitute an impediment in the realization of the long-term objective. In India total subsidies for agriculture and rural development programmes have risen from about Rs. 10,000 million in 1974-75 to about Rs. 25,000 million by 1980-81 *at constant prices* (see Table 2.2, columns [2] and [3]). It is interesting that the ratio of public expenditures on direct assistance (input subsidies) to public expenditures on crop production programmes and irrigation declined from 0.45 in 1974-75 to 0.22 by 1976-77, but again rose sharply to 0.48 by 1980-81.

TABLE 2.2 : Growth in Subsidies for Agriculture and Rural Development

Years	*Total Subsidies for Agriculture and Rural Development*		*Direct Subsidies for Agricultural Inputs (fertilizer, irrigation power, in mlns of Rs.*[a]	*Ratio of Agricultural Input Subsidies to Total Subsidies (col. 4 ÷ col.2)*	*Public Investments in Crop Production and Irrigation (in mlns of Rs.)*	*Ratios of Direct Input Subsidies to Public Investments (col. 4 ÷ col. 6)*
	at current prices (in mlns of Rs.)	*at constant prices (in mlns of Rs.)*				
(1)	(2)	(3)	(4)	(5)	(6)	(7)
1974-75	7,643	10,630	7,203	0.94	15,861	0.45
1975-76	7,852	11,385	6,772	0.86	19,757	0.34
1976-77	6,591	8,942	5,148	0.78	23,429	0.22
1977-78	7,951	10,369	5,917	0.74	27,186	0.22
1978-79	14,808	19,090	10,435	0.71	32,129	0.33
1979-80	22,640	25,288	14,638	0.65	31,479	0.47
1980-81[b]	25,576	25,576	15,838	0.62	32,915	0.48

a. Includes (i) subsidies on domestic and imported fertilizers, and on freight and distribution; (ii) revenue expenditures on public major, medium and minor irrigation systems, including returns on capital, less revenues from water charges; and (iii) operating losses of the State Electricity Boards for rural electrification (operating costs, including interest less revenues).

b. These are revised Budget estimates, except for the fertilizer subsidy which is preliminary. The rural electrification subsidy is a World Bank estimate.

Sources : Accounts and budget documents of the Central and State governments (1974/75-83), as compiled in Government of India.
Budget of the Central Government.
Bulletin of the Reserve Bank of India (various issues).
State Electricity Board (annual) reports, published in the state captials
Annual Reports of the Food Corporation of India, 1974-82 (New Delhi).

Price Supports and Regional Benefits

Price supports have now become important policy instruments for inducing farmers to adopt new technologies. Furthermore, full average cost has now been more or less recognized as the least unsatisfactory criterion for fixing support prices (Krishna 1983).

From the viewpoint of regional equity, price supports need to be analyzed against the backdrop of the wide variations between regions in the costs of producing rice and wheat per hectare and per quintal, and the growing regional concentration of procurement noted earlier. The cost differences between the technologically advanced Punjab and the less developed states are especially striking.[2] For example, the total cost per hectare for paddy was 9 per cent higher in the Punjab than in Andhra Pradesh, but the cost per quintal was 32.4 per cent lower, owing to much higher yields caused by a more favourable agro-climatic environment in the Punjab for the growth of high-yield paddy (see Table 2.3). The cost differences for wheat between the highly developed Punjab and the less developed state of West Bengal are similar (Subbarao, 1982). There are also interesting differences in the structure of costs: for example, the labour costs per hectare were much lower in the Punjab than in other states for both rice and wheat, reflecting the widespread use of labour-saving cultivation techniques in the Punjab.

The distribution of benefits from support prices over the different states is a function of (a) support prices exceeding the unit cost of production, and (b) the share of each state in the total procurement for all of India. Published cross-section data on the costs of rice and wheat are only available for the year 1975-76 for a large number of states. The amounts by which support prices exceeded the cost of production in different states for these two products are given in Table 2.4, columns (1) and (5). Multiplying these columns by the actual procurement of the product in each state, I arrived at the absolute benefit accruing to each state (not shown in the table). To highlight inter-state disparities, I computed the relative shares of different states in the benefits for all of India; these are shown in columns (2) and (6), along with benefits per hectare and per unit of output (columns [3], [4], [7] and [8]).

Benefits per hectare under rice and wheat cultivation, as well as per unit of output, largely accrued to the states of Punjab, Haryana, Uttar Pradesh and Tamil Nadu. Producers in the rest of the states did not

TABLE 2.3 : Inter-State Differences between Andhra Pradesh and the Punjab in Average Costs of the Production of Paddy, 1978-79

States	*Total Cost of Paddy*[a]	*Purchased Material Inputs*[a]	*Labour Inputs (hired plus family)*[a]	*Rent (owned) Plus Actually Paid*[a]	*Other Fixed Costs*[a]	*Yield of Paddy*[b]	*Average Cost of Production*[c]
Andhra Pradesh							
Costs	3,083	1,289	849	846	99	30.7	100.4
% of Total	100.0	41.8	27.5	27.4	3.2	—	—
The Punjab							
Costs	3,361	1,911	418	763	269	49.5	67.9
% of Total	100.0	56.8	12.4	22.7	8.0	—	—
% of Differences in Cost between Andhra Pradesh and the Punjab	9.0	48.2	–50.7	–9.8	171.7	61.2	–32.4

a. In Rs. per hectare.
b. In quintals per hectare.
c. In Rs. per quintal.
Source : Subbarao, 1982.

TABLE 2.4 : Inter-State Distribution of Benefits from Support Price Policy, 1975-76

State	Paddy				Wheat			
	Amount by which Support Prices Exceeded Cost of Production per Quintal (in Rs.)	*Relative Share in Aggregate Benefits*	*Aggregate Benefit per Hectare of Area Under Paddy (in Rs.)*	*Aggregate Benefit per Unit of Output Paddy Produced (in Rs. per tonne)*	*Amount by which Support Prices Exceeded Cost of Production per Quintal (in Rs.)*	*Relative Share in Aggregate Benefits*	*Aggregate Benefit per Hectare of Area Under Wheat (in Rs.)*	*Aggregate Benefit per Unit of Output Wheat Produced (in Rs. per tonne)*
	(1)	(2)	(3)	(4)	(5)	(6)	(7)	(8)
Andhra Pradesh	–4.98	nil	nil	nil	—	—	—	—
Bihar	+4.83	1.27	0.78	0.90	+12.16	1.53	3.38	2.70
Haryana	—	—	—	—	+21.35	22.78	62.12	37.94
Karnataka	+4.64	7.3	16.09	7.90	—	—	—	—
Madhya Pradesh	—	—	—	—	+13.15	4.47	5.14	6.60
Orissa	+6.14	8.6	4.93	4.76	—	—	—	—
Punjab	+6.10	—	104.36	76.05	+5.55	32.99	48.15	22.96
Rajasthan	—	43.6	—	—	+20.03	9.13	18.62	16.19
Tamil Nadu	+6.53	34.6	32.06	14.82	—	—	—	—
Uttar Pradesh	+0.94	4.3	2.10	2.46	+15.31	29.80	16.64	14.86
West Bengal	–5.63	nil	nil	nil	–8.99	nil	nil	nil

Source : Computed by the author. The procedure for deriving columns (2), (3), (4), (6), (7) and (8) is described on p. 23

benefit much from the support price policy either because their unit costs were higher and (or) because their share in public procurement was negligible.

Regional Benefits from Input Subsidies

(i) *Fertilizer.* In terms of treasury cost, fertilizer subsidies accounted for the massive amount of Rs. 9,000 million in 1983-84 (Financial Express 1984). This amount consisted of the difference between the domestic cost of fertilizer and the actual (subsidized) price at which it was supplied to the farmer.[3]

That the prevailing fertilizer subsidies violate regional equity is evidenced by Table 2.5, which confirms the disproportionately large share of the states with high irrigation ratios (see columns [3] and [6]). Four states (Uttar Pradesh, Punjab, Andhra Pradesh and Tamil Nadu), which accounted for about half of the nation's gross irrigated area, claimed 55 per cent of the fertilizer subsidies but accounted for only 30 per cent of all of India's gross cropped area. Five states (Madhya Pradesh, Rajasthan, Bihar, Orissa and West Bengal) accounted for about 40 and 20 per cent of all of India's gross cropped and gross irrigated area respectively, but their share in fertilizer subsidies was only 16 per cent.

The pre-empting of fertilizer supplies by the states which are better off has led to low fertilizer use in the rain-fed eastern part of India and in the semi-arid tropics, where the potential is least tapped. The National Council of Applied Economic Research (1979) estimated the potential demand for fertilizer in each state for 1986-87. A comparison of actual consumption in 1981-82 with the potential estimated by NCAER (see Table 2.6) shows that the gap between the two is substantial for the eastern and central states. In the southern and northern states, nearly 50-60 per cent of the potential for 1986-87 was realized by 1981-82; for the Punjab, 81 per cent of the potential was realized.

(ii) *Irrigation.* The subsidies on irrigation consisted of the difference between the revenue from the expenditures (including a nominal return on capital invested) on the publicly operated major, medium-sized and minor irrigation systems. A substantial part of the subsidies is on the canal irrigation system.

Water charges collected from farmers differ among all the major canal systems depending on the age of the canal system. An accurate

TABLE 2.5 : Relative Shares of the States in Input Subsidies

States	*Relative Shares in Input Subsidies (average) for the Period 1979-80 to 1981-82*				*Relative Shares in 1978*		*Total Input Subsidies*		*Irrigation Water Charges as a % of Gross Value Added (in Rs. per hectare)*
	Irrigation subsidy	*Power subsidy*	*Fertilizer subsidy*	*Total subsidies on agricultural inputs*	*Gross cropped area*	*Gross irrigated area*	*Per hectare of gross cropped area (in Rs.)*	*Per unit of food grains (in Rs. per tonne)*	
	(1)	(2)	(3)	(4)	(5)	(6)	(7)	(8)	(9)
Central									
Madhya Pradesh	12.9	2.49	3.5	6.07	12.60	5.07	44.35	81.26	3.6 (W)
Rajasthan	11.8	4.46	2.7	6.16	10.14	7.26	53.00	142.93	2.7 (W)
Uttar Pradesh	12.0	20.67	20.8	18.04	14.08	22.24	119.25	120.49	3.8 (W)
East									
Assam	1.8	0.04	0.1	0.61	1.92	1.20	29.49	37.17	NA
Bihar	7.0	3.03	3.8	4.53	6.59	7.79	59.00	78.75	1.91 (P)
Orissa	3.9	0.22	1.4	1.76	4.79	3.33	37.55	50.00	NA
West Bengal	3.1	0.65	4.6	2.75	4.56	3.24	72.49	55.06	3.7 (P)
North									
Haryana	2.4	6.64	4.2	4.53	3.20	6.26	143.13	115.99	0.77 (W)
Jammu & Kashmir	0.5	0.30	0.3	0.36	0.57	0.85	61.37	43.16	NA
Punjab	2.7	10.76	12.2	8.75	3.84	11.58	234.18	103.34	1.1 (W)
South									
Andhra Pradesh	8.8	8.16	11.7	9.55	7.60	9.86	115.18	135.47	2.33 (P)
Karnataka	7.6	3.63	7.0	6.02	6.45	3.61	83.18	128.35	0.60 (P)

Table 2.5 (*Contd*).

	(1)	(2)	(3)	(4)	(5)	(6)	(7)	(8)	(9)
Kerala	1.0	0.77	1.9	1.23	1.67	0.74	113.77	135.38	NA
Tamil Nadu	4.9	17.72	10.3	11.27	4.45	8.03	233.26	240.54	NA
West									
Gujarat	6.8	10.08	6.9	8.03	6.02	4.07	116.25	259.42	NA
Maharashtra	12.6	10.37	8.1	10.34	11.51	4.85	77.17	147.71	1.6(P)

(W) = wheat; (P) = paddy; NA = not available.

Source : Computed by the author. For the procedure adopted see the text.

TABLE 2.6 : Actual and Potential Fertilizer Use in Different States

State	*NCAER Estimated Potential Demand for Fertilizer, 1986-87*		*Actual Fertilizer Consumption, 1981-82*		*Actual (1981-82) as % of Potential (1986-87)*
	Total demand	*Relative share*	*Total consumption*	*Relative share*	
Central					
Madhya Pradesh	789.7	5.77	236.3	4.02	29.9
Rajasthan	490.0	3.58	138.0	2.35	28.2
Uttar Pradesh	3,391.5	24.76	1,269.6	21.62	37.4
East					
Assam	123.5	0.90	10.8	0.18	8.7
Bihar	1,678.8	12.26	205.3	3.50	12.2
Orissa	182.0	1.33	82.0	1.40	45.1
West Bengal	1,036.1	7.56	258.4	9.40	24.9
North					
Haryana	429.5	3.14	251.6	4.29	58.6
Jammu & Kashmir	77.5	0.57	21.6	0.37	27.9
Punjab	1,009.7	7.37	820.5	13.98	81.3
South					
Andhra Pradesh	1,067.5	7.79	655.5	11.16	61.4
Karnataka	652.0	4.76	383.6	6.53	58.8
Kerala	162.3	1.18	94.8	1.61	58.4
Tamil Nadu	874.3	6.38	512.6	8.73	58.6
West					
Gujarat	742.7	5.42	401.4	6.84	54.0
Maharashtra	989.2	7.22	529.1	9.01	53.5
Total	13,696.3	100.00	5,871.0	100.00	

Source : NCAER (1979): *Passim,* and fertilizer statistics for India, 1982-83.

interregional distribution of subsidies would require data on the canal-irrigated areas, along with data on the age composition of different systems. In the absence of data on the latter, I have distributed the subsidies in proportion to the area, listed state-wise, under canal irrigation for the total area (for all of India) under canal irrigation. The relative shares of the states in subsidies thus derived are shown in column (1) of Table 2.5.

The states with predominantly older irrigation systems (Punjab, Tamil Nadu and Andhra Pradesh) accounted for a small proportion of the subsidies in comparison with Uttar Pradesh, Rajasthan, Madhya Pradesh and Maharashtra, where relatively newer canal irrigation projects were located (see column [1] in Table 2.5). However, it needs to be stressed that water rates as a proportion of gross value added per hectare are extremely low in the states which historically have been endowed with canal irrigation for several decades (see column [9] in Table 2.5). To the extent that these states (Punjab, Tamil Nadu and Andhra Pradesh) are the richest, in terms of both per capita income and per capita agricultural output, there appears to be a strong case for upward revision of the water rates charged in them, with a view to reducing the *overall* subsidy on irrigation.

(iii) *Power.* The subsidies on power represented the operating losses of the State Electricity Boards in the course of rural electrification (operating costs, including interest *less* revenues), Gallerson (1979) estimated the revenue and cost of the electricity supply for each region for different uses. His estimates show that out of all the sectors, the agricultural sector is the most heavily subsidized so far as power tariffs are concerned. I have used Gallerson's regional data for the states falling within each region, along with the state data on power consumption for agricultural operations, to arrive at the relative distribution of subsidies over the different states.

Five states (Punjab, Tamil Nadu, Maharashtra, Gujarat and Uttar Pradesh) accounted for about 70 per cent of all of India's subsidies on power consumption by the agricultural sector (see column [2] in Table 2.5). Four major eastern states (Bihar, West Bengal, Assam and Orissa) accounted for only 4.3 per cent of all of India's subsidies on agricultural power consumption.

Total input subsidies per hectare of gross cropped area were higher for the Punjab and Tamil Nadu (see column [7] in Table 2.5) but much lower for the eastern and central states, with the exception of Uttar

Pradesh (which, as is well known, is a large and heterogeneous state). Table 2.7 summarizes how government input subsidies were distributed over the different states. Even when one considers the subsidy per unit of foodgrain produced, which takes account of the productivity effect, it is interesting that the output produced was heavily subsidized in Tamil Nadu, Gujarat, Andhra Pradesh, Karnataka, Haryana, Punjab, Uttar Pradesh and Rajasthan. The output produced in the poorer, eastern states was the least subsidized.

TABLE 2.7 : The Changing Position of Different States in Terms of Input Subsidies : A Summary

Range of Subsidy (in Rs.)	*Subsidy per Hectare of Gross Cropped Area*	*Subsidy per Tonne of Food Grain Output Produced*
0-50	Madhya Pradesh, Assam, Orissa	Assam, Orissa, Jammu and Kashmir
51-100	Rajasthan, Bihar, West Bengal, Jammu and Kashmir, Karnataka, Maharashtra	Madhya Pradesh, Bihar, West Bengal
101-150	Uttar Pradesh, Gujarat, Kerala, Haryana, Andhra Pradesh	Rajasthan, Uttar Pradesh, Haryana, Punjab, Andhra Pradesh, Karnataka, Kerala, Maharasthtra
151-200	—	—
201+	Punjab, Tamil Nadu	Tamil Nadu, Gujarat

Source : Compiled from Table 2.5

Subsidized Institutional Credit

Finally, the provision of subsidized institutional credit has been an important component of state policy for indirectly promoting agricultural growth by improving the financial ability of farmers to purchase modern inputs. The share of institutional credit going to the agricultural sector for purchasing modern inputs has risen sharply over the last ten years. Expansion of commercial banks and setting up regional rural banks has contributed to easing the flow of institutional credit for agriculture. In particular, farmers are supplied with production credit at rates of interest which are much lower than the market rates of interest.

Table 2.8 summarizes the past and present position of different states in respect of *total* institutional credit (from cooperative as well as commercial banks). A number of states (including Haryana) were receiving a credit per hectare of less than Rs. 50 in 1967-68. Of these,

TABLE 2.8 : Summary Statement of the Changes in the Relative Position of the States in Institutional Finance of the Agricultural Sector, 1967-68 to 1977-78

Range of Institutional Credit per Hectare of Gross Cropped Area (in Rs.)	*1967-68*	*1977-78*
0-50	Andhra Pradesh, Assam, Bihar, Haryana, Jammu and Kashmir, Karnataka, Madhya Pradesh, Orissa, Rajasthan, Uttar Pradesh, West Bengal	Assam
51-100	Gujarat, Maharashtra, Punjab	Bihar, Jammu and Kashmir, Madhya Pradesh, Orissa, Rajasthan
101-150	Tamil Nadu, Kerala	Uttar Pradesh, West Bengal
151-200	—	Andhra Pradesh, Maharashtra
201-300	—	Gujarat, Karnataka, Haryana and Punjab
300+	—	Tamil Nadu, Kerala

Note : The Institutional credit per hectare that was supplied to the agricutural sector in 1967-68 and 1977-78 is estimated in Subbarao 1984. This table forms the basis for deriving the distribution of the states over the credit categories in a summary form.

Andhra Pradesh, Haryana and West Bengal improved their position substantially, while the others (including Assam) moved only to the next higher credit category (Rs. 50-100 per hectare). Punjab, Haryana, Gujarat, Tamil Nadu, Kerala and Karnataka are now in the two top categories of high credit per hectare. Considering the extremely slow upward mobility of the eastern and central states (with the exception of West Bengal), it appears that the credit per hectare *in real terms* may have virtually remained the same in these states over the period 1967-77.

The evidence examined above suggests that the benefits emerging from a combination of the state's incentive policies—comprising price supports, input subsidies and subsidized credit—have accrued disproportionately to the prosperous states, which *already* have access to water, power and fertilizer. This has left fewer public resources for new public investments designed to spread these critical inputs in a regionally more equitable manner. The impact of such a regionally inegalitarian incentive framework has been quite serious so far as the eastern states are concerned. Let us now turn to a discussion of them.

Extraction of Surplus, Dependence and Poverty in the Eastern States

The backward eastern states, in sharp contrast to the Green Revolution belt, have fallen into a low-level equilibrium with stagnating yields, rising poverty and dependence. Although explicit taxation of the agricultural sector—an important feature during the colonial period—diminished after independence, the impoverishment of the rural poor in these states continued as a result of food-price management. A combination of two factors may be identified: (a) a rise in the real price of food grains (at least until the mid-1970s) and (b) the eastern states' increasing dependence for food grains on other states, owing to the slow growth of agricultural output in relation to the explosive rise in population-induced demand. Thus, the proportion of food grain imports in total imports into Bihar, for example, was less than 10 per cent in 1961, but it rose to over 50 per cent by 1975.[4] During the same period, the real price of food grains rose substantially.

The principal exports of the eastern states (Orissa and Bihar in particular) continued to be minerals. The price index for coal and metallic minerals rose by about 110 per cent over the period 1961-75, whereas the consumer price index for foodgrains rose by 257 per cent in Bihar, 264 per cent in Assam, 284 per cent in Orissa and 252 per cent in West Bengal. Clearly, the terms of exchange between mineral-surplus states and foodgrain-surplus states deteriorated for the former during the period 1961-75. It should be stressed that the rural poor in the eastern states were virtually out of the purview of publicly distributed subsidized food, which was confined to urban areas in general (with the exception of Kerala).

It might be argued that the rising food prices may have contributed to the rise in the *supply* of foodgrains in the eastern states as much as in other states. Available studies, however, have categorically shown that the technological lag in the eastern states is entirely due to the non-availability of technology suited to the agro-climatic environment of this region (Rao 1976), which has not been changed or controlled by appropriate comprehensive water management through either private or public investments.

Private investments in agriculture have been tardy in the eastern states, essentially because of the low investment capacity of the farmers, the small size of farms and their non-viability and high weather uncertainty. At the same time, returns to private investments in trading

TABLE 2.9 : Irrigation in Selected Backward and Advanced States

States	*Major and Medium-sized Irrigation*	*1,000 Hectares of Irrigation*		*Minor Irrigation*			*Energization of pumpsets in Sixth Five Year Plan (in 1,000 nos.)*			
							1980-81		*1981-82*	
	Ultimate potential	*Potential created upto 1980-81*	*Potential created as a % of ultimate potential*	*Ultimate potential*	*Potential created upto 1980-81*	*Potential created as a % of ultimate potential*	*Target*	*Achieve-ment*	*Target*	*Achieve-ment*
(1)	(2)	(3)	(4)	(5)	(6)	(7)	(8)	(9)	(10)	(11)
Bihar	6,500	2,525	38.85	5,900	2,464	41.76	35.0	7.7	25.2	4.2
Orissa	3,600	1,468	40.78	2,300	708	30.78	5.0	3.2	11.0	2.0
Uttar Pradesh	12,500	6,281	50.25	13,200	9,595	72.69	—	—	—	—
West Bengal	2,310	1,559	67.50	3,800	1,495	39.34	1.0	0.8	13.1	0.8
Punjab	3,000	2,337	77.90	3,550	2,956	83.27	—	—	—	—
Tamil Nadu	1,500	1,197	79.40	2,400	1,915.5	79.81	—	—	—	—
All India	58,475	27,455	46.95	54,927	31,400	57.17	410.00	381.3	422.5	324.7

Source : Government of India, Planning Commission 1982: 50-51, columns (2)-(9); and Rao 1983; 16-9.

and money-lending are high, essentially due to massive poverty and fluctuating weather conditions which necessitate frequent consumption loans. Thus, the overall situation increases the prospects of high profits from speculative trading in food grains during bad years and trading in scarce inputs during good years.

The nature of the development problem in the eastern states as outlined above clearly suggests that the basic growth impetus must come from non-price policies related to massive *public* investments in irrigation and flood control, agro-distributional infrastructure, structural reforms and institutional changes. Unfortunately, the record of state policies in these areas has so far been disappointing, as can be seen from the low levels of utilization of irrigation potential in these states (see Table 2.9). It is interesting that while the infrastructure necessary for extracting minerals, tea and jute *out of* these states has grown enormously because the necessary massive public investments have been made, the infrastructure needed for transferring agricultural technology, resources and basic needs *into* these states has *not* been built through public investments. As a consequence, even the remaining resources—labour and capital—are migrating out of these states, thus creating further problems in raising the overall productivity and levels of living in them.

Since state policies played such a crucial role in the continued backwardness of the eastern states, I shall now briefly turn to a discussion of the class character and role of the state in the context of this growing regional differentiation in agriculture.

The Role of the State

The discussion of the state's role will admittedly be very sketchy. There are several views on the class character of the Indian state, one of which is that since independence, the dominant political coalition has been between landlords and commercial producers on the one hand and the monopolistic industrial bourgeoisie on the other (Mitra, 1977). This alliance ensures that the former receives benefits in the form of protected agricultural prices, input subsidies and other institutional rents, apart from postponed land reforms. The latter enjoy being completely independent of industrial, trade and licensing policies as well as fiscal, monetary and foreign exchange management.

The rise of another dominant class, the bureaucracy, has been noted by many scholars (Bardhan, 1983). This class was already quite large at the time of independence and has increased manifoldly over the three decades since then. Rao observes:

> The economic ladder contributed by Independence and governmental support has led to a spurt of money power to the upper and middle classes and given them a place in the society and influence that far exceeds that exercised by the professional classes in the pre-Independence period. The growth in number and influence and power of this commercial, industrial, bureaucratic and political class is an outstanding feature of post-Independence development. (1984: 4)

The economic interests of these dominant classes—rich farmers, the industrial bourgeoisie and the bureaucracy—may often conflict, but there are reasons to believe that the rich farmer class has strengthened its position in recent years. The proportion of farmers in the Central Parliament rose from about 15 per cent in the early 1950s to about 40 per cent in the 1980s. There has also been a sharp regional concentration of the rich farmer class after the Green Revolution. Thus in the Punjab, 36 per cent of all rural households owned more than Rs. 20,000 in assets, while in Orissa this proportion was about 5 per cent (Bardhan). The changing qualitative difference in the ruling classes also should be mentioned: in the early 1950s (the Nehru era), the Central Parliament included several leaders with a certain national consciousness and vision, whereas in the period following the Green Revolution, the number of such leaders dwindled, the power bloc increasingly became atrophied, and the state came under the influence of the dominant classes. Although the rich farmers acquired a position of dominance, the state became a rent-seeking instrument not only for this class but also for the other dominant classes in various ways (the industrial bourgeoisie enjoyed access to licenses, export subsidies, raw material subsidies, sheltered markets etc., while the bureaucracy enjoyed the power to grant these).

Whether or not the situation described above reflects a well articulated 'alliance' between the dominant classes, the net result of the instrumentalist demands on the state has been the slow growth of central revenues due to the complete absence of taxation on rising agricultural incomes, as well as tax evasion and avoidance; a sharp escalation in public subsidies; and a fall in public expenditures in critical areas such as flood control and irrigation. The chief victims of this rent-seeking phenomenon have been the rural poor in the eastern and central states.

Summary

This article shows that ever since the introduction of new seed varieties in the mid-1960s, there has been a sharp accentuation of regional disparities in Indian agriculture. A substantial proportion of the incremental production originated in five or six advanced states, and the poorer eastern states and the semi-arid zone participated little in this impressive national performance. This growth in the advanced states was aided by state-induced market price distortions resulting in heavy subsidies and the easy availability of subsidized public funds. The few prosperous states which *already* had access to water, power and fertilizer cornered a disproportionate share of public subsidies, leaving fewer public resources for new public investments designed to spread these political inputs in a regionally more equitable manner. The net result has been a deficiency of public investments in critical areas such as flood control and water management in the poorer eastern and central states. The prevailing atmosphere of institutional rent-seeking from the state by the dominant classes is beginning to severely limit the state's capacity to invest in the very items that contribute to accelerated agricultural growth and regional equity. The consequence of this policy failure to halt regional disparities has been massive poverty, malnutrition and a shrinkage of effective demand for wage goods in the poorer states, alongside the accumulation of food stocks with the government and a rise in regional tensions.

Note: Reproduced from *Development and Change,* Vol. 16, No. 4, October, 1985, pp. 523-545.

NOTES

1. A recent upsurge of regional tensions led to a situation in which the traditional Marxian-type class conflict was complemented if not supplanted by a conflict between prosperous regions and lagging regions *vis-à-vis* the Central Government.
2. The cost concept adopted includes the imputed value of family labour as well as the imputed value of owned land.
3. It should be pointed out that these subsidies are in addition to the subsidies

given to the fertilizer industry for the naphtha which goes into producing fertilizer. During 1981-2, naphtha was imported at the rate of Rs. 2,900 per ton but supplied to the fertilizer plants at the administered price of Rs. 1,723. The electricity supplied to fertilizer plants is also subsidized; furthermore, the freighting of both inputs by rail to fertilizer plants and the movement of fertilizer from the plants to the farmers are also heavily subsidized. Thus, practically everything associated with the manufacture and sale of fertilizer is subsidized.

4. See Government of India, Department of Commercial Intelligence and Statistics 1975 for the twelve months ending in March 1975.

REFERENCES

Bardhan, P. (1983), The State, Classes and Economic Growth in India (unpublished Radhakrishnan Memorial Lecture, All Souls College, Oxford).

Bhaduri, A (1983): *The Economic Structure of Backward Agriculture,* Delhi Macmillan.

Bhalla, G.S. & Y.K. Alagh (1979), *Performance of Indian Agriculture,* Delhi Sterling Publishers.

Bulletin of Food Statistics (various years) (New Delhi Government of India, Ministry of Agriculture, Directorate of Economics and Statistics, Controller of Publications).

Bulletin of the Reserve Bank of India (1974-83), various issues (Bombay).

Financial Express, (New Delhi) 2 April 1984.

Food Corporation of India (1974-82), *Annual Reports* (New Delhi, Food Corporation of India).

Gallerson, M.W. (1979), Marginal Cost-Based Electricity Tariffs. Theory and Case Study of India, *Indian Economic Review,* XIV (2) 169-76.

Government of India (1974/5-83), *Budget of the Central Government*, (New Delhi, Government of India.

Government of India, Department of Commercial Intelligence and Statistics (1975, *Accounts Relating to be Indian (Rail and River-Borne) Trade Consignments of India* (Calcutta, Government of India, Department of Commercial Intelligence and Statistics).

Government of India, Planning Commission (1982 *Annual Plan, 1982-83* (Delhi).

Grewal, S.S. & P.S. Rangi (1983), 'An Analytical Study of Growth of Punjab Agriculture, *Indian Journal of Agricultural Economics,* XXXVIII(4) 509-19.

Krishna, R. (1983), 'Some Aspects of Agricultural Growth. Price Policy and Equity in Developing Countries. *Food Research Institute Studies,* XVIII(3) 239-61.

Mitra, A. (1977), *Terms of Trade and Class Relations,* (London Frank Case).

National Council of Applied Economic Research (NCAER) (1979), *Projection of Fertilizer Demand; Final Report* (New Delhi, National Council of Applied Economic Research).

Rao, C.H.H. (1975), *Technological Change and Distribution of Grains in Indian Agriculture*, Delhi, Macmillian.

——(1976), 'Factor Endowments, Technology and Farm Employment. A Comparison of East Uttar Pradesh with West U.P. and Punjab' *Economic and Political Weekly,* 11-39 (25 September): A117- 23.

——(1983), 'Consumption of Fertilizers in Indian Agriculture: Emerging Trends and Policy Issues', Pusa, New Delhi, Indian Agricutlrual Research Institute Sastri Memorial Lecture, 4 February.

Rao, V.K.R.V. (1984), T.A. Pai Memorial Lecture, Syndicate Bank, Manipal, May.

Sawant, S.D. (1983). 'Investigation of the Hypothesis of Deceleration in Indian Agriculture', *Indian Journal of Agricultural Economics,* XXXVIII 4 (October-December) 475-508.

Subbarao, K. (1982), 'Food Production, Prices and Income Distribution. A Review of Recent Trends in India (paper presented at a Conference on 'Independent India: The First Thirty Five Years', University of California Santa Cruz).

——(1984), 'Incentive Policies and India's Agricultural Development: Some Aspects of Regional and Social Equity' (paper presented at the Institute of Economic Growth Silver Jubilee National Seminar, Delhi, April-May);

Whitcombe, E. (1983) 'Irrigation and Railways on Dharm Kumari (ed.) *Cambridge Economic History of India,* Vol. 2, Cambridge, University Press, Cambridge.

3

Some Aspects of Wheat Price Policy in India

G.S. RAYCHAUDHURI & RAJ KRISHNA

I. INTRODUCTION

The main purpose of this paper is a rigorous statistical analysis of:

1. The factors influencing the determination of procurement prices;
2. The influence of the procurement price, and other important factors, on the volume of procurement;
3. the effects of price movements and other shifter variables on the output of wheat; and
4. The effect of zoning on inter-State price dispersion.

The paper is organized as follows. The estimated procurement price functions are presented in Section 3, the procurement (volume) function in Section 4, and the wheat supply function in Section 7. The procurement function relates to the whole country, but procurement price functions have also been estimated for individual States, Supply functions have been estimated both for the whole country and the Punjab.

Section 2 contains a brief, descriptive review of trends in Indian wheat price and procurement policy, particularly since 1965 when the Agricultural Prices Commission was established and official support/procurement prices began to be fixed on a regular basis. In particular, the relationship between the cost of production, the harvest price, the procurement price, and the market wholesale price, in each State, is analysed in detail. Trends in the volume of procurement and the procurement-output ratio are also examined. This review suggests hypotheses about factors determining procurement and procurement prices which are tested in Sections 3 and 4.

There are two new features in the supply function estimated in Section 7. First, the supply response elasticity is estimated directly instead of being approximated by the acreage response elasticity as in many earlier studies. Second, the relative price used in the specification is the wholesale price of wheat deflated by an input price index rather than the wholesale price deflated by an index of the prices of competing crops. With the increasing role of purchased inputs the allocation of land and other inputs is likely to be strongly influenced by variations in input prices as well as output prices.

Section 5 is devoted to zoning. The hypothesis exmphasized some years ago (Krishna, 1965) that zoning increases inter-State price dispersion is tested with an equation in which a measure of price dispersion is regressed on a scale variable and a dummy variable representing the zoning policy every year.

The policy implications of estimated relationships are indicated in each Section and summarized in the concluding Section.

II. PROCUREMENT PRICE, COST OF PRODUCTION, HARVEST PRICE, AND WHOLESALE PRICE

The most important quantitative indicator of official policy for a crop is the procurement price[1] at which the government stands ready to purchase grain, subject to the annual procurement targets. This price is fixed for each State on the basis of the recommendations of the Agricultural Prices Commission; but the prices fixed by the States have not been the same as the prices recommended by the Commission.

An examination of the ratio of prices actually fixed in seven States, to the prices recommended by the Commission, during the eleven-year period 1966-67 to 1976-77, shows (Table 3.1) that the actual prices have usually exceeded the recommended prices. The average of the excess in different years has varied between 1 per cent and 9 per cent. In 1967-68 the mean excess was as high as 34 per cent. And all the important wheat-producing States except Gujarat and Madhya Pradesh have been fixing procurement prices in excess of recommended prices by 6 to 9 per cent.

The phenomenon of States fixing prices higher than the recommended prices is partly explained by the influence exercised by farmers' interest on policy at the State level. But it is also due to the realistic economic reasoning that the procurement targets recommended by the Centre could not be realized with unattractive purchase prices.

In its numerous reports the Agricultural Prices Commission (APC) has repeatedly mentioned the considerations which have influenced the prices recommended: (1) the cost of cultivation; (2) the trend in the open market wholesale price reflecting overall shortage; (3) the need for securing a balanced growth of the output of related crops; (4) the need to reduce inter-State price dispersion; and (5) the need to limit the rate of inflation in the interest of the consumers of grain, industrial development and overall stability.[2] The record shows, however, that the various criteria listed above were applied and emphasized in an uncoordinated way. Some of them were stressed and used for some decisions and others on other occasions. They were never integrated into an objective model to compute the price to be recommended.

TABLE 3.1 : Mean Excess of the Procurement Price Fixed by the State Governments over the Price Recommended by the Agricultural Prices Commision*

Average of eleven years		*Average of seven years*	
State	%	*Year*	%
Bihar	7	1966-67	1
Gujarat	3	1967-68	34
Haryana	9	1968-69	7
Madhya Pradesh	4	1969-70	9
Punjab	6	1970-71	6
Rajasthan	8	1971-72	3
Uttar Pradesh	8	1972-73	6
		1973-74	3
		1974-75	0
		1975-76	1
		1976-77	6

*Calculated from primary data in Reports of the Agricultural Prices Commission and issues of the *Bulletin of Food Statistics*.

Therefore, the effective influence which different criteria did have on the recommendations of the Commission can only be analysed by reviewing actual empirical relationships. The effect of the criterion of inter-crop balance cannot even be analysed empirically because prices for related crops were not worked out systematically and simultaneously. Also, the Commission's concern with inflation only limited the recommended increases in purchase prices when increases were warranted on other grounds. But the use of the other three criteria—cost, market price trends, and inter-regional dispersion—can be examined in some detail.

It has been argued that the procurement price, which acts as the effective support price (except in abnormally good crop years), should cover the full cost of production, including the imputed value of family resources, in order that it may provide sufficient incentive to farmers to maintain and increase production. A price fixed on this basis would ensure commercial farmers against losses in good harvest years; and provide to subsistence and semi-commercial farmers a surplus above paid-out (cash) costs which they can use to finance innovations (see Krishna, 1967). Thus, it is necessary, first of all, to compare the procurement prices fixed with the available estimates of the cost of production.

Second, it is necessary to compare the procurement price with the open market wholesale price because, on the demand side, the distribution of the marketed surplus between government procurement and sale in the open market should be expected to depend on the ratio of the procurement price to the open market price, if the procurement price falls short of the open market price by a substantial margin, procurement is likely to remain unsatisfactory, even if administrative pressure is applied in the form of compulsory levies and other penalties.

The results of a comparison between the procurement price and the full cost of production, for the years and States for which cost observations[3] are available, are presented in Table 3.2. The table shows that in five States (U.P., Punjab, Haryana, Bihar and Rajasthan) accounting for more than three-fourths (Table A. 1) of national wheat output, the procurement price was clearly *below* the full cost of production in the fifties and sixties for which cost data are available. But in the case of *irrigated* wheat in the Punjab, the deficiency of procurement price in the fifties was negligible (only 1.5%) because the yield per hectare was high and cost per quintal was low due to irrigation. In two districts of western U.P. in the fifties, the procurement price even *exceeded* the cost of production by a small margin in the case of irrigated wheat, for the same reason.

The relationship between procurement price and cost of production appears to have changed significantly in the first four years of the green revolution period beginning 1966-67. All the available observations for this period (Table 3.2, lower panel) show that the procurement price *exceeded*[4] cost by a comfortable margin varying between 11 per cent

TABLE 3.2 : Procurement Price and Cost of Production of Wheat

State (district)	*Period*	*Number of observations*	*Average excess of procurement price over cost of production (%)*
Haryana* (Karnal, Rohtak and Jind Tehsil of Sangrur)	1961-62 to 1963-64	3	–23.3
Haryana** (Karnal, Rohtak and Jind Tehsil of Sangrur)	1961-62 to 1963-64	3	–28.6
Bihar (South Monghyr)	1957-58 to 1959-60	3	–27.0
Bihar (Shahabad)	1960-61 to 1962-63	3	–37.9
Rajasthan (Pali)	1962-63 to 1964-65	3	–29.2
Punjab** (Amritsar and Ferozepur)	1954-55 to 1956-57	3	–14.0
Punjab* (Amritsar and Ferozepur)	1954-55 to 1956-57	3	–1.5
Uttar Pradesh* (Meerut and Muzaffarnagar)	1955-56 to 1956-57	2	–21.0
Uttar Pradesh** (Meerut and Muzaffarnagar)	1955-56 to 1956-57	2	+8.0
Punjab*** (Ferozepur)	1967-68 to 1969-70	3	+11.0
Punjab† (Ferozepur)	1967-68 to 1969-70	3	+30.0
Eastern Uttar Pradesh* (Deoria)	1966-67 to 1968-69	3	+39.6
Eastern Uttar Pradesh** (Deoria)	1966-67 to 1968-69	3	+21.0
Uttar Pradesh* (Muzaffarnagar)	1966-67 to 1968-69	3	+41.0
Uttar Pradesh** (Muzaffarnagar)	1966-67 to 1968-69	3	+66.0

* = Unirrigated wheat; ** = Irrigated wheat; *** = Desi wheat; † = Mexican wheat.

Soruces : Various Issues of *Bulletin of Food Statistics*, and *Studies in the Economics of Farm Management.*

in the case of desi wheat in the Punjab and 66 per cent in the case of irrigated wheat in western U.P.

This remarkable change was due to the fact that the cost of production increased less than the procurement price as new technology was adopted. In Table 3.3, for example, we can see that between the fifties and sixties the procurement price increased in western U.P. by Rs. 33 per quintal and the cost of production by only Rs. 19 per quintal for irrigated wheat and Rs. 11 for unirrigated wheat. In the Punjab also the procurement price increased by Rs. 42 per quintal and the cost by only Rs. 22 to Rs. 38 per quintal (for irrigated, unirrigated, and Mexican wheat).

We can conclude from this evidence that the wheat procurement price did not cover the full cost of production, (or exceeded it only by a small

TABLE 3.3 : Changes in Procurement Price and Cost of Production, Punjab and Uttar Pradesh

	Increase in procurement price	*Increase in cost*
	Average 1967-68/1969-70 Minus Average 1954-55/1956-57 (Rupees per quintal)*	*Average 1967-68/1969-70 Minus Average 1954-55/1956-57 (Rupees per quintal)*
Punjab	42.33	
Ferozepur (Local Irrigated)		37.98
Ferozepur (Local Unirrigated)		32.67
Ferozepur (Maxican Irrigated)		27.12
Ferozepur (Maxican Unirrigated)		21.81
	Average 1967-68/1968-69 Minus Average 1955-56/1956-57 (Rupees per quintal)*	*Average 1966-67/1968-69 Minus Average 1955-56/1956-57 (Rupees per quintal)*
Uttar Pradesh	32.64	
Muzaffarnagar (Irrigated)		19.23
Muzaffarnagar (Unirrigated)		11.38

* The procurement prices relate to 1951-52/1952-53 but are assumed to be the same even for 1954-55/1956-57 for the purpose of this comparison, because no revisions were announced and there was no procurement in the latter years (Table 17.8).

Sources of basic data : *Farm Management Reports* and issues of the *Bulletin on Food Statistics*.

margin in some irrigated tracts) in the fifties, but it provided a comfortable margin above full cost in the early years of the Green Revolution.

In contrast with the procurement price the farm harvest price, namely, the price prevailing during six to eight weeks immediately after the harvest in each State, has been above the cost of production in all important wheat producing States in the fifties as well as the sixties. The few exceptions to this relationship pertain to unirrigated or inadequately irrigated areas of Uttar Pradesh, Punjab, Bihar and Rajasthan (Table 3.4).

Tables 3.5 and 3.6 document the additional fact that in all important wheat producing States, without exception, the procurement price has been less than the harvest price and the annual average wholesale price.

The two main price hierarchies prevailing in the fifties, sixties, and seventies revealed by the evidence in Tables 3.2, 3.4, 3.5, and 3.6 can be neatly summarised as:

Fifties : Procurement Price < Cost of Production < Harvest Price < Annual Average Wholesale Price.

Sixties & Seventies: Cost of Production < Procurement Price < Harvest Price < Annual Average Wholesale Price.

The latter ranking implies that even when the procurement price has covered the cost of production, it has been consistently below the two market prices.

TABLE 3.4 : Farm Harvest Price and Cost of Production of Wheat

State (district)	*Period*	*Number of observations*	*Average excess of farm harvest price over cost of production (%)*
Bihar (South Monghyr)	1957-58 to 1959-60	3	–6.0
Rajasthan (Pali)	1962-63 to 1964-65	3	–16.0
Uttar Pradesh* (Meerut and Muzaffarnagar)	1955-56 to 1956-57	2	–16.0
Punjab* (Amritsar and Ferozepur)	1954-55 to 1956-57	3	–2.0
Haryana* (Karnal, Rohtak and Jind Tehsil of Sangrur)	1961-62 to 1963-64	3	+30.0
Eastern Uttar Pradesh* (Deoria)	1966-67 to 1968-69	3	+78.0
Bihar (Shahabad)	1960-61 to 1962-63	3	+38.0
Punjab** (Amritsar and Ferozepur)	1954-55 to 1956-57	3	+12.0
Uttar Pradesh* (Muzaffarnagar)	1966-67 to 1968-69	3	+69.0
Haryana** (Karnal, Rohtak and Jind Tehsil of Sangrur)	1961-62 to 1963-64	3	+21.0
Punjab*** (Ferozepur)	1967-68 to 1969-70	3	+9.0
Eastern Uttar Pradesh** (Deoria)	1966-67 to 1968-69	3	+54.0
Uttar Pradesh** (Meerut and Muzaffarnagar)	1955-56 to 1956-57	2	+15.0
Uttar Pradesh** (Muzaffarnagar)	1966-67 to 1968-69	3	+92.0
Punjab† (Ferozepur)	1967-68 to 1969-70	3	+28.0

* = Unirigated wheat; ** = Irrigated wheat; *** = Desi wheat; † = Mexican wheat.

Sources : Various Issues of *Farm (Harvest) Prices of Principal Crops, Agricultural Situation in India,* and *Studies in the Economics of Farm Management.*

TABLE 3.5 : Procurement Price and Farm Harvest Price of Wheat

State	*Period*	*Number of observations*	*Average excess of procurement price over farm harvest price (%)*
Bihar	1951-52 to 1953-54	3	–22.0
	1966-67 to 1975-76	10	–20.9
Gujarat	1966-67 and		
	1969-70 to 1972-73	5	–23.0
Haryana	1966-67 to 1972-73	7	–12.0
Madhya Pradesh	1952-53 to 1953-54	2	–16.0
	1965-66 to 1975-76	11	–14.9
Punjab	1951-52 to 1952-53	2	–12.0
	1966-67 to 1975-76	10	–18.9
Rajasthan	1951-52 to 1953-54	3	–12.0
	1967-68 to 1975-76	9	–5.7
Uttar Pradesh	1951-52	1	–6.0
	1966-67 to 1975-76	10	–14.1

Sources : Various issues of *Bulletin on Food Statistics, Farm (Harvest) Price of Principal Crops,* and *Agricultural Statistics in India.*

TABLE 3.6 : Procurement Price and Annual Average Wholsale Price of Wheat

State	*Period*	*Number of observations*	*Average excess of procurement price over wholesale price (%)*
Bihar	1951-52 to 1953-54	3	–28.7
	1965-66 to 1975-76	10	–29.3
Gujarat	1951-52 to 1952-53	2	–32.4
	1965-66 to 1974-75	10	–27.3
Haryana	1965-66 to 1975-76	11	–11.6
Madhya Pradesh	1952-53 to 1953-54	2	–9.1
	1965-66 to 1975-76	11	–17.2
Punjab	1951-52 to 1952-53	2	–5.3
	1966-67 to 1975-76	10	–11.4
Rajasthan	1951-52 to 1953-54	3	–16.1
	1967-68 to 1975-76	9	–19.9
Uttar Pradesh	1951-52 to 1952-53	2	–16.1
	1966-67 to 1975-76	10	–18.5

Sources : Various issues of *Bulletin on Food Statistics, Agricultural Prices in India, Statistical Abstract of Punjab,* and *Statistical Abstract of Haryana.*

III. THE EMPIRICAL BASIS OF PROCUREMENT PRICE DETERMINATION

Hypotheses to explain the behaviour of procurement during the period 1965-66 to 1975-76 will be tested in the next Section. But we investigate here the empirical basis of the procurement prices actually fixed. The notion suggested by the evidence is that the procurement price is strongly influenced by the trend of the wholesale price in the recent past. We have defined this past period as two preceding years and estimated the elasticities of the procurement price (PP) with respect to the lagged average wholesale price in two preceding years (PWL) for six States and for the whole country. These elasticities are tabulated in Table 3.7. They range from 0.61 in Madhya Pradesh to 0.86 in Haryana. The all-India elasticity[5] turns out to be 0.68. The squared correlation coefficient (r^2) is satisfactory in the case of five of the six States and in the all-India regression, and all the elasticity coefficients in the estimated log-linear regressions have significantly high t-values. The synchronous movements of the all-India PP and PWL can be seen in Figure 3.1. In the eleven-year period 1965-66 to 1975-76, the procurement price was increased five times following an increase in the lagged wholesale price (1966-67, 1967-68, 1973-74, 1974-75, and 1975-76). The year 1968-69 was the only one in which the all-India procurement price was slightly reduced although the lagged wholesale price increased. In the remaining five years, 1968-69 to 1972-73, the procurement price was kept stable because the wholesale price also varied within a narrow interval.

Therefore, we can confidently infer that the procurement price has simply followed the trend of the market wholesale prices with a time lag. In other words, the market price has been the leader price and the procurement price a reluctant follower. Regardless of the principles of price fixation, the prices actually fixed by the States seem to have been based on the simple rule of thumb that if the market price rises for one or two years, the procurement price should be raised by about two-thirds of the percentage increase in the market wholesale price.

IV. PROCUREMENT, OUTPUT, AND PROCUREMENT PRICES

Before estimating a procurement function, it is useful to review the trend of procurement as a proportion of production. The time-series of

TABLE 3.7 : Estimated Elasticity of the Wheat Procurement Price with Respect to the Average Wholesale Price in two Preceding Years in Different States (1965-66/1975-76)

State	*Elasticity*	*'t' value*	R^2
Bihar	0.83	(5.17)	0.75
Haryana	0.86	(6.22)	0.81
Madhya Pradesh	0.61	(5.10)	0.74
Punjab	0.78	(3.35)	0.56
Rajasthan	0.67	(5.92)	0.80
Uttar Pradesh	0.62	(2.83)	0.47
All India	0.68	(5.11)	0.74

the all-India procurement/output ratio (PQR) and the ratio in seven major States for the 26-year period (1951-52/1976-77) is tabulated in Table 3.8. In the early fifties (1951-52 and 1952-53) about 10 to 12 per cent of output was procured. For the next fifteen years (1953-54/1967-68) the all-India procurement/output ratio remained negligible, evidently because the availability of concessional imports led the government to relax domestic procurement effort. In six of these fifteen

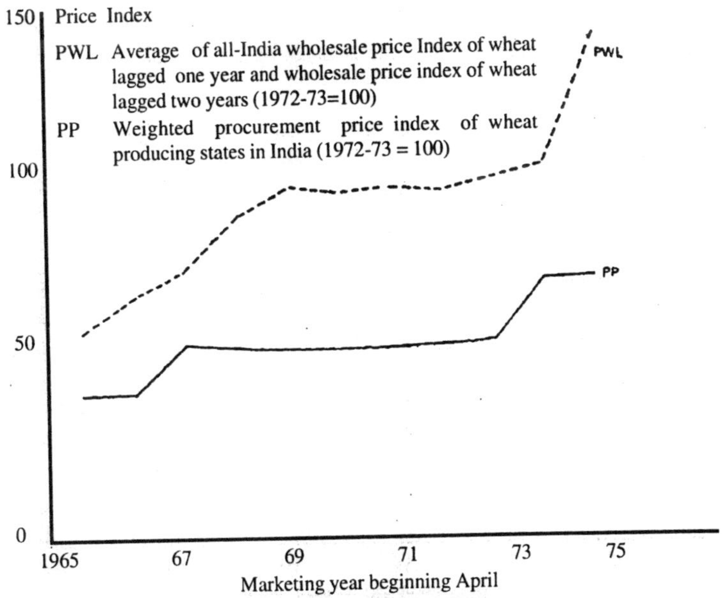

FIG. 3.1 : Procurement Price and Lagged Wholesale Price of Wheat

TABLE 3.8 : Ratio of Wheat Procurement to Production, Indian States (1951-52/1976-77)

Year	*Bihar*	*Gujarat*	*Haryana*	*Madhya Pradesh*	*Punjab*	*Rajasthan*	*Uttar Pradesh*	*All India*
1951-52	—	—	—	2.81	23.08	14.63	8.39	12.35
1952-53	—	—	—	2.73	16.30	6.73	11.49	10.02
1953-54	—	—	—	1.46	2.80	2.83	—	3.43
1954-55	—	—	—	—	—	—	—	—
1955-56	—	—	—	—	—	—	—	—
1956-57	—	—	—	—	—	—	—	—
1957-58	—	—	—	—	—	—	—	—
1958-59	—	—	—	—	—	—	—	—
1959-60	—	—	—	2.58	7.61	1.24	1.78	2.98
1960-61	—	—	—	5.06	9.21	—	1.31	3.53
1961-62	—	—	—	—	—	—	—	1.82
1962-63	—	—	—	—	—	—	—	—
1963-64	—	—	—	—	3.05	—	—	0.84
1964-65	—	—	—	—	3.07	—	—	0.95
1965-66	—	—	—	3.46	10.65	0.46	0.78	3.78
1966-67	—	—	3.68	6.78	3.97	—	—	1.90
1967-68	—	—	5.69	8.54	23.36	2.18	3.69	7.84
1968-69	—	—	14.59	5.47	40.51	5.31	0.83	13.89
1969-70	0.79	—	17.81	1.05	41.46	0.42	3.27	12.80
1970-71	—	—	22.74	0.36	49.33	0.08	5.10	15.89
1971-72	1.03	5.64	30.27	1.93	57.10	7.89	14.89	21.40
1972-73	0.08	0.11	34.05	2.04	55.69	2.44	11.17	18.96
1973-74	1.59	—	26.27	8.40	50.43	8.33	10.85	18.32
1974-75	0.12	2.32	13.75	4.88	20.90	7.09	5.43	8.98
1975-76	2.25	2.31	22.16	5.87	44.55	10.17	11.01	16.80
1976-77	4.40	11.81	37.23	9.97	51.57	20.71	21.04	23.30

Source : Various annual issues of the *Bulletin on Food Statistics*.

years no procurement operation was undertaken at all. The average procurement/ output ratio for the whole fifteen-year period turns out to be only 1.8 per cent. (If the years 1965-66 and 1967-68 are excluded, the average sinks to only 1.2%). The procurement ratio recovered to a significant level only in 1968-69, after the big increase in procurement price in the previous marketing season and a record harvest in the rabi season of 1968. This was the first normal season after the introduction of the new HYV technology. In the following three years (1969-70/ 1971-72) the procurement/output ratio rose from 14 to 21 per cent. Punjab and Haryana farmers sold about one-half and one-third of their wheat crop to the government. In 1972-73 and 1973-74 the government introduced the "socialization" of the wheat trade involving monopoly purchase of market arrivals by State agencies. The PQR remained high in these two years (18% to 19%) but lower than the peak reached in the pre-socialization year (21%). In the season 1974-75 the procurement price was raised by 38 per cent to a record level of Rs. 105 per quintal (from Rs. 76 in the preceding year), but PQR suddenly sank from 18 per cent to a mere 9 per cent. Two reasons seem to explain this decrease: the decline in output by about 3 million tonnes and the withdrawal of the socialization of the wheat trade in April 1974 after two years of experimentation. In the next two years (1975-76 and 1976-77) the PQR recovered to 17 per cent and then rose to the record level of 23 per cent.

Apart from price policy the growth factors which account for the satisfactory trend in procurement since 1968-69 is that all-India wheat procurement is dominated by the Punjab, Haryana, and (Western) Uttar Pradesh, where the irrigation ratio is very high, the new wheat technology has been rapidly adopted, and the wheat crop is raised on medium and large-scale farms as a commercial crop (Table 3.A-1).

In view of the record, the hypothesis set up to explain the volume of procurement from year to year (PR) is simply that it depends on (a)

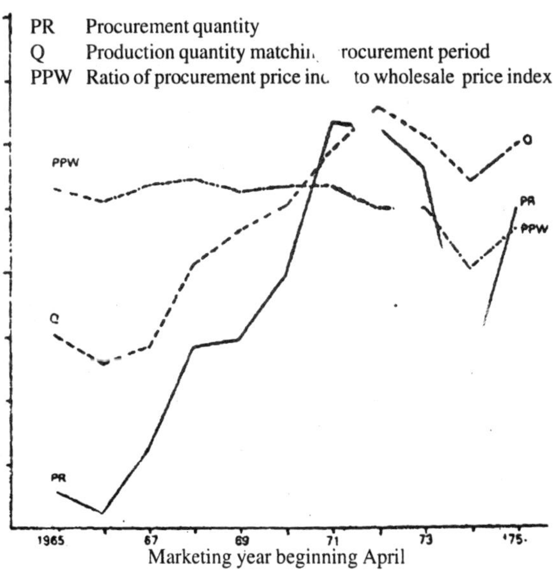

FIG. 3.2 : Wheat Procurement in India (1965-66/1975-76)

the level of output (Q) and (b) the ratio of the procurement price and the open market wholesale price (PPW):

$$PR = f\,(Q, PPW) \qquad \ldots (1)$$

In this equation, the price ratio is expected to catch the substitution effect: the division of the farmers' marketed surplus between sales to the government and sales in the open market. The output variable captures the scale effect.[6] The movement of the three variables can be seen in Fig. 3.2.

Equation (1) has been estimated with all-India data for the period 1965-66 to 1975-76:

$$\begin{aligned} PR \;=\; & -\,11.050 + 0.332\ Q + 7.324\ PPW \qquad \ldots (2) \\ & (-7.168) \quad (17.429) \quad (5.488) \end{aligned}$$

$$\bar{R}^2 = 0.968$$

(t-values are given in parentheses.)

The coefficients of output and price ratio are highly significant with expected signs. The elasticity of procurement with respect to output turns out to be as high as 2.32; and the elasticity of procurement with respect to the procurement price/wholesale price ratio is even higher (2.72).

The magnitude of the price elasticity of procurement points to the importance of the procurement price/wholesale price ratio as a stimulant of the quantity of procurement. And it refutes the view of some policy-makers that the volume of procurement can or should be raised substantially by mere administrative pressure, without an improvement in the price offered to the farmers. Our equation shows that the supply of any crop to the government is responsive to relative prices just like total output or the total marketed surplus. Therefore, the importance of an appropriate procurement price policy for raising the level of procurement cannot be overemphasized.

The estimated equation also underscores the obvious truth that the propitious seasons for a stock build-up are not the bad ones but those in which output levels are comfortable.

It remains to review the recent evidence on overall supply response. But a brief study of inter-State price dispersion is in order here.

IV. INTER-STATE PRICE DISPERSION AND ZONING

The zoning policy under which inter-State or inter-zone movement of grain under private auspices is restricted has been a controversial subject in India. The supporters of zoning have argued that it facilitates procurement at relatively low prices by restricting the export of grain from surplus States. The opponents have pointed out that it hurts producers in the surplus States and consumers in the deficit States by pushing prices down for surplus State producers below the free trade level and pushing them up for deficit States consumers above this level. The total imports of deficit States and the total exports of surplus States are likely to be less with zoning than without it (see Bhagwati and Chakravarty [1969] and Krishna [1965]). it is difficult to estimate what procurement levels and inter-State grain transfers would be under hypothetical movement policies other than the ones actually followed. But the hypothesis that the actual zoning policy increases inter-State price dispersion can be tested with the equation:

$$CVP_t = f(Z_t, S_t, C_t) \qquad \ldots (3)$$

where *CVP* is the coefficient of inter-State variation of wholesale wheat prices in year t; *Z* is a dummy variable having the value zero in years of free trade and value 1 in years of large (multi-State) or single-State zones; *S* is total wheat supply defined as all-India net wheat output plus net imports plus the initial government stock; and *C* is an index of the concentration of wheat output.[7]

The variable Z reflects zoning policy and is expected to have a positive sign, for when the inter-State movement of wheat is restricted the coefficient of inter-State price variation is likely to be higher. The supply variable is included for the obvious reason that, given the zoning policy, the larger the overall availability of wheat, the greater will be the volume of inter-State price dispersion. However, data show that this negative effect of availability on inter-State price variation is effective only after availability exceeds a critical level. In other words, the relationship between the CPV and availability, *ceteris paribus,* is parabolic.

The inclusion of the concentration index became necessary because the geographical distribution of wheat production in India is highly skewed. The share of Punjab and Haryana in total all-India output has increased from about 20 to 30 per cent over the 25 years (1950-51/1974-75). The share of Uttar Pradesh has come down from about 42 to 30 per cent. And the share of other States has changed marginally. But the overall index of concentration has declined. It is obvious that, apart from other factors, a higher concentration of production should be associated with greater inter-State price variation. Therefore, the concentration index is expected to have a positive sign.

The second-degree equation estimated with 24-observations (1951-52 to 1974-75) turned out to be:

$$CPV = -112.65 + 5.39Z + 3.70S - 0.08S^2 + 1.93\,C \quad \ldots (4)$$
$$(-2.67) \quad (2.44) \quad (3.24) \quad (2.31) \quad (2.38)$$
$$R^2 = 0.563$$

(t-values are shown in the parentheses.)

In the estimated equation, the coefficients of all the parameters are significant with the expected signs of the supply (< 0) concentration (> 0) and zonal restriction (> 0) variables.

The coefficients of S and S^2 signify that the inverse relationship between supply and price dispersion becomes effective only at a high supply level of 23.42 million tonnes. The equation supports the proposition that inter-State price dispersion increased systematically with the intensity of zonal restrictions. The large-zone system prevailed during 1951-52/1954-55 and then in 1967-68, and the single State zone system in four periods (1957-58 to 1960-61, 1964-65 to 1966-67, 1968-69 to 1969-70, and 1972-73 to 1974-75). In seven years of the period (1955-56, 1956-57, 1961-62 to 1963-64, 1970-71 and 1971-72) there were no zonal restrictions. it is interesting to note that the coefficient of variation averaged only 11.7 per cent in the seven years of unrestricted movement, 14.93 per cent in the five large-zone years and 19.42 per cent in the twelve single-State zone years (Table 3.9). This relationship is faithfully captured by the estimated equation.

The equation provides strong endorsement for the view that with zone restrictions surplus producers earn less and deficit area consumers pay more than they otherwise would; and both lose more in bad crop years than in normal years. This proposition would hold even if it is true that the government can procure more grain with zoning; for price dispersion clearly increases in spite of the transfer of this ''extra'' grain from the surplus to the deficit areas. The transfer must be less than it would be without zoning; otherwise dispersion could not increase. Whatever zoning may be achieving, it is in all probability not increasing the sum of producer and consumer welfare as it is measured in welfare economics.

TABLE 3.9 : Wheat Zoning and Price Variation : India (1951-52/1974-75)

Year	*Coefficient of variation (CPV)*	*Average*	*Zoning (Z)*	*Supply (million tonnes) (S)*	*Concen-tration index (C)*
(1)	(2)	(3)	(4)	(5)	(6)
1951-52	19.26		1.00 L	8.71	51.54
1952-53	13.68		1.00 L	7.78	52.07
1953-54	8.30		1.00 L	8.57	48.68
1954-55	13.10	13.59	1.00 L	8.10	49.38
1955-56	10.46		0.00 F	8.30	48.00
1956-57	9.14	9.80	0.00 F	8.95	47.08
1957-58	14.46		1.00 S	10.59	47.39
1958-59	15.55		1.00 S	10.05	48.34
1959-60	19.99		1.00 S	12.12	46.07
1960-61	12.15	15.54	1.00 S	12.39	47.36
1961-62	13.14		0.00 F	12.78	48.48
1962-63	10.09		0.00 F	13.90	47.42
1963-64	7.63	10.29	0.00 F	13.36	45.87
1964-65	30.39		1.00 S	15.58	46.41
1965-66	32.43		1.00 S	16.49	48.11
1966-67	23.27	28.70	1.00 S	17.18	45.47
1967-68	20.31	20.31	1.00 L	16.66	47.06
1968-69	21.12		1.00 S	18.07	45.50
1969-70	10.72	15.92	1.00 S	19.13	43.61
1970-71	16.94		0.00 F	20.13	44.35
1971-72	15.03	15.99	0.00 F	20.84	43.71
1972-73	14.97		1.00 S	25.99	41.77
1973-74	22.33		1.00 S	24.88	43.19
1974-75	15.64	17.65	1.00 S	23.18	41.62

Sources : Various issues of *Bulletin on Food Statistics, Agricultural Prices in India,* and *Statistical Abstracts* (different States).

L = Multi-State zone.

S = Single-State zone.

F = Free movement.

VI. PROCUREMENT PRICES AND PRODUCTION

As we have seen, procurement prices have a significant direct influence on procurement. But the effect of procurement prices on production (or acreage) is indirect. Farmers' decisions about acreage allocation, and input levels, for a crop depend on the Weighted Average realized price (WARP) per unit of output. This can be defined as the procurement price weighted by the procurement/output ratio (PQR) plus the market harvest price weighted by (1 – PQR).

The time series of the weighted average realized price (WARP) for the 11 years 1965-66/1975-76 have been estimated in Table 3.10 and plotted in Figure 3.3 for the Punjab, Haryana, and Uttar Pradesh. The movements of WARP and the market price can be seen to be similar because the ratio of procurement to output is low (less than 8%) in all States except the Punjab, Haryana, and U.P. where it averaged about 52 per cent, 29 per cent, and 12 per cent respectively in the 6 years 1970-71/1976-77 excluding 1974-75 (Table 3.8). The market price dominates WARP and the two move together everywhere. In the Punjab alone the market price and the procurement price carry an equal weight in the determination of WARP.

TABLE 3.10 : Weighted Average Realised Prices (WARP) and Wholesale Prices (PW) of Wheat : Indian States (1965-66/1975-76)

(Rupees per quintal)

Year	*Haryana*		*Punjab*		*Uttar Pradesh*	
	WARP	*PW*	*WARP*	*PW*	*WARP*	*PW*
1965-66	—	55.44	65.07	58.91	70.40	76.94
1966-67	97.25	84.92	79.63	87.88	116.35	98.49
1967-68	72.25	86.52	78.21	80.46	78.43	105.37
1968-69	82.12	85.03	72.00	91.54	78.98	72.21
1969-70	84.75	87.18	76.57	93.92	88.15	82.54
1970-71	78.93	77.29	76.42	91.38	75.35	73.18
1971-72	82.55	82.00	76.07	90.58	76.11	70.00
1972-73	82.96	88.23	76.53	79.96	81.53	97.14
1973-74	105.28	90.55	94.99	82.33	131.48	92.75
1974-75	111.62	132.34	109.54	116.65	141.18	200.64
1975-76	106.29	107.90	106.30	113.04	105.12	116.70

Sources : Annual issues of *Bulletin on Food Statistics, Farm (Harvest) Prices of Principal Crops,* and *Agricultural Prices in India.*

In most States, therefore, the influence of the procurement price on acreage/production decisions, mediated through WARP, should be low. But in the Punjab which provides about one-fourth of national wheat output and more than half of total wheat procurement, farmer incentives should depend critically on the procurement price. Therefore, in fixing this price its effect on wheat acreage/output in the Punjab needs to be constantly kept in view. In average crop years excess demand usually

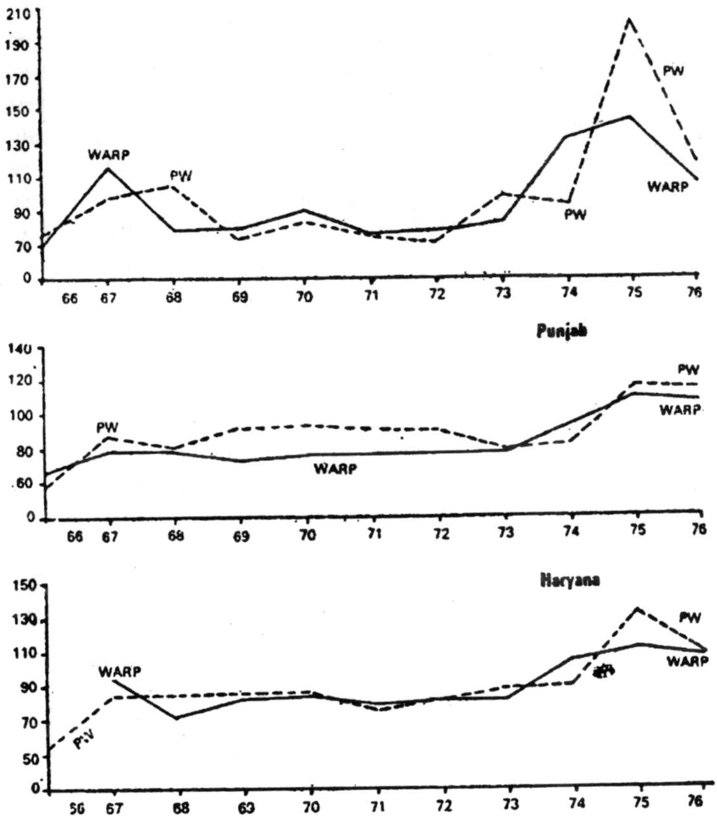

WARP = Weighted average realised price (rupees per quintal)
PW = Market Wholesale Price (rupees per quintal)

FIG. 3.3 : Wheat Price Movements in States (1965-66/1975-76)

keeps WARP above the cost of production. But in good harvest years the procurement price has to be high enough so that a decline in the wholesale price should not depress WARP below cost. And the government should be prepared to purchase any quantity the farmers wish to offer at the procurement price. Otherwise farmers are likely to suffer losses in good crop years.

VII. PRODUCTION RESPONSE TO PRICES & IRRIGATION

The fact that the acreages of individual crops in India vary systematically in response to inter-crop price movements is now widely accepted on the basis of numerous acreage response studies (Table 3.11). Therefore, we did not consider it necessary to run another set of routine acreage response regressions. On acreage response it is sufficient to recall the evidence from a few recent studies. The early papers (Krishna [1963] and Dharam Narain [1965]) demonstrated the presence of positive acreage response with pre-independence data. But in recent work post-Independence data have been used. In some cases, the period covered by the studies includes the early years of the Green Revolution.

TABLE 3.11 : Wheat Acreage Elasticity Estimates

	Region	*Period*	*Elasticity*		*Significance*	*Source*
			SR	*LR*		
1.	Punjab	Pre-Ind.	(.22)		S	Krishna (1963)
2.	Punjab	Pre-Ind.	.08	.14	S	Krishna (1963)
3.	Punjab	1950-67	.10	.13	NS	Cummings (1975)
4.	Punjab	NA	.002	.004	NA	NCAER (1969)
5.	Punjab	Post-Ind.	(.898)		S	Sud & Kahlon (1969)
6.	Punjab	NA	.08	.86	NA	Ramesh (1965)
7.	Haryana	1961-73	(.01 to 0.5)		NS	Singh & Kumar (1976)
8.	U.P.	1939-58	(.06)		NA	NCAER (1962)
9.	U.P.	NA	.06	.06	NA	NCAER (1969)
10.	M.P.	NA	.09	.20	NA	NCAER (1969)
11.	Rajasthan	1951-68	.02	.03	NS	NCAER (1969)
12.	Rajasthan	NA	.04	.12	NA	NCAER (1969)
13.	Bombay	1939-58	(.64)		NA	NCAER (1962)
14.	Gujarat	1954-67	.93	1.00	S	Cummings (1975)
15.	Maharashtra	1955-67	.24	.24	S	Cummings (1975)
16.	Mysore	1954-67	.23	.33	S	Cummings (1975)
17.	Delhi	1948-67	.17	.25	S	Cummings (1975)
18.	H.P.	1949-66	.02	.01	NS	Cummings (1975)
19.	West Bengal	1946-67	.23	.20	NS	Cummings (1975)
20.	India	1939-58	(.16)		NA	NCAER (1962)
21.	India	1949-66	(.67)		S	Maji *et al.* (1971)
22.	India	NA	.11	.72	NA	Ramesh (1965)

NS : Not significant.
( ) : Short-run and long-run not distinguished.
NA : Not available.
S : Significant.
SR : Short-run.
LR : Long-run.

The most numerous elasticity estimates based on an identical model have been generated recently by Cummings [1975]. The model includes lagged price. lagged acreage, rainfall, and trend. Cummings reports short-run and long-run elasticities of wheat acreage for 8 States estimated with post-Independence data. For the Punjab the short-run estimate (0.1) is close to the estimate made with pre-Independence data. Other estimates are given in Table 3.11. But very few of them are statistically significant. (The long-run elasticity estimates are particularly unreliable.) The significant short-run estimates from well-specified equations lie mostly in the interval 0.08 to 0.24. Only the Cummings equation for Gujarat yields a significant higher estimate of the order of 0.9. This estimate is out of line with most other significant estimates. The reasons for this unusual estimate need to be investigated.

Since wheat is a major subsistence crop with a large acreage in the growing States it is reasonable that the short-run acreage elasticity should be low (0.08 to 0.24). But even a 2 per cent shift in a large acreage, induced by a 10 per cent change in the realized relative price of wheat, can cause a substantial variation in production.

While acreage response has been the subject of considerable study, relatively little work has been done on output response. The elasticity of output with respect to the appropriate relative price is the sum of the elasticity of acreage and the elasticity of yield per acre; therefore, the elasticity of acreage can be a good approximation to the elasticity of output only if yield is stagnant or insensitive to price changes. In the pre-Independence period, the yield per acre of wheat was not only stagnant but declining at a low rate. During the 30-year period 1914 15/1943-44, for example, wheat acreage increased at the rate of 0.64 per cent per annum, but yield per acre decreased at the rate of –0.14 per cent. In the post-Independence period (1950-51/1973-74) the acreage growth rate escalated to 2.78 per cent and the yield growth rate escalated, even more, to 3.08 per cent. For this period the elasticity of output with respect to relative price should be much higher than the elasticity of acreage with respect to relative price. In response to price changes (and other exogenous variables such as the overall irrigation capacity created by the government, the availability of new technology, etc.) farmers would have adjusted not only the acreage in wheat but also the quantities of other inputs applied from year to year. In other words, a yield response on their part would have been added to acreage response to generate the recorded response of output.

In order to test these hypotheses, we have estimated a wheat output response equation for the whole country and for the Punjab for the period 1957-58/1969-70 (Figure 3.4). The exercise was designed not only to test the additivity of acreage response and yield response but also to incorporate input prices in the explanatory mechanism.

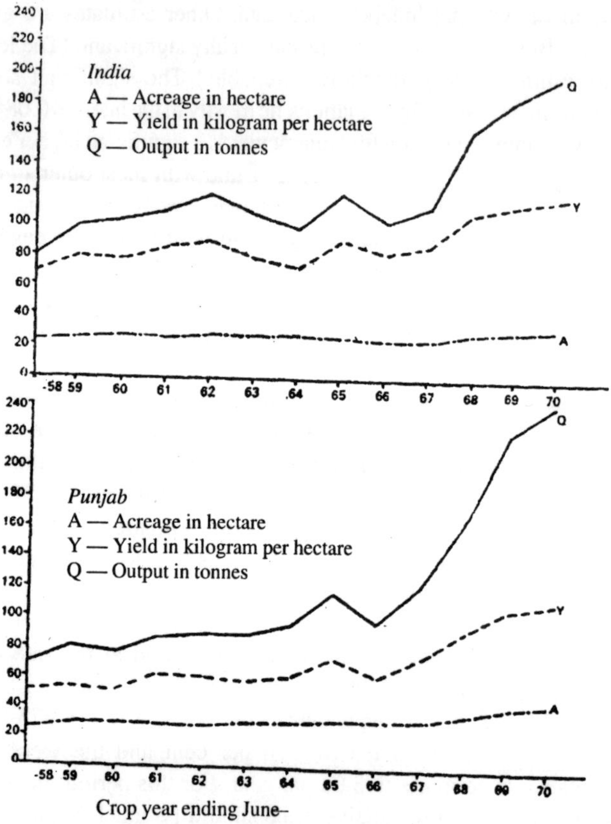

Crop year ending June–

FIG. 3.4 : Wheat Average, Yield and Output in India and Punjab (1957-58/1969-70)

Earlier studies have focused mainly on product substitution in response to the movement of intercrop price ratios. These ratios were, therefore, specified as the price variables in acreage response equations. Sensitivity to input prices was ignored on the implicit assumption that they have a relatively negligible effect on acreage or output. This assumption could have some legitimacy in a period when the use of modern, purchased inputs was minimal and/or their prices were invariant. But this assumption cannot be assumed to hold in a period covering the late 60s and the early 70s when the use of modern inputs in wheat cultivation has recorded a phenomenal growth and input prices have registered substantial variations. In fact, in this period, intercrop price variations seem to have had a negligible effect on wheat acreage and output compared to the effect of variations in input-use and input-prices. Accordingly, the relative price variable we have used in our output response function is not the intercorp price ratio index but an index of the ratio of the wheat price index to an input price index (with a one-year lag). The non-price shifter variable is the lagged irrigation ratio. The specification is:

$$Q_1 = a + bP_{t-1} + cV_{t-1} + u_{t'} \qquad \ldots (5)$$

where Q is wheat output in year t, P is the index of the wholesale price of wheat deflated by the input price index, and V is the ratio of the irrigated area in wheat to the total area in wheat. From the estimated equation 12.3 in Table 3.12, the elasticity of output with respect to the output-input price ratio turns out to be 0.59 and the elasticity with respect to the irrigation ratio 0.81.

TABLE 3.12 : Wheat Supply Response Regressions (1975-58/1969-70)

Eq. No.	*Dependent variable*	*Regression coefficients*			R^2	*Elasticity*	
		Constant	P_{t-1}	V_{t-1}		P_{t-1}	V_{t-1}
			INDIA				
12.1	A_t	7.764	3.286	0.078	0.673	0.221	0.211
		(5.821)	(2.248)	(1.482)			
12.2	Y_t	0.779	3.281	0.139	0.839	0.338	0.575
		(0.696)	(2.675)	(3.146)			
12.3	Q_t	–5.028	7.996	0.273	0.820	0.594	0.812
		(–1.927)	(2.796)	(2.648)			
			PUNJAB				
12.4	A_t	0.735	0.498	0.715	0.840	0.276	0.269
		(3.255)	(2.914)	(1.266)			
12.5	Y_t	–0.305	0.690	1.840	0.891	0.432	0.781
		(–0.967)	(2.889)	(2.329)			
12.6	Q_t	–1.865	2.207	3.770	0.874	0.824	0.954
		(–2.023)	(3.164)	(1.635)			

A_t : Current acreage.
Y_t : Yield in quintals per hectare.
Q_t : Current output.
P_{t-1} : Index of wholesale price (base 1965-66 = 100) in pre-sowing period (October-September), lagged one year, deflated by the input cost index for current year.
V_{t-1} : Ratio of irrigated wheat acreage to total wheat acreage, lagged one year.
The equations are linear in form.
t-values are shown in parentheses.

In order to verify additivity, acreage and yield elasticities have been estimated from equations 12.1 and 12.2 in Table 3.12. These equations have the same explanatory variables as the output equation. The price elasticity of acreage given by equation 12.1 is 0.22 and the price elasticity of yield given by equation 12.2 is 0.34. These elasticities add up to 0.56 which is very close to the price elasticity of output (0.59) estimated from equation 12.3.

It is interesting to note that the price elasticity of acreage (0.22) for the period covering the late 50s and the 60s is comparable with the elasticities estimated in earlier studies (ranging between 0.08 and 0.24).[8] But, as we hypothesized, the output elasticity is much higher than (almost three times) the acreage elasticity. The difference between the output elasticity and the acreage elasticity is accounted for by the elasticity of yield.

The elasticity of output with respect to the irrigation ratio (0.81) is also roughly equal to the sum of the acreage elasticity (0.21) and the yield elasticity (0.58) with respect to the irrigation ratio. The irrigation coefficient is clearly significant in the yield and output equations but not in the acreage equation. This result suggests that yield now responds to an increase in the irrigation ratio much more than (almost three times as much as) acreage. Thus, the effect of the growth of irrigation capacity on output accrues mainly via the resulting yield increases.

Since the Punjab is the most important wheat-growing State, equations similar to the all-India equations have been estimated for the State (Table 3.12). The price elasticities of acreage, yield, and output are, respectively, 0.28, 0.43, and 0.82, compared with 0.22, 0.34, and 0.59 for the country as a whole. The additivity condition is again fully verified. But every elasticity is somewhat higher in the Punjab than in the country as a whole, because of the higher technological level and the greater commercilization of the Punjab agriculture.

The hypothesis that input-price variations have a significant effect on acreage, yield, and output is also verified, for the coefficients of the output-input price ratio are clearly significant in all equations.

We have used lagged irrigation ratio as the non-price variable, rather than the lagged dependent variable derived from a Nerlovian adjustment or expectation mechanism. The reason is that irrigation ratio is the most important proxy for all technological changes which shift the supply function. It is the critical precondition for the use of all other modern inputs such as hybrid seeds, chemical fertilizers, pesticides, equipment, etc. A real causal variable which obviously has a strong influence on output from year to year is preferable to the lagged dependent variable which merely reflects a hypothetical adjustment or expectation process.

All the estimated equations bring out an important fact relevant for policy; namely, that the responsiveness of output to the irrigation ratio is usually higher than the responsiveness to relative price changes. The irrigation elasticity of all-India output (0.8) is about 33 per cent higher than the price elasticity (0.6). This finding shows the pointlessness of the superficial controversy which often turns up about the relative price or technological variables alone being important determinants of output. It is clear that both are important but the responsiveness of output to technological change is, at the margin, higher than the responsiveness of output to relative price movements—though one can legitimately argue that the rate at which farmers adopt technological innovations, other than the expansion of public sector irrigation capacity, which is independent of farmer decisions and price movements, is itself a function of the prevailing output-input price ratio regime. The policy implication is that for sustaining a satisfactory rate of growth of output, policy-makers must maintain both a satisfactory price regime and a rapid growth of irrigation capacity, input supplies, and technical knowledge.

VIII. CONCLUSIONS AND POLICY IMPLICATIONS

The main findings of the preceding analysis can be summarised as follows. First, it is clear that the procurement price actually fixed by the government has been influenced dominantly by the open market wholesale price in the recent past, except in abnormal years, and not by the cost of production. The elasticity of the procurement price with respect to the lagged wholesale price has been about 0.6. In the fifties the procurement price did not cover the cost of production in important wheat-producing States but in the late sixties it allowed a margin of profit above full cost. The procurement price almost always remained below the market harvest price and the market wholesale price. And whereas the market prices did cover the cost of production in most States and periods, the procurement price did so only in recent years.

Second, it is established that the volume of procurement has been systematically sensitive to output (with elasticity 2.3) and to the procurement price/market price ratio (with elasticity 2.7). The policy implication of this finding is that a satisfactory procurement price is as important for government obtaining the desired quantity as the trend of output.

Third, zoning restrictions have been shown to increase inter-State price dispersion and thus to hurt the producers of surplus States and the consumers of deficit States. But this perverse effect of zoning is moderated in good crop years when greater open market availability tends to reduce price dispersion.

TABLE 3A.1 : Proportion of All-India Wheat Output and Procurement in Different States (Average 1972-73 to 1974-75)

State	*Production of output (%)*	*Proportion of procurement (%)*
Uttar Pradesh	28.9	17.1
Punjab	22.7	59.5
Madhya Pradesh	10.3	4.0
Bihar	9.6	0.4
Haryana	8.5	14.0
Rajasthan	7.6	3.5
Maharashtra	3.4	0.4
Gujarat	3.1	0.4
Andhra Pradesh	0.1	0.0
Others	5.8	0.7
Total	100.0	100.0

Source : *Bulletin on Food Statistics*

Finally, it is shown that the sensitivity of wheat output to relative price movements has increased in the post-Independence period to 0.6 from about 0.2 in the pre-Independence period. This is due to the fact that in the early period acreage responded to price movements but yield remained stagnant; but in recent years yield response has been added to acreage response. Production is now sensitive not only to intercrop price ratios but also, and much more, to the output-input price ratio. And, as in the past, the effect of the technological supply shifter, of which irrigation capacity is the most basic, on output, remains much stronger than the effect of price movements. Thus, for maintaining satisfactory output growth, steady growth of irrigation capacity and input supplies is as essential as a favourable output/input price regime.

NOTES

1. The "support" price which is lower than the procurement price is not discussed in this paper for it is a notional price which is supposed to become operational only when the market price sinks much below the procurement price. This has happened only rarely.
2. See Commission reports on wheat.
3. Cost data relate only to samples in districts selected as representative of State/regions for Farm Management Studies.
4. The big increase in the (weighted average) all-India procurement price came in a single marketing season beginning April 1967. After this increase the procurement price was kept nearly stable for six years.
5. The all-India procurement price is the weighted average of States' procurement prices, the weights being the States' shares in all-India procurement in 1972-73.
6. Both the PR and Q figures relate to the period April-March every year.
7. The index is $\sqrt{\sum_{i=1}^{n} r_i^2}$ where r_i is the percentage share of State i in all-India output.

REFERENCES

Bhagwati, J. and Chakravarty, S. (1969), "Contributions to Indian Economic Analysis: A Survey", *American Economic Review*, Vol. 58. September 1969, (A Supplement).

Cummings, J.T. (1975), "The Supply Responsiveness of Indian Farmers in the Post-Independence Period: Major Cereal and Cash Crops", *Indian Journal of Agricultural Economics*, Vol. 30, January-March 1975, pp. 25-40.

Government of India, Ministry of Agriculture (1965), *Report of the Agricultural Price Commission on Price Policy for Kharif Cereals for 1965-66*, Delhi, 1965.

Government of India, Ministry of Agriculture & Irrigation (1978), *Agricultural Price in India*, Delhi.

Government of India, Ministry of Agriculture & Irrigation, Directorate of Economics & Statistics: *Bulletin on Food Statistics* (Annual issues).

Government of India, Ministry of Agriculture & Irrigation, Directorate of Economics & Statistics: *Farm (Harvest) Price of Principal Corps* (Various issues).

Government of India, Ministry of Agriculture & Irrigation, Directorate of Economics & Statistics: *Studies in the Economics of Farm Management* (Various issues).

Krishna Raj, (1963), "Farm Supply Response in India-Pakistan: A Case Study of the Punjab Region", *The Economic Journal*, Vol. 73, September 1963, pp. 477-487.

Krishna Raj, (1965), "Rice Zone Policy: A None of Dissent" in Ministry of Food and Agriculture, Government of India, *Report of the Agricultural Prices Commission on Price Policy for Kharif Cereals for 1965-66 Season*, May-July 1965.

Krishna Raj, (1967), "Agricultural Price Policy and Economic Development", in Herman M. Southworth and Bruce F. Johnston, (eds.), *Agricultural Development and Economic*

Growth, Cornell University Press, Ithaca, New York, 1967.

Maji, C.C. *et al.,* (1971), "Dynamic Supply and Demand Models for Better Estimation and Projections: An Economic Study for Major Foodgrains in the Punjab Region", *Indian Journal of Agricultural Economics,* Vol. XXVI, No. 1, January-March 1971.

Narain, Dharam, (1965), *Impact of Price Movements on Areas Under Selected Crops in India: 1900-1939,* Cambridge, U.K. 1965.

National Council of Applied Economic Research (1962), *Long-Term Projections of Demand and Supply of Selected Agricultural Commodities, 1960-61 to 1975-76,* New Delhi, 1962.

National Council of Applied Economic Research (1969), *Projections of Demand and Supply of Agricultural Commodities,* New Delhi, 1969.

Ramesh, D., (1965), "Long and Shortrun Price Elasticities of Acreage Under Crops", *Agricultural Situation in India,* Government of India, Ministry of Food & Agriculture, August 1965, pp. 319-324.

Singh, I.J., and Kumar, P., (1976), "Impact of Price and Price Variability on Acreage Allocation in Haryana", *Indian Journal of Agricultural Economics,* Vol. 31, April-June, pp. 31-37.

Sud, L. and Kahlon, A.S., (1969), "Estimation of Acreage Response to Price of Selected Crops in Punjab State", *Indian Journal of Agricultural Economics,* Vol. 24, July-September, pp. 46-50.

4

Agricultural Price Policy

A Survey with Reference to Indian Foodgrain Economy

N. KRISHNAJI

I. INTRODUCTION

The Context

Before the mid-nineteen sixties the activities of the government in the foodgrain economy were limited in scope to the import of grain and its distribution—mainly in the urban areas under various forms of rationing—during periods of food scarcity following poor harvests.[1] Thereafter, however, with the adoption and promotion of the 'new agricultural strategy', based on the cultivation of high yielding varieties of seeds with the use of 'modern' inputs—fertilisers, pesticides, etc.—in areas of assured water supply, the government began to play a major role in the transformation of the agricultural sector. This led to the involvement of the state in diverse activities such as the development of infrastructural facilities, provision of subsidies of different kinds, supply of inputs and credit, and the promotion of agronomic research for the adaptation of exotic crop varieties to Indian conditions.[2]

The agricultural price policy which evolved subsequently must thus be seen as an essential part of a larger 'package' of policies designed to promote rapid growth in a few regions endowed well with irrigation facilities, and to encourage private investment in the necessary means (modern inputs, tubewells, farm machinery, etc) for the cultivation of the new varieties. It can be argued that, given the highly skewed distribution of land an assets and a marked regional concentration of irrigated areas, the policy package was bound to promote both inter- and intra-regional inequalities. For this reason, one can perceive a build-in regressive character in the price policy and locate it within the overall policy of promoting growth through various production incentives, a policy not based on sufficient consideration of its likely distributional consequences.

Agricultural Prices Commission (APC)

Both before and after independence the Government of India had been constituting 'committees' and 'commissions' with a fair regularity to investigate the 'food problem' and suggest ways to solve it. But all these were bodies formed *ad hoc* in response to periodic scarcities. The work and recommendations of these various bodies probably had no lasting impact on the Indian agrarian economy.[3]

It was only in 1965 that a permanent body, viz., the Agricultural Prices Commission, was set up with presumably long-run goals in view. The APC, like the Planning Commission (a national body concerned with growth and development in general), is only an advisory body, with all the decision-making powers resting unquestionably with the government. The APC was charged with the responsibility of evolving a balanced and integrated price structure "in the perspective of the overall needs of the economy and with due regard to the interests of the producer and the consumer". The terms of reference of the commission refer not only to the need for providing price incentives for promoting agricultural growth but also need to "ensure rational utilisation of land and other production resources" and to the "likely effect of the price policy on the rest of the economy, particularly on the cost of living, level of wages, industrial cost structure, etc".[4]

Wide-ranging as these terms of reference were, they have since been modified and expanded in 1980 in response to the changes in the agrarian economy that have taken place during the late sixties and the seventies. An important addition is the directive that price fixation should "take into account the changes in the terms of trade-between agricultural and non-agricultural sectors": the original terms of reference did not cover any principie of pricing in relation to costs or parity between sectors, barring a reference to marketing costs and margins. We shall discuss the specific political context of this directive later but note here that the underlying concern with agricultural input costs in relation to output prices, and, more generally with parities of different kinds, have been given further and explicit recognition in 1985 with the Commission being renamed as the Commission for Agricultural Costs and Price (CACP).[5]

The composition of the commission has undergone some change in character over the years. It was headed in succession by several distinguished economists and originally composed of economists and civil servants with experience in matters relating to the working of the food economy in the country. However, after 1974, one or more representatives of the farmers have been functioning as full-time members in the commission, with obvious implications for the evolution of policy.[6]

Objectives

The objectives as laid down in the terms of reference of the APC are to some extent mutually conflicting, and difficult to realise for other reasons as well. One may thus say that the APC was given an impossible task. Let us elaborate.

(a) If the sole objective is to promote agricultural growth—whether or not broad-based with respect to crops and regions—through subsidies and price incentives, there is a clear possibility that the implied policy would hurt the interests of the buyers of food and agricultural raw materials. A resolution of the conflict of interests between producers and consumers—or between net sellers and net buyers, and between 'surplus' and 'deficit' areas, as some have described it in the case of foodgrains—thus becomes a necessary element of overall policy-making. It is a recognition of this necessity that is reflected in the expressed concern about the impact of policy on cost of living, etc., and has led to the conversion of price policy into an approach which seeks to integrate in some measure production objectives with those of distribution. In the case of foodgrains this meant that price policy had to be combined with the operations of a Public Distribution System (PDS) to ensure supply of foodgrains at subsidised prices to at least some of the poorer sections of the population. However, to the extent that subsidising both the producer and consumer is severely limited by financial constraints, it cannot be sustained on either a short- or long-term basis without being supplemented by compensating measures of taxation. All this makes the design of price policy difficult, especially by a body with no decision-making powers; more importantly, the difficulty leads to politically expedient measures which exacerabate rather than resolve conflicts.

(b) The difficulty is compounded by two other obvious factors. First, if relative prices are to play an efficient allocative role and not lead to undesirable cropping pattern shifts, policies must be designed to cover all the major crops on a country-wide basis, an impossible task to accomplish.[7] Second, since public distribution of subsidised grain cannot hope to meet the entire market demand (of both the rural and urban consumers), the impact of policies on market prices assumes importance. Thus, with the government handling only a fraction of the marketed surplus, the pricing objectives may in the end be defeated by market forces. The design of price policies with only a limited control over the market thus not only involves an appropriate combination of procurement, stock accumulation and depletion policies but must necessarily be based on a sound knowledge of the implied working of dual markets. The biggest hindrance in this respect is not lack of knowledge about production trends and demand elasticities but our virtual ignorance about the operations of private traders at the different tiers of the grain markets and their impact on prices.

(c) Finally, it must be noted that the objectives are not only conflicting—the fact that there is no indication in policy formulations whether it is only in the short-run but also in the long-run these conflicts could be left unresolved is also important—they are also incomplete and imprecise in some respects. For example, income and price stabilisation are common objectives of price policy; maintenance of some form of parity between the agricultural and industrial sectors and prevention of the erosion of purchasing power of agricultural incomes is a related concern, and all these are adequately discussed in the literature. Indeed they can be stated with some precision: unlike the vague 'protection' of producer and consumer 'interests'; but it is doubtful if multiple objectives are realisable.

Instruments

Given the all-embracing but vaguely set-out nature of the objectives of price policy, the APC had to recommend not one but several sets of prices; its recommendations had also to cover issues relating to the procurement and distribution of foodgrains by governmental agencies. It had to develop notions about: 'support' prices to prevent farmer distress,[8] 'incentive' prices to promote investment and growth in agriculture, 'procurement' prices—possibly involving an element of tax—at which grain could be procured by the government under varying degrees of compulsion (or none) and 'issue' prices—involving an element of subsidy—at which grain could be distributed under the public distribution system whatever be its coverage.

The following discussion attempts to cover the different instruments of policy adopted by the APC, more factually by the government, their rationale, as well as how they have worked in terms of the stated objectives.

II. PRINCIPLES

Support and Incentive Prices

Given large output fluctuations arising out of the vagaries of the monsoon (see below) and given the low price elasticity of demand for foodgrains in the country, prices tend to fluctuate a lot from year to year: changes in the supply induce disproportionately large changes in the market prices.[9] A notion of a support price can be seen to acquire meaning in this context.[10] Such a price may be regarded as an offer price at which the government is willing to buy any amount of grain from the farmers in years of good harvest when, in the absence of a support operation, the market price may fall below the cost of production; it has therefore to be above that hypothetical market price, and for that reason has a built in element of subsidy to the producer. Some economists have argued, however, that support prices conceptualised in this manner need bear no relation to costs, at least in the short-run: as long as it is above the low market price that will be realised in the absence of intervention, there is an element of subsidy, and the government absorbs a part of the loss and neutralises in some measure the effects of output fluctuations.

A logical corollary to the concept of the 'minimum' support price is that of a maximum or ceiling price the rationale for which lies in two factors.[11] First, protecting farm incomes in years of abundance through purchases by the government at minimum prices implies building up of stocks; and for viability—at least over the long run—such a policy requires stock depletion as well. Second, this is a related factor—a maximum price would imply protection of consumer interests in years of crop failure, without it a support programme unilaterally favours the farmer and is clearly undesirable because the effects of agricultural output fluctuations in India are not restricted to the farm sector alone but are quickly transmitted to the other sectors of the economy, especially in periods of scarcity, through high food price. In a sense the food price acts as a base price in the determination of both agricultural and industrial costs.

We thus see that the two notions, one of price stabilisation—or, rather, price control within a certain range—and the other of support of farm incomes along with some degree of protection of non-farm interests are interrelated. This, however, still leaves some asymmetry, for the question remains whether in poor crop years corresponding to a disproportionate price rise, farmers and traders should be left free to sell all the surplus on the market to earn super-normal profits. Of course the policy of stock depletion by the government tends to depress the market prices to some extent and holds these profits in check but it can be obviously more efficacious in combination with a policy of compulsory procurement, at lower-than-market prices in lean years.

Thus the logic of a support price leads to the notion of a price range, especially in relation to the objective of dampening price and income fluctuations arising out of unpredictable monsoons. Supporting agriculture in this context is clearly different in nature in India from, for example, that in the US, where it has more to do with the preservation of the purchasing power of farmers in relation to manufactured goods under conditions of abundant and growing farm supplies than with weather-induced fluctuations in farm output. (The US has been able to successfully pursue this policy because of her dominance in the international grain market).

In contrast to the support price, it is not easy to settle what constitutes an incentive price. The rationale of a price incentive lies in the promotion of growth. In the case of a single crop such as wheat—hindsight enables us to cite this as an example—a support price covering costs and some profit during a period of rapid growth in production is in effect an incentive price, with the government standing by to lift surpluses and ensure profit margins irrespective of output growth. Clearly such a policy cannot be adopted without undesirable consequences (including political tensions bound to arise because of the regional concentration of surpluses except in the short-run; on the other hand, a price incentive policy is not viable if the objective is to promote a balanced growth with respect to all regions and crops because there are obvious limits to the provision of incentives.

It must also be noted that incentives need not cover output prices alone but may be extended to include subsidies in the supply of inputs and credit. It is possible to analyse the growth and welfare effects of the different underlying choices, but it is generally agreed that input price subsidisation may be cheaper only in the very short-run—an immediate increase in food and raw material prices being avoided; in the long run, with price adjustments taking place across the board, the effects may be similar to those of producer price subsidies.[12] More importantly, as it has been repeatedly stressed in the literature, the promotion of technological improvement through, for example, the development of irrigation, may at all times be a better alternative than the provision of price incentives to secure agricultural development. The belief is firmly grounded in the empirical finding that generally the elasticities of supply (in terms of output) with respect to such technological change variables as their irrigation ratio exceed price elasticities: greater gains can be secured through upward shifts in the production function than from movements along the supply curve. In other words, if price policy is to be the only instrument for agricultural development, the low supply response would require a tilting of the terms of trade in favour of agriculture to a degree the economy cannot bear except over a very short period.[13] Finally, given the uneven spatial distribution of irrigation and other resources, the scope for achieving desired levels of overall growth through even marginal increases in per-hectare yields in the unirrigated tracts of the country (covering about two-thirds of cultivated area) must indeed be enormous; the means for such a raising of the yields are without doubt technological in nature.

Procurement Prices, Costs of Production and Parity

Given the structure of the Indian economy and the decisive role of drought-induced agricultural, output fluctuations within it, the operation of a support programme is inextricably linked with that of a public distribution system for the supply of grain at subsidised prices. It is clear that a system of subsidies at both ends (production and consumption) can be financially viable only if there are compensatory measures of taxation within the agricultural sector itself or elsewhere; subsidies beyond a certain quantum without taxes will cripple the economy in the long run.

This leads to the concept of procurement as a tax on surplus producers especially in years of poor harvest when market prices and hence profit margins tend to be very high.[14] And it is obvious that a degree of compulsion is inevitable if producers are to be made to surrender a part of the surplus at prices lower than those prevailing in the market. While both the support and procurement operations involve the government in the acquisition of stocks, they can thus be seen to be analytically quite distinct in character—one, a more of subsidisation and the other, a form of taxation.

This important analytical distinction was however, lost sight of in the academic debates as well as in policy implementation, after some initial consideration.[15] The reasons are many. Soon after the procurement and distribution programmes began to gather momentum by the late sixties, the procurement price emerged as a guarantee price, announced well ahead of harvest time, at which the government was committed to make purchases. More decisive, of course, was the concern of the government to promote growth in the production of wheat through the new technology. All in all the price guarantee scheme was a hasty response to a combination of several factors : the need to recover from an environment of shortages resulting from two consecutive years of severe drought (1965-67), the possibility of such a recovery at a rapid rate through the cultivation of the newly emerging high-yielding varieties and finally the purely localised and limited nature of the support operation which made it eminently feasible. However, neither the subsequent emergence of surpluses on a wider scale nor the greater involvement of the government in the markets it entailed led to any basic revision of the policy with regard to the rationale of procurement and its pricing.

The policy had predictable consequences. While the support operation was satisfactorily served by the levels of procurement prices whenever and wherever support was needed in response to bumper harvests, procurement of grain became increasingly difficult under conditions of scarcity, local or global. This meant that procurement prices had to remain close to, even if slightly below, market prices at all times. Since procurement prices were being announced in advance of harvest and could be based at best on insufficient knowledge of the crop prospects and market conditions expected to prevail in the coming season, this in turn required the upward revision of prices[16]—secured in practice through diverse means—for meeting procurement targets with varying degrees of success. (Section IV-Price Markets and Relations discusses further details).

Returning briefly to the question of compulsion which could have been the basis of procurement under conditions of relative scarcity, we may note that the government did employ a number of instruments incorporating a coercive element. These range from direct compulsory levies on producers, traders and millers to the zoning or cordoning off supplies to facilitate procurement. But the measures proved to be ephemeral in character and ineffective in execution, whether it is the lack of 'political will' or some other factor that could be held responsible for the failure of policy. Take, for example, the constitution of zones comprising one or more regions. The usefulness of zoning has been much discussed in the literature. Let us note however that movement restrictions generally amplify the changes in prices in any region as they occur in response to variations in domestic production. Hence, to the extent that a normally surplus region can be kept immune to short-falls occurring elsewhere in order to create conditions favourable to the procurement agency, zoning may be seen to be consistent with the principle of procurement below market prices in lean years. However, the net impact of zoning on prices, both in terms of all-India average and the range of regional variation would crucially depend not only on the coverage of zoning but also on how much, where and at what prices grain is procured and subsequently distributed. In practice, movement restrictions have been imposed and relaxed in an arbitrary manner, sometimes in dissonance with stated objectives.[17]

Let us now consider the relevance of costs of production to the fixation of prices. The general consensus is that minimum support prices—now indistinguishable from procurement prices—must fully cover the average cost of production, but the question what elements of cost should be included in the calculation of the average total cost continues to be debated.[18] Specifically, the debate refers to the appropriateness of the inclusion of not only the so-called paid-out costs but also the imputed values of the services of family land and labour inputs in the calculation of the 'complete' average cost. A price covering such a complete cost has been described as a 'forward-looking' floor because it ensures cash incomes to farmers over and above the actual money expenditure incurred. Moreover, it is seen to incorporate at least one principle of parity viz., input return parity, since the family inputs are given the same remuneration that they could notionally ear outside the family farming activity.[19]

(Even if one accepts this complete cost principle of pricing, it is beset with many operational difficulties arising from the fact that even for a single crop the cost of production varies a lot not only from region to region but also within regions across different farms.[20] A solution suggested in this respect recommends the use of a basic line cost, the minimum cost that covers the costs of farmers producing a major part of the output, weeding out the very 'inefficient' farmers. But then the derived price would cease to have the same meaning as that based on the average cost in standard economic theory. It is equally difficult to grapple with the question of regional variation in costs. The actual practice of fixing a uniform price for all regions is obviously defective. For example, the adoption of a price that covers the costs in a high cost region, where, say traditional methods of cultivation are used, results in abnormal profits to farmers of other regions using the cost-reducing new technologies. Thus the principle may be consistent with the objective of growth but not with that of the reduction of interregional disparities).

Two important points have been made in relation to the imputation of values at market prices to the use of family land and labour resources. Since the price of produce is raised by the inclusion of these elements in the total cost, agricultural labourers and other net buyers of food for whom wages constitute a major element of income—lose because wages actually paid generally tend to be below the statutorily fixed minimum levels (which in principle could be in line with output prices fixed by the complete costing principle) while employers of hired labour earn an abnormal profit for the same reason. In simpler terms money incomes of net buyers (especially of wage earners in rural areas) generally lag behind food prices and the total cost principle reinforces the tendency. Regarding the imputation of market rental values to the use of owned land, it has been argued that there has been a continuous increase in these values arising from increasing productivity as well as demographic pressure. Regions well endowed with, say, public irrigation resources are also in general regions of high productivity and it is debatable whether such regions should be allowed to enjoy high economic rents through the enhancement of product prices resulting from the inclusion of rental values in the total cost.

Similarly, it is arguable whether increases in rental values arising from exogenous factors such as demographic pressure or insufficient expansion of non-agricultural employment should be automatically transmitted as additional costs to the product price. Thus the debates in the literature demonstrate that even a well-defined cost based principle for price fixation can become meaningful and reasonably operational only in relation to a clearly specified policy goal (such as the reduction of spatial disparities in incomes earned from the cultivation of the same crop).

The parity with input costs is only one among different kinds of parity one can think of in the context of product pricing. For example, there is the parity between farm incomes and incomes earned in other sectors of the economy. A little reflection shows that in India, with large output fluctuations the rule rather than the exception, it is difficult to maintain such a parity through a product pricing policy alone: here again a system of subsidies and taxes is needed to limit the range of fluctuations in farm incomes and ensure their parity with incomes in other sectors. Also, because of the stickiness of nominal wage incomes over the short period, product price adjustment can at best protect incomes from cultivation only. Thus, upward adjustments in agricultural prices made in response, for example, to increase in the prices of some manufactured consumer articles may fail to maintain the purchasing power of all rural classes uniformly; wage earners may in fact suffer a loss in real income following the rise in the price of food because food absorbs most of their expenditure. A uniform output price consistent with the income parity criterion cannot be calculated simply because the commodity baskets of consumption vary as between the different rural classes.

III. POLITICS

Political factors behind the making of economic policy are generally ignored in what may be described as 'mainstream' economics but constitute an important element of discussions of economic problems in the growing literature on Indian political economy. The general concern in the alternative stream is with production relations (such as those between landowners and labourers) and how they influence processes of change in the spheres of production and accumulation. However, the politics of price-fixing in the Indian context has received a lot of attention from both the camps.[21]

In the agrarian economy, the importance of politics with respect to policy-making arises from two basic factors. First, policies are formulated by governments at the state and the central levels, the Constitution and working of the government depend very much on the configuration of class forces. (For example, the failure of land reforms for imposing ceilings on landholdings and for redistribution of land, despite much legislation and endless rhetoric in day-to-day politics and plan documents, cannot be explained without a consideration of the balance of class forces and the manner in which it has influenced the implementation process). Second, and more readily seen, is the very large disparity in landholdings in all regions of India, with big landowners constituting a small proportion of rural households but controlling the major share not only of land and assets in general but also, consequently, marketed surpluses.

Rural disparities have grown further with the introduction of the new agricultural technology which has strengthened the economic and political power of the class of surplus producers. Described as the rural oligarchy this class had appropriated a disproportionately large share of the benefits generated by the policy of incentives and subsidies to the agricultural sector since the late nineteen-sixties. The political strength of the class is seen to lie, moreover, in its control over the rural vote in the electoral process, which enables it to further its own interest through bargaining with the other major ruling group, the industrial (urban) bourgeoisie. As a result there is in existence a politically powerful 'farm lobby' which influences policy-making in agriculture with a fair degree of success.[22]

The success is derived from two factors: First, given the nature of the electoral process and the decisive role of the rural vote in its outcome, there is a heavy representation of the 'farmer interest' in the state legislatures and governments. We should note that in reality the farmers interest relates to the amassing of incomes and wealth by the surplus producing big farmers, although the political rhetoric refers always to the 'peasant cause' in general, as if the peasantry constitute an undifferentiated mass; small, poor farmers and agricultural labourers have hardly any representation in any tier of the decision-making process. Second, practically all political parties accord support to the 'peasant cause' and to rural agitations on a variety of issues such as demands for 'remunerative' prices for the products of agriculture; for provision of inputs like water and electricity, and credit, at cheap rates; for cancellation of rural debt in hard times; and in general for all-round support at all times.

For most political parties, slogans about the rural poor area necessary means to secure political power through the electoral process. The bias of leftist support for the farmer interest has another dimension however. The Communist parties, for example, recognise landlords and capitalist farmers as constituting an exploiting class but seek to mobilise not only poor peasants and labourers but also the middie and richer sections of the peasantry in order to secure peasant unity considered necessary to carry forward the struggle against exploiting classes in general. These parties recognise the fact, for example, that the benefits from increases in agricultural prices flow in a disproportionately large manner, if not wholly, to the surplus-producing big farmers (and therefore, suggest measures of taxation on the rich and for relief for the poor to counteract the growth in inequality that prices increases entail) but support the struggles for 'remunerative prices' perhaps because of the priority accorded to the necessity for securing peasant unity from a long-term perspective.[23]

The course of events relevant to foodgrain price policy since the mid-sixties can now be described in brief. The heralding of the green revolution with the introduction of new seed varieties of wheat and the unequivocal commitment of the central government to the new agricultural strategy led to the support of wheat prices at levels prevailing during 1966 and 1967—a period of extreme scarcity—without any reference to principles of cost or parity or to the demand and supply conditions in a period of abundance resulting from the wheat surpluses in the north-west. The sole aim was to promote growth at any cost. During this period policy-making was wholly the prerogative of the chief ministers of the surplus states. Such attempts as were made by the APC to introduce some principles into the setting of relative prices and procurement targets proved to be futile.

The most important concern of the farm lobby till about 1973 appears to be the prevention of a fall in the price of wheat which the bumper harvests of 1968-72 would have normally entailed. It must be noted in this context that during this period rice surpluses and quantities procured were marginal; further some cost comparisons show that the margins of the procurement prices over the costs of production were far higher in the case of wheat than in that of rice (see below). The parity between rice and wheat seems to have been established only from about the mid-seventies,[24] following both the greater quantum of rice being procured and the conditions of relative scarcity prevailing in 1973 and 1974.

At any rate, the terms of trade moved generally in favour of agriculture during 1956-73. The trend has reversed since then, with the prices of manufactured articles and agricultural inputs rising at rapid rates and stocks of foodgrains piling 'up with the government as a consequence of a succession of good harvests. This adverse movement in the terms of trade is the background to numerous farmers agitations for remunerative prices. As incidental factor behind the struggles is the emergence in 1977 of peasant leaders like Charan Singh at the Centre. The eighties saw not only a reconstitution of the APC (to allow for explicit representation of the farmer interest) but also to its renaming as the Commission of Agricultural Costs and Prices, enshrining the cost principle in price-fixing. The farm lobby has in recent years been growing further with the mushrooming of self-styled 'non-party' 'peasant' organisations such as the one led by Sharad Joshi. The rhetoric of Joshi's movement refers to the exploitation of agriculture (Bharat) by industry (India) and the goals may therefore look broad-based but the programme has been aptly described as a one-point programme for securing higher agricultural prices in all seasons.[25]

IV. PRICE MOVEMENTS AND RELATIONS

Growth and Fluctuations in Output

In per capita terms the output of foodgrains in India as a whole has stagnated since the beginning of the sixties, implying thereby that the growth rate in foodgrain production just about equalled that of population.[26] However, different crops exhibit different trends; for example, there has been an impressive increase in the output of wheat but a distinct decline in that of coarse cereals and pulses (both in per capita terms). The impact of these differential trend on relative price movements as well as consumption levels, especially of the poor, has still to be studied in detail.

But more important perhaps is the fact that both at the specific crop level and in the aggregate there have been considerable year-to-year fluctuations in yield per acre as well as production. Since the production of foodgrains as a whole has stagnated in per capita terms, output fluctuations are clearly relevant to price movements in a period characterised by a growth—albeit modest—in real incomes per capita. Some early studies show that the price flexibility coefficient trends to be very high (more than two in absolute value) for cereals, such as, rice and wheat.[27] Of course, availability is not determined by levels of production alone; the operations of the government (through imports and distribution) and of surplus holding farmers and traders at the different tiers in the market economy are equally relevant. A good part of the price flexibility may be due to the speculative activities of the traders (singularly absent in studies of price trends), but output fluctuations contribute also to the quantum of grain that could be procured by the government at given prices, to amounts that have to be distributed through the public distribution system, etc., and thus are of central importance to the interaction of private and public agencies in the foodgrain economy, i.e., to the dual market economy.

From this perspective, the recently well-documented increase in fluctuations—with much inter-regional and inter-crop variation—in productivity and production of foodgrains in the post-green revolution period assumes importance[28] but is not adequately reflected in policy-making with reference to pricing, building-up of stocks and distribution. The studies which demonstrate this increase in instability argue that the improved technologies are not responsible *per se* for this trend in variances and point to a number of concomitant factors such as the sub-optimal use of irrigation and fertilisers, simultaneous increases in year-to-year variability in areas sown as well as an increase in the covariance between yields obtained and areas sown and so on; all this requires more detailed study at a greater level of disaggregation with respect to both crops and regions than has been attempted in the literature so far. A recent study on output fluctuations comes to the conclusion that "aggregate production instability is an inevitable consequence of rapid agricultural growth and there is little that can be effectively done about it" and suggests that policies addressed to the stabilisation of consumption through storage and trade policies may be more promising.[29] This conclusion is, however, contestable because a broad-based strategy for the improvement of yields and stability in output in the dry and rainfed tracts of the country is a clear alternative, admittedly not yet spelt out in detail on the basis of existing knowledge.

The supply of grain is more relevant to price movements than is domestic production. If one takes net production augmented by changes in stocks with the government as a first approximation to supply (designated usually as availability), this variable exhibits characteristics similar to those of production; since the early sixties net availability of foodgrains in per capita terms has virtually stagnated but year-to-year fluctuations have been fairly wide—although distinctly less wider than in output, with the government distributing somewhat higher amounts in years of crop failure.[30] The good harvests of the late seventies and early eighties have been utilised more for stock-building than for improving levels of availability. Indeed, levels of stocks which were about 3 million tonnes in the early seventies grew close to 30 million tonnes by 1985 (more on this later). The fluctuations in actual availability in the market might have been wider than these first approximations indicate, because of the possible speculative hoarding by the private trade.

Market Prices, Procurement Prices and Costs

In this and the following sub-sections we shall discuss the working of the dual markets in the foodgrain sector, especially in relation to the period after the mid-sixties marked by a significant intervention by government agencies in the market. Apart from procurement and distribution of certain quantities of grain at set prices, the government was involved also in stock-building operations. This is the basis for dual market models which seek to capture the interaction between the operations of the government and the functioning of the open market.[31] It must be noted, however, that a realistic formulation of such models in their fullest complexity is not an easy task, for it would require apart from the specification of supply and demand functions under the dual price structure, an explicit consideration of the role of exogenous output fluctuations and of the complexity introduced by different policy instruments such as zoning and stock-piling. Perhaps, it is for this reason that most available analytical models are addressed to narrowly focused issues such as the impact of price and other instruments on levels of procurement.

Let us begin with a descriptive account of the trends in prices and costs. Although procurement prices have generally been lower than market prices, both series exhibit a strong upward trend since 1960-61, characterised, in the case of foodgrains as a whole in India, by a clearly visible staircase type of movement, with prices rising significantly in poor crop years but remaining steady or increasing moderately in years in good harvest. The only exceptions to this type of movement—with prices declining in response to good crops—were in 1968-69, following the initial spurt in wheat production, and in 1975-76, the period of Emergency and all round 'discipline'.[32]

Two factors appear to underline this type of price behaviour: first, weather-induced output fluctuations (especially in per capita terms) of a wide order, combined with high price flexibility coefficients; and second, the bounty of good harvests being used more for stock-building than for price reductions. (The asymmetry between the protection of producer interests through price support—inherent in withholding supplies in good years—and the protection of consumers is perhaps most clearly seen in this aspect of government operations. The question of optimal level of stocks is discussed below).

It must be added however, that the nature of both the trends in prices and the order relations between procurement and market prices, portrayed above in the case of foodgrains as a whole for all-India, may not hold uniformly across regions and crops.

Turning now to a comparison between procurement prices and costs to see whether procurement has operated as a disincentive to production, let us recall that the comparison is beset with difficulties arising from the conceptualisation of cost. Studies in this regard rely on the average cost at the state level (estimated on the basis of sample surveys) with cost C as the relevant concept. This cost includes the imputed value of family labour as well as the imputed rental value of owned land. The conclusions are that in the case of wheat, since the early years of green revolution, procurement prices have been higher than—well above, in the initial part of this period—costs of production and in some years close to the open market prices. For rice, until the mid-seventies support prices did not cover costs in West Bengal and the southern states but they did in the case of northern states. Thereafter, however, procurement prices of rice have been raised substantially so as to cover costs uniformly in all the producing regions.[33]

It is possible to argue that uniform purchase prices (generally fixed in relation to costs prevailing in a high cost region) are likely to have promoted inter-regional inequalities because of the wide variation in costs (howsoever defined). For example in 1978-79, while the cost per hectare for paddy was higher in Punjab than in Andhra Pradesh (by about 11 per cent), the cost per quintal of paddy was lower (by about 22 per cent) because of the higher yields per unit area more favourable conditions for the cultivation of high yielding paddy varieties in Punjab.[34] It must be noted in this context that in the Punjab region paddy is produced almost exclusively as a commercial crop and the region contributes significantly to the rice procured by the government.

To the extent that farmers are free to sell on the open market, prices received by farmers (a weighted average of procurement and market prices, with the relevant quantities as weights) are more relevant than procurement prices alone to a judgement about disincentives. At a hypothetical but more sophisticated level one might compare costs with the prices that would have ruled in the open market in the absence of purchases and distribution by the government. The determination of such prices would require a properly specified supply-demand-prices model but crude calculations can be done on the basis of available demand elasticities and price flexibility coefficients under the assumption that output levels are exogenously given. At any rate, the few studies involving such price-cost comparisons show once again that procurement prices have in general been 'remunerative' in relation to cost.[35]

As noted before, the parity in relation to cost is only one among the different kinds of parity which can form the basis of price fixation. Although the APC has from time to time taken into account, apart from costs of production, prices of specific inputs such as fertilisers or the general price level in the price fixation process, presumably in conformity with some principle of parity, it is difficult to discern from data the working of any particular parity principle (such as that between crops, between input and output prices, or between prices paid and received) possibly because of the arbitrary manner in which elements relevant of different types of parity have in practice entered price fixing in different combinations over the years.[36]

The arbitrariness has made it easy for those who argue for price increases to invoke this or that principle of parity as the occasion demands. For example, if the prices of fertilisers are raised by the government, that alone is sufficient to ensure political mobilisation in favour of the demand for remunerative prices regardless of the *total* cost of production or market conditions. If in the following season the prices of fertilisers are reduced but the general price level rises (for whatever reasons), the basis for the demand for price increases may change; for example, it may shift to the cost of living or the real income of poor farmers. The point is important because of the differential trends in prices of commodities covering different crops, agricultural inputs and articles of consumption; it is easily illustrated; the demands for high support (procurement) prices of wheat were based on an appeal to the level and changes in market prices till about the mid-seventies but later, with the prices of some inputs and manufactured articles increasing in general at rates faster than those of agricultural commodities (including wheat), slogans for political mobilisation began to be centered more and more around input prices and costs. It will be interesting to analyse in this light the shifting ground of the farm lobby in response to the changing economic environment.

At a different level, among the academics, questions relating to the terms of trade (TOT) between agriculture and industry have received much attention.[37] They include questions about the relevance of TOT to price policy and mobilisation of rural surpluses in the specific Indian context as well as those relating to the actual movements of relative prices. In contrast to the first set of these questions, the second has generated much controversy in the literature. This is probably a fall-out of the political process in which slogans based on relative prices have, as remarked before, assumed much significance. In any case, the shift in the focus is unfortunate because what is relevant to policy-making concerned with welfare is not so much the actual movement in the net barter TOT as its impact on aggregate farm output, farm incomes and their distribution. It must be pointed out in this context that generally the supply response is low and, therefore, the question to what extent agricultural growth can be stimulated through changes in TOT brought about by policies with a limited coverage of crops and regions becomes crucial to our understanding of such policies.

The picture of actual movements in the TOT (whether in net barter or in net income terms) depends on how the relevant indices are constructed, the commodity compositions taken, the weights assigned, the price data employed, the base year chosen and so on. Indeed, the studies in this area demonstrate, if anything at all, that constructed series are extremely *sensitive* to methods of construction, yielding as they do sometimes opposing conclusions about the direction in the movements. A continuation of this debate can be fruitful only if this sensitivity is examined more closely and in relation to the impact of the TOT on the different aspects of the farm sector.

Imports and Stocks

One of the objectives underlying the promotion of high yielding varieties and the new technology was the attainment of self sufficiency in foodgrain production, i.e., a drastic reduction in the dependence on imports. Indeed, the role of imports in maintaining supplies has become insignificant from about the mid-seventies.

Both imports and internal procurement enable the government to build up stocks needed to maintain supplies at given levels and stabilise the prices, i.e., to curb price movements arising out of the output fluctuations characterizing Indian agriculture. Two questions arise in this context, the first about the optimal level of stocks to be held by the government and the second relating to the possibility of choice between imports and domestic procurement for the purpose and the conditions governing such a choice. It is obvious that for meaningful analysis the questions have to be set within a framework that includes not only the dual domestic markets but also the international market for foodgrains.

But let us begin with a brief review of the facts. There were broadly two phases, covering 1968-72 and 1975-85, during which stocks held by the government have been allowed to increase, from 2 to 8 million tonnes during the first and from 8 to roughly 30 million tonnes during the second phase respectively. Both these phases cover strings of good harvest, with peak and record levels of per capita production during the early eighties. While both imports and procurement contributed to this stock piling during the first phase, it was almost wholly the internal efforts that enabled the government to accumulate stocks during the second phase. (The droughts of 1986 and 1987 have led to a depletion of stocks). As noted before, the government has generally been releasing somewhat larger amounts, in per capita terms, for distribution through the PDS under distress conditions, but these amounts have only marginally augmented the fluctuating production levels. Generally, the levels in per capita availability have fluctuated along with those in production, although the year-to-year variations were smaller in the former. It is clear that the government's stock management has not been successful in bringing about a stability in supplies, a fact readily seen from the low levels of per capita net availability in years such as 1975 and 1980 corresponding to poor harvests (406 and 410 gms per day respectively, in contrast to an average of 447 gms over the whole period 1961-84). More importantly, the precedence accorded to the policy of accumulation of stocks over that of depletion inadequate amounts led to the expected price changes: large increases in years of crop failure and moderate changes during times of good harvest following the use of procured grain for accumulating stocks to prevent prices falling. It can be seen thus that the stock management policy of the government has contributed in some measure to the inflation in grain prices and to the disadvantage of consumers.[38]

The question of optimal level of stocks assumes importance in the context.[39] The gross output of foodgrains per capita per day has over the period 1961-84 stagnated around an average of 494 gms with a coefficient of variation of 8 per cent. (The computation includes the worst ever years, 1966 and 1967). This means that actual output levels in any year are unlikely to deviate from 494 gms by more than 12.8 per cent, i.e., 63 gms per capita per day (1.6 times the standard deviation corresponding to a 5 per cent chance for larger deviations, under the assumption of a normal distribution). For a population of 800 million this corresponds to a predictable shortfall not exceeding 18.4 million tonnes (with a probability of 95 per cent) in a single year. The technical committees constituted by the government during the sixties and seventies have recommended buffer stocks in the range of 10 to 15 million tonnes (for lower levels of total population) on a basis of probability calculations similar to the one above.[40]

An insurance cover against two consecutive poor harvests would of course require a much larger buffer stock but the policy is likely to be extremely costly and economically crippling. In any case, it is difficult to argue in favour of a stock holding in excess of 20 million tonnes at the current population level, especially if the possibility of imports (during a second consecutive year of poor crops) is kept in mind. Indeed, some studies show that the government's stock policies have been sub-optimal in many ways: they demonstrate that higher level of per capita supplies than in the past could be maintained with appropriately designed and coherent policies for fixing prices, accumulating and depleting stocks, determining import levels and so on.

Above all, if the range of price variation is to be limited so as to benefit both producers and consumers, any sensible policy must include the depletion of stocks in adequate amounts when necessary as well as accumulation to the desired extent. The point is important because the increase in inventories beyond optimal levels not only imposes a crippling burden on the economy but is also regressive in character. For example, the so-called food subsidy—mounting and accounting for a big proportion in the budget deficits of the Central Government in recent years, and usually interpreted as a subsidy to the poor covered by the PDS—includes the costs of holding stocks; it can be argued that the poor would be better off if prices are allowed to decline in response to good harvests especially when the stocks held by the government are already adequate to meet contingencies such as drought. Accumulation of stocks beyond the level needed to cover contingencies benefit the surplus producers only at the cost of the consumers, especially the rural poor who have no access to the PDS.

Analytical Issues

The highly correlated movements in costs, procurement prices and market prices, both contemporaneously and with lags in the variables, are not very useful for assessing the impact of state intervention in foodgrain markets. Given the limited coverage of government operations, with respect to crops and regions on the one hand and quantities procured and distributed on the other, it would appear *prima facie* that the impact of these operations, for example, on market prices, would also be limited with regard to regions and crops. On this understanding, the impact of operations in the wheat market would be limited to regions in the north-west where the procurement effort is concentrated and the regions—mainly large urban areas—where subsidised wheat is distributed in large enough quantities to influence the open market prices. This is questionable however, because to an unknown extent the markets for different cereals are integrated, with demand, supply and price conditions in the market for one cereal influencing those in the markets for other cereals. The explanation for a strong correlation between market prices and support prices in the case of foodgrains as a whole may lie in such an integration, not addressed adequately in the literature. Our ignorance in this respect is attributable in part to the lack of the relevant data; for example, on prices in the rural areas of both the 'surplus' and 'deficit' regions; but also to the absence of analyses of the required nature, for example, of how prices of inferior cereals are influenced (in the different regions of the country) by those of rice and wheat; of trends in the regional dispersion in price levels and so on.

In contrast, the phenomenon of increases in market prices leading on, with some lags, to increases in procurement and issue prices, analysed in some detail is perhaps easier to understand.[41] In general a rise in foodgrain prices can be expected to lead to a rise in wage costs, albeit with time lags in adjustment. More importantly, the burden imposed on the government by its own operations in the foodgrain markets, is relieved in practice, although only in part, through increases in the so-called administered prices, for example, fertiliser prices, and these increases have a direct bearing on the movement of agricultural costs. Finally, increases in costs of production lead to pressure on the government through the political process to raise support prices. Observed correlations between procurement prices and lagged market prices (and costs) notwithstanding, the chain of causation in this respect requires further study on the support of available documentation on the disposal of subsidies and the budgetary transactions of the central government.

In sum, the elaboration of the means for achieving some form of inter-crop parity and reducing inter-regional disparities in prices, in consistence with a set of well-defined welfare goals, remains elusive in Indian economic literature.

V. NET IMPACT

What is the net impact of foodgrain price policy on the economy in general, on the processes of inflation, investment, growth and distribution, and more particularly on growth rates in agricultural production, the levels of prices and consumption of foodgrains of different types in the different parts of the country?

The question cannot be answered easily, and not merely because it encompasses many interrelated issues. Three other reasons are pertinent; first, the quantitative significance of the limited nature of state intervention in the foodgrain markets is still to be assessed in meaningful analytical terms. (For example, we know practically nothing in this context about rural urban price differentials) Second, while there is no doubt that relative prices, irrespective of the extent to which they have been influenced by actually implemented price policies, have played a significant role in shaping the economy, it is necessary to take into account the working of taxes and subsidies of different kinds for understanding this role. And third, there is a need to incorporate into the analysis the experience with non-price instruments such as direct public investments in irrigation development, etc. and so on.

What follows covers a limited ground only.

Output

Numerous studies deal with supply response—grist to the agricultural economics mill.[42] Barring a few exceptions, they are concerned with the response in terms of the allocation of areas of different competing crops corresponding to relative prices and profitability. These seek to demonstrate, in general, that the Indian farmer is a rational economic animal. The exceptions consider also the response in terms of the increase in input use, especially of modern inputs such as fertiliser, relevant to the generation of higher profits through increases in the productivity per unit land area. We thus have some plausible estimates of the overall output response (combining both the area and input responses). The long-run elasticity of output with respect to the lagged output-input price ratio is in the range of 0.4 to 0.6 for the major cereals: rice and wheat, and the supply response, reckoned thus, is lower than the response of output to increases in the levels of irrigation.[43] These estimates based on past data summarise the Indian experience in some fashion and show therefore that agricultural growth has been promoted more by the development of irrigation than by the price factor.

This is not to deny the very significant role support prices and subsidies (and hence the terms of trade) have played in the promotion of agricultural growth in certain area of the country. In general only some crops and some regions have benefited from the mix of policies (pricing, fiscal and other) and this has promoted both inter-crop differences in growth and inter-regional inequalities in agricultural performance. The impressive growth in the production of superior cereals has been brought about to some extent through shifts in the cropping pattern away from inferior cereals and pulses, and in some areas at the expense of commercial crops too. These shifts have been made possible by the policy-mix referred to earlier. In the overall, a substantial part of the increment in production associated with the new technology has originated in five or six advanced states with the eastern region and semi-arid parts of the country performing poorly. Both price policies and the easy availability of subsidised funds have promoted these developments.[44]

Income Distribution and Poverty

The exacerbation of inter-regional inequalities in foodgrain production and the decline in per capita production of inferior cereals and pulses are strongly suggestive of a general widening of disparities in income and consumption. Let us consider in brief some relevant analytical factors. Procurement prices, farm harvest prices and wholesale market prices are all relevant to the determination of the price producers receive for their surplus crop output. In general and in the simplest case price increases are regressive in their impact on the income distribution among the producers because the bigger producers account for the lion's share in the marketed part of the output and therefore appropriate the gains, if any, accruing from a price rise in a larger proportion than do the small farmers. Of further relevance in this context is the fact that small farmers and agricultural labourers are in general net buyers of foodgrain in contrast to the big farmers who are net sellers, so that by virtue of their capacity for retention in full measure of grain for home consumption the latter are not hurt by the rise in market prices. Likewise, the seasonal movements in prices, especially in years of poor harvest, have a regressive impact since the poor as net buyers may depend wholly on market purchases in the lean season when prices tend to be higher.

However, for a more complete assessment of the impact of price increases on income distribution it will be necessary to take into account incomes earned through non-firm activities by wage earners and small farmers. In the absence of detailed research of this type, some analysis have sought to directly relate estimated poverty levels to changing relative prices to see, for example, whether rising food prices hurt the interests of the poor. This effort has also led to some controversy because the same data sets analysed in different ways produce contradictory results. The underlying methodological difficulties have received much attention in the recent literature.[45]

But it is difficult to contest some relevant empirical findings. It has been demonstrated convincingly that at the all-India level poverty ratios (the proportion of people below the so-called poverty line) exhibit fluctuations with no discernible trend—positive or negative.[46] The variations in the ratio are, however, correlated—enough to attract much academic attention—to both real income (inversely) and food prices (directly), suggesting thereby that while agricultural growth contributes to a reduction in poverty rising food prices may act independently of levels of output and lead to decreases in the consumption of food and hence to higher levels of poverty.

Regional disparities assume importance in this context. Inter-state differences in the per capita production of foodgrains have widened. But through its operations of procuring grain from the pockets of surplus and distributing it through the PDS the government has ensured that regional disparities in terms of availability have not increased since the mid-sixties. However, except in Kerala and Gujarat (and more recently Tamil Nadu and Andhra Pradesh) and PDS hardly covers the rural areas. An early study dealing with inter-state differences in food intake, covering both the rural and urban parts, is relevant in this context.[47] It suggested that given the magnitudes of regional disparities in incomes and prices, the markets work in such a way that a part of the surplus tends to remain in the surplus-producing regions and deficits left uncovered in other regions. A reason could be that the rural areas in the deficit regions do not attract market supplies, given the range of price differences and the organisation of markets. Therefore, the implied lack of adequate price quantity adjustments through market forces has generally a greater impact on rural than on urban consumption levels. Of course, the public distribution system reinforces this tendency so that generally while the urban poor are ensured of supplies at given levels, the rural poor eat better only if the local harvest is good.

Concluding Remarks

Price policy refers to the doings of the government, but prices are not determined by policies alone; the markets play a part and so do some non-market factors un-related to policy. It is not easy then to unscramble observed price trends in relation to the different aspects of the economy to arrive at unambiguous judgements about the correctness or otherwise of this or that policy. Relatively high and rising agricultural prices favour surplus-producing big farmers and may indeed be necessary to promote agricultural growth at a satisfactory rate. But they have adverse effect on the costs and standards of living of the rural and the urban poor, and on a wider stretch, as some have argued convincingly, may also lead to industrial stagnation.[48]

The subject remains controversial and the debate on each specific underlying question is unresolved despite the participation in the debate of several eminent economists and political leaders. To some extent there is in all this a loss of perspective resulting from scholarly affinity to hair splitting. But it has led to a polarisation of sorts.

One school denounces 'cheap food' policies and argues for price-support programmes that give adequate incentives to stimulate supply. On the other hand, it can argued that aggregate supply response is low, and that more importantly, price-support programmes imply high private and social costs, price incentives fundamentally result in income transfers to the large farmers. Consequently, a well-balanced package of government instruments including promotion of technology, infrastructure investments and production subsidies is preferable to a reliance on the price instrument alone. The literature suggests that public investment, say, in irrigation is better than the provision of price incentives for not only promoting growth but also for 'cost-efficiency' and distributing gains from growth uniformly across the different social and economic classes in India.[49]

These conclusions emerge in part from some well known facts; agricultural growth has remained fairly stable since independence at about 3 per cent per annum; the rates of growth in foodgrain production have been somewhat lower and there is no evidence of an improvement in the levels of consumption; the green revolution has exacerbated inter-regional inequalities in the production of foodgrains, productivity increases having been largely confined to wheat in the north-west; the Public Distribution System (PDS), restricted mainly to the urban area, has failed to protect the rural poor; and the social costs of public intervention in this sector, including direct subsidies to agriculture and house involved in the operation of the PDS have heavy and rising. The point is that these subsidies cut into the resources meant for development expenditure, apart from increasing rural urban inequity since the vast mass of the rural poor are denied access to the PDS.

But the conclusions have been, and continue to be disputed. In some measure the controversies arise from the fact that while the issues can be stated in fairly simple terms (such as whether price policy has contributed to agricultural growth), they cannot be satisfactorily resolved (through, for example, a regression of production on some price variable to measure the supply response) because of the complexity of the structure of economic and social relations within which the question have to be addressed for a meaningful analysis.[50]

Note : Reproduced from *Economic and Political Weekly,* June 30, 1990, pp. A54-A64 with kind permission of the publisher.

NOTES

1. Internal procurement of foodgrains was generally not high till the mid-sixties; thereafter it gradually increased in quantum from about 4 million tonnes to the current level of about 15 million tonnes a year. For a discussion of the evolution of food policy and data relating to the government operations during the earlier period see Chopra (1981). Comprehensive data on the foodgrain economy are also to be found in the annual publication : *Bulletin on Food Statistics.*
2. The general economic setting for the adoption of the new technology as a part of overall growth strategy is explained in Chaudhuri (1978).
3. Apart from these bodies, the Americans have also contributed to our wisdom about the food problem. The Ford Foundation compiled a report in 1959 entitled *India's Food Crisis and the Steps to Meet it.* For details see Chopra (1981).
4. The terms of reference are reprinted in the Appendix to the *Report of APC on the Price Policy for Kharif Cereals, 1965-66,* p. 47. For some discussion of these terms see Kahlon and Tyagi (1983).
5. The revised terms of reference are contained in a notification of the Government of India, No 14011/2/78, Ministry of Agriculture, dated March 5,1980.
6. de Janvry and Subbarao (1986) p 17. They add that there is no representation for consumers and on the commission.
7. Raj Kishna, who was closely associated with the APC, had written : "A price support programme is unavoidably a disequilibrium programme. The case for it is only that, if effective, it may convert a diseqilibrium of shortages into a disequilibrium of supplies in the markets for *selected* commodities". Krishna (1967) p 525. Emphasis added.
8. It must be noted that there are two kinds of distress. Small farmers sell a part of their output, sometimes at price disadvantage, although they may in general be net buyers of grain. This form of distress is clearly distinct from farmers—big and small—having to sell at low prices arising from bumper harvests. An appreciation of this distinction is absent in discussions on

support prices.

9. The price flexibility coefficient has been estimated at—2.38 for foodgrains as a whole and—2.24 for cereals. See Ray (1972) for details. Earlier contribution in this respect are those of Thamarajakshi (1970a and 1970b). These also demonstrate high price-flexibility coefficients for rice and wheat.
10. The principles for price fixation by the government for its limited intervention in foodgrain markets have been discussed threadbare in the literature published since the mid-sixties. Some relevant references are : Dandekar (1965), Dantwala (1967). Krishna (1967 and 1982) and Thamarajakshi (1977). Attempts to sum up the dabtes have been made, for example, in Kahlon and Tyagi (1983), Subbarao (nd.) and Rath (1985). No such attempt is made in the following text. It should be noted, however, that much of the literature is concerned with the efficacy of stimulating growth through price incentives. Issues relating to distribution are addressed somewhat tangentially.
11. Dandekar (1965).
12. Krishna (1967), pp 526-27.
13. Assuming a realistic estimate of 0.4 for the long-run elasticity of output with respect to the terms of trade, Raj Krishna says : "...16 per cent growth over five years (in output) would require a one-shot 40 per cent increase in the real terms of trade of agriculture. This is equivalent to 7 per cent annual increase over this period, which will also, of course, spread out the resulting growth. This order of terms-of-trade increase is hardly a practical proposition, even assuming that a government can fix terms of trade". In the sequel, he concludes "...a unit percentage change in the important shifter variable (technology) will yield much greater growth than a unit percentage price shift", Krishan (1982) pp 235-36.
14. Procurement as a tax on farm produce has a long history in the development phase of many countries. This has a bearing on the larger issue (not treated in this survey) of using surpluses from the farm sector to promote industrialisation. See in this context Krishna (1982) and Mitra (1979).
    However, in India this has a narrower, more immediate context as recognised, for example, by the APC itself. An early APC report says: "It should be a social obligation on the part of cultivators to surrender a portion of the produce to the state in the wider national interest...While direct taxation involves no *quid pro quo,* in the case of procurement the transactions would still be in the natural of buying and selling and the producers would receive a price, even though this price may be lower than the prevailing market price. The case for such as impost becomes particularly strong in view of the extremely light incidence of agricultural taxation in the country". *Report of the APC on Price Policy for Kharif Cereals (Procurement Prices) for 1967-68 Season,* September 1967, p 5.
15. See Krishnaji (1973) and Nadkarni (1987), p 203.
16. The prices fixed by the State Governments were generally higher than

those recommended by the APC and the Central Government. A study making a detailed demonstration of this fact says : "The phenomenon of states fixing prices higher than the recommended prices is explained by the influence exercised by farmer interests on policy at the state level. But it is also due to the realistic economic reasoning that the procurement targets recommended by the centre could not be realised with unattractive purchase prices, especially in bad crop years." Krishan and Raychaudhuri (1980) p 4.

17. Subbarao (1978) summing up the literature on zoning says : "That food zones actually accentuated regional price differences is an empirically well established fact." However, there is some evidence which indicates that the relaxation of zoning in the north-west during the initial spurt of wheat production is a response to the fear that prices would fall and impede growth in that region. Krishnaji (1973).
18. Subbarao (nd). See also Kahlon and Tyagi (1983).
19. Krishna (1987). pp 517-21.
20. What follows is largely based on Subbarao (nd).
21. The debates on price policy are not restricted to professional academic journals, they are conducted also, fairly regularly, newspapers, at public meetings and within the states and central legislatures.
22. For a cogent exposition of this view, which attracted a good deal of controversy sea Mitra (1979).
23. Representative views of the different parties are explained, for example, in (CPI(M)(1973 and 1981), Nathan (1982) and Sinha (1980).
24. See Krishnaji (1975) and Krishan and Raychaudhuri (1980).
25. The growth of farmers' lobbies and movements has been traced in Nadkarni (1987).
26. The growth rate calculated separately for aggregate output shows a higher value than that of population but the growth in per capita output is found to be statistically insignificant. This is a statistical artefact. The data are given by Roy (1984) and Krishnaji (1988). See in particular the graphical representation of the data in Roy.
27. Ray (1972).
28. Mehra (1981), Hazell (1982) and Ray (1970), 1971 and 1983).
29. Hazell (1982), p 10.
30. Roy (1984).
31. For example, Krishna and Chibber (1983) and Gulati (1987).
32. For the data see Roy (1984) and Krishnaji (1988).
33. For cost-price comparisons see Krishna and Raychaudhuri (1980).
34. de Janvry and Subbarao (1986) pp 20-27.'
35. See in this context the discussion in Roy (1984); also for a case study of rice, Subbarao (1978a).
36. A review of the actual process of price fixation by the APC is contained in Kahlon and Tyagi (1983).

37. For a review of the debate see Vittal (1966). See in particular Mitra (1979) and Tyagi (1979).
38. See Roy (1984) for the data and Krishnaji (1988) for further discussion.
39. Questions relating to stock polices have been addressed in Ray (1970). Well (1978) Krishna and Chibber (1983). Krishnaji (1984) and Reutlinger (1978).
40. For details of these recommendation see Acharya (1983).
41. Krishna and Raychaudhuri (1980) discuss the empirical determinants of procurement prices.
42. Askari and Cummings (1976).
43. Krishan and Raychaudhuri (1980).
44. Subbarao (1985). See also Mahendra (1987).
45. For a debate on the poverty price relationship see Millor and Desai (1986) and for a comment on the statistical methodology see Krishnaji (1987).
46. See Bhattacharya *et al* (1985).
47. See Chapter 1 in United Nations (1975).
48. Mitra (1979). This important aspect of relative prices is not dealt with in this survey .
49. This section draws heavily from de Janvry and Subbarao (1986).
50. de Janvry and Subbarao (1986 attempt to grapple with this complexity through a Computable General Equilibrium (CGE) model, which enables them in carryout several simulation exercises. They find that technological change in agriculture and irrigation development *with flexible prices* are essential for alleviating poverty. Price flexibility benefits both rural and urban workers through a combination of employment and deflat is very effects in general it benefits the net buyers of food through relatively low prices. Generally, they find price incentive policy an inferior alternative to technological development, especially in relation to welfare goals.

REFERENCES

Acharya, K. C. S., (1983), *Food Security System in India,* Concept, New Delhi.

Askari, H. and Cummings, J. T. (1976), *Agricultural Supply Response : A Survey of the Econometric Evidence,* Praeger, New York.

Bhattacharya, Nikhilesh *et. al.* (1985), *Relative Price of Food and the Rural Poor. The Case of India,* Indian Statistical Institute (mimeo).

Chaudhuri, Pramit, (1978), *The Indian Economy : Poverty and Development,* Vikas, New Delhi.

Chopra, R.N. (1981) *Evolution of Food Policy in India,* Macmillan India.

Communist party of India (Marxist), (1973), *Central Committee Resolution on Certain Agrarain Issues and An Explanatory Note by P. Sundaryya* (party document).

—1981, *New Peasant Upsurge : Reasons and Remedies,* Documents and Resolutions at the AIKC meeting at Trichur.

Danadekar, V. M., (1965), 'Minimum Support Prices for Foodgrains : Guidelines for a Policy and a Programme', *Artha Vijnana,* Vol. 7, No 4, pp 273-83.

Dantwala, M. L., (1967), 'Incentives and Disincentives in Indian Agriculture', *Indian Journal of Agriculture Economics,* Vol. 22, No 2, pp 1-25.

—(1976), 'Agricultural Price Policy in India since Independence,' *Indian Journal of Agricultural Economics,* Vol. 31, No 4, October-December. 9, 37.

de Janvry, Alain and Subbarao, K, (1986), *Agricultural Price Policy and Income distribution in India* Oxford, Delhi.

Government of India, Reports of the Agricultural Prices Commission, various issues.

—*Bulletin of Food Statistics,* various issues Gulati, Ashok, 1987, *Agricultural Price Policy in India An Econometric Approach,* Concept, New Delhi.

Hazell, Peter B. R., (1982), *Instability in Indian Foodgrain Production,* International Food Policy Research Institute, Report No 30, Washington.

Kahlon, A. S., and Tyagi, D. S., (1980), 'Intersectoral Terms of Trade, *Economic and Political Weekly,* Vol. 15, Review of Agriculture, pp A173-84.

—1983, *Agricultural Price Policy in India,* Allied Publishers, New Delhi.

Krishna Raj, (1967), Agricultural Price-Policy and Economic Development' in Herman M. Sourthworth and Bruce F. Johnston (eds.), *Agricultural Development and Economic Growth,* Cornell, New York.

Krishna, Raj and Raychaudhuri, G. S., (1980), *Some Aspects of Wheat and Rice Price Policy in India,* World Bank, Staff Working Paper, No 381.

Krishna, Raj and Chibber, Ajay, (1983), *Policy Modelling of Dual Grain Market : The Case of Wheat in India,* Report No. 38, International Food Policy Research Institute.

Krishna, Raj (1982), 'Some Aspects of Agricultural Growth, Price Policy and Equity in Developing Countries', *Food Research Institute Studies,* Vol. 18, No. 3, pp. 219-60.

Krishnaji, N., (1973), 'Wheat Price Movements: An Analysis', *Economic and Political Weekly,* Vol. 8, Review of Agriculture, June.

—(1975), 'State Intervention and Foodgrain Prices', *Social Scientists,* Vol. 3, Nos. 6/7.

—(1987), 'Agricultural Growth, Prices and Poverty', *Economic and Political Weekly,* Vol. 22, No. 6, June.

—(1988), 'Foodgrain Stocks and Prices' in Amiya Kumar Bagchi (ed.), *Economy, Society and Polity: Essays in the Political Economy of Indian Planning,* Oxford, Calcutta.

Krooneman, P.J. (1990), 'Managing Towards Sustainability: The Common Property Resource Approach' (a paper presented to the International Seminar organised by the Commission on duration and Training, Bangalore).

Ladejinsky, Wolf (1977), *Agrarian Reform as Unfinished Business,* Oxford University Press.

Mahendra Dev, S. (1987), 'Growth and instability in Foodgrains Production: An Inter-State Analysis', *Economic and Political Weekly,* Review of Agriculture, September, pp. A82-92.

Mehra, Shakuntala, (1981), *Instability in Indian Agriculture in the Context of the New Technology,* International Food Policy Research Institute, Report No. 25.

Mellor, John W and Desai, Gunvant, U (eds), (1986), *Agricultural Charge and Rural Poverty,* Oxford, Delhi.

Mishra, G. P. (1988), 'Technological Change and Agricultural Wages in Uttar Pradesh' in *Development and Change in India,* R. T. Tiwari and A Joshi (eds.), Ashish, New Delhi.

Mitra, Ashok, (1979), *Terms of Trade and Class Relations,* Rupa and Co., Calcutta, also published by Frank Cass, London, 1977.

—(1989), 'Agrarian Reform, Output Growth, Wages and Employment in UP's

Agriculture' (paper prepared for the Study Group on Land Reforms and Common Property Resources), Giri Institute of Development Studies, Lucknow.

Nadkarni, M.V., (1987), *Farmers' Movements in India,* Allied Publishers.

Nadkarni, M. V. and R. S. Deshpande (1983), 'Growth and Instability in Crop Yields—A Case Study of Agriculture in Karnataka, Regional Studies' *Journal of the Regional Studies Association,* Cambridge University Press, Vol. 17, No. 1, February, 1983.

Nathan, Dev, (1982), 'On Agricultural Prices', *Economic and Political Weekly,* Vol. 17, No. 36, pp. 2101-04.

National Commission on Rural Labour (1990), 'Background Paper for National Seminar on Agricultural Labour held in Delhi in March, 1990.

Pandey, P. S. (1990), 'Land Reforms in UP—Problems and Prospects' (paper prepared for the Study Group on Land Reforms and Common Property Resources), Academic Staff College, Gorakhpur University, Gorakhpur.

Pasha, S. A. (1990), '*Field Notes* on the Status and Use of Common Property Lands' (unpublished), Institute for Social and Economic Change, Bangalore.

Patel, B. B. (1990) 'Land, Land Reforms and Dimensions of Survival Struggle of the Rural Labour in Gujarat' (paper prepared for the Study Group on Land Reforms and Common Property Resources), Gandhi Labour Institute, Ahmedabad.

Planning Commission (1989), 'Land Reforms A Retrospect and Prospect' (papers and proceedings of a seminar), Planning Commission (Rural Development Division).

—(1989), 'Background Note for Seminar on Land Reforms, Planning Commission (Rural Development Division).

—(1990a), 'Employment: Past Trends and Prospects for 1990s' (Working Paper).

—(1990b), *Towards Social Transformation: Approach to the Eighth Five-Year Plan 1990-95.*

Prasad, Pradhan H. *et al* (1989), *The Dynamics of Employment and Poverty in Bihar* (mimeo), A. N. Sinha Institute of Social Studies and International Labour Organisation, Patna.

Raj, K. N. *et al* (1983), 'Agrarian Reform in Kerala and its impact on the Rural Economy—A Preliminary Assessment' in *Agrarian Reform in Contemporary Developing Countries,* Ajit Kumar Ghose (ed.). Croom Helm, London.

Rajapurohit, A. R. (ed.) (1984). *Land Reforms in India,* Ashish, New Delhi.

Rao, V.M. (1986), Agricultural Growth in India : A Review of Experiences and Prospects', *Economic and Political Weekly.*

—(1988). 'Interventions for the Poor: Critical Dimensions, Potentialities and Limitations', *Economic and Political Weekly,* (Special Number).

Rath, Nilakantha, (1985), 'Prices, Costs of Production and Terms of Trade of Indian Agriculture', *Indian Journal of Agricultural Economics,* Vol. 40, October-December, pp. 451-81.

Ravindranath, H. H. *et al* (1990), 'Guidelines for Micro-level Planning for Integrated Development of Wastelands', Centre for Ecological Science, Indian Institute of Science, Bangalore.

Ray, S.K., (1970), 'Imbalances, Instability and Government Operations in Foodgrains', *Economic and Political Weekly,* Vol. 5, Review of Agriculture, September, pp. A115-24.

—(1971), 'Weather and Reserve Stocks of Foodgrains', *Economic and Political Weekly,* Vol. 6, Review of Agriculture, September, pp. A131-42.

—(1972), 'Effects of Consumption, Availability, Fluctuations on Foodgrain Prices',

Economic and Political Weekly, Vol. 7, Review of Agriculture, June, pp. A86-92.

—(1983), *Growth and Instability in Indian Agriculture,* Institute of Economic Growth (mimeo), Delhi.

Reddy, Atchi M. *et al* (1990), 'Economics of Dry Farming : A Study of Cotton Crop in Rayalaseema 1800-1980 (paper presented to the Eighth Annual Conference of Andhra Pradesh Economic Association, Hyderabad).

Reddy, V. R. (1990), 'Drought and Land Utilisation in Andhra Pradesh' (paper presented to the Eighth Annual Conference of Andhra Pradesh Economic Association, Hyderabad).

Reutlinger, Shlomo, (1978), 'The Level of Stability of India's Foodgrain Consumption', *Indian Occasional Papers,* World Bank, Staff Working Paper No. 279.

Roy, Shyamal, (1984), 'Government Management of Foodgrains Supplies : A Critical Appraisal', *Towards Continuing Education,* Vol. 5, pp. 1-19.

Saha, Kalyani (1989), *'Impact of Land Reforms on Rural Labour: A Case Study of West Bengal 1981-85'* (mimeo) Indian Institute of Management, Bangalore.

Sharma, Usha S. (1990), 'A Study of Impact of Agricultural Land Ceilings Act on Economic Conditions of Surplus Land Allottees in Gujarat (mimeo), Sardar Patel Institute of Economic and Social Research, Ahmedabad.

Sidhu, M. S. *et al* (1984), *A Study of Migrant Labour in Punjab*, Punjab Agricultural University, Ludhiana.

Singh, A. K. (1981), *Patterns of Regional Development,* Sterling, New Delhi.

Sinha, Indradeep, (1980), *The Changing Agrarian Scene, Problems, and Tasks,* Peoples Publishing House, New Delhi.

Singh, J. P. (1990), 'Land Reforms: Policy, Legislation, Implementation and Impact in Northern Region of India', (paper prepared for the Study Group on Land Reforms and Common Property Resources), Agricultural Economic Research Centre, University of Delhi, Delhi.

Sivanandan, P. (1989), 'Caste and Economic Opportunity: A Study of the Effect of Educational Development and Land Reforms on the Employment and Income Earning Opportunities of the Scheduled Castes and Scheduled Tribes in Kerala', Ph. D. thesis, Centre for Development Studies, Trivandrum.

Sivanandan, P. *et al* (1986), 'Land Hunger and Deforestation : Case Study of Cardamom Hills in Kerala', *Economic and Political Weekly*, March, 29.

Sub-Committee of the Parliamentary Consultative Committee (1987), *Report on the Problems of Unorganised Workers in Agriculture Sector.*

Swamy, M. C. K. (1990), 'Conserving Common Property Resources: Changes in Control and Management' (paper prepared for the Study Group on Land Reforms and Common Property Resources).

Subbarao, K., (1978), 'Rice Price Behaviour and Public Procurement: An Analysis of the Experience of Andhra Pradesh', *Indian Journal of Agricultural Economics* Vol. 33 No. 3, pp. 1-20.

—(1978a), *Rice Marketing Systems and Compulsory Levies in Andhra Pradesh,* Allied Publishers, New Delhi.

—(1979), 'Producers Levy Evasion and Income Loss', *Economic and Political Weekly,* Vol. 14, Review of Agriculture, March, pp. A2-7.

—(1985), 'State Policies and Regional Disparity in Indian Agriculture', *Development and Change,* October.

—(nd), Farm Prices: A Survey of the Debate, Institute of Economic Growth (mimeo).

Thamarajakshi, R. (1970a), 'Determinants of Rice Prices' *Agricultural Situation in India,* March, pp. 1075-1979.

—(1970b), 'Determinants of Rice Prices' *Agricultural Situation in India,* May, pp. 129-36.

—(1977), 'Role of Price Incentives in Stimulating Agricultural Production in a Developing Economy' in D. E. Ensminger (ed.), *Food Enough or Starvation for Millions,* Tata McGraw Hill, pp. 376-90.

Technical Committee, Department of Rural Development (1989), *Computerisation of Land Records: Review of On-going Pilot Projects and Suggestions for the Future.*

Thimmaiah, G. and Abdul Aziz (1984), *Political Economy of Land Reforms,* Ashish, New Delhi.

Tyagi, D.S., (1979), 'Farm Prices and Class Bias in India', *Economic and Political Weekly,* Vol. 14, Review of Agriculture, September, pp. A111-24.

United Nations, (1975), *Poverty, Unemployment and Development Policy.*

Verma, M. S. (1985), *Consolidation of Agricultural Holdings: A Case Study of Bihar and Orissa,* University of Lucknow.

Vittal, Nalini, (1986), 'Intersectoral Terms of Trade in India : A Study of Concept and Method', *Economic and Political Weekly,* Vol. 21, Review of Agriculture, December, pp. 147-66.

Wadhwa, D G (1989), 'Registration of Title to Land : A Preliminary Study', Committee for Records-of-Rights in Land, Planning Commission.

5

Agricultural Holdings Tax

A Modified Scheme

AMARESH BAGCHI

I. INTRODUCTION

Notwithstanding persistent criticism of the Indian tax system on account of inadequate taxation of agriculture, the level of agricultural taxation in India continues to decline. For the year 1978-79, receipts from land revenue and agricultural income tax—the two main direct taxes on agriculture[1]—are estimated at little over Rs. 200 crores (Table 5.1). With growing tendency to enlarge the area of exemptions the receipts are unlikely to exceed Rs. 200 crores. As a proportion of NDP from agriculture, this works out to less than 0.5 per cent.[2] In the late sixties (1967-68 to 1969-70) the proportion was around 1 per cent.[3] Fifty years ago as much as 10 to 13 per cent of the gross produce from land used to be taken away by the government through land revenue,[4] and, as every student of Indian history knows, in ancient times usually one-sixth of the produce was the king's share.

The main reason for such dramatic decline of agricultural taxation in India in recent years is that the land revenue of the British days continues to be the mainstay of direct taxation of agriculture,[5] even though the rates were fixed long ago and bear no relationship to present day productivity and prices in agriculture. Attempts have been made by some states to raise the yield of land revenue by adding on cesses and surcharges, sometimes in a graduated form. But these have not gone beyond *ad hoc* improvisation of the existing system of land revenue and, as the figures just mentioned would show, have evidently

TABLE 5.1 : Receipts frcm Agricultural Income Tax and Land Revenue (Budget Estimates for 1978-79)

(Rs. Crores)

	States	Agricultural Income Tax	Land Revenue				Total of Agricultural IT and Land Revenue (Col 2 + 6)	Remarks
			Land Revenue	Rates & Cesses	Other Receipts	Total (Col. 4, 4 and 5)		
	(1)	(2)	(3)	(4)	(5)	(6)	(7)	(8)
1.	Andhra Pradesh	—	24.00+	2.98	0.99	27.97	27.97	+ Includes Rs. 9.64 crore from irrigation projects.
2.	Assam	18.85	2.70	0.70	0.60	4.00	22.85	
3.	Bihar	0.19	6.24	11.15	0.20	17.59	17.78	
4.	Gujarat	—	6.50	0.45	0.55	7.50	7.50	
5.	Haryana	—	5.55+	—	0.11	5.66	5.66	+ Land Tax.
6.	Himachal Pradesh	—	0.41	—	0.29*	0.70	0.70	* Includes sale proceeds of waste land and redemption of Land Tax of Rs. 0.17 crore.
7.	Jammu and Kashmir	—	0.51	—	0.13	0.64	0.64	
8.	Karnataka	6.50	4.38	—	1.62	6.00	12.50	
9.	Kerala	6.50	1.58*	0.86**	1.09	3.53	10.03	* Land Tax ** Tax on plantations.
10.	Madhya Pradesh	—	10.48	1.10	2.18	13.76	13.76	
11.	Maharashtra	0.08	8.54	10.00	2.70	21.24	21.32	
12.	Meghalaya	—	0.04	—	0.02	0.06	0.06	

TABLE 5.1 (*Contd.*)

(1)	(2)	(3)	(4)	(5)	(6)	(7)	(8)
13. Manipur	—	0.47	—	—	0.47	0.47	
14. Nagaland	—	0.23	—	—	0.23	0.23	
15. Orissa	0.08	0.24	2.73	0.66*	3.63	3.71	* Land Development Tax Rs. 0.04 crore; Village Development Tax Rs. 1.06 crore.
16. Punjab	—	2.12+	0.01	0.07	2.30	2.30	* Includes sale proceeds of Waste lands and redemption of Land Tax of Rs. 0.03 crore.
17. Rajasthan	Neg.	6.37	0.01	5.12*	11.50	11.50	
18. Sikkim	—	—	—	—	—	—	
19. Tamil Nadu	2.50	0.27*	—	1.85	2.12	4.62	+ Includes receipts by way of income from jagirs resumed under Land Reform and Resumption of Jagirs Act estimated at Rs. 3.23 crore.
20. Tripura	0.01	0.25	—	—	0.25	0.26	
21. Uttar Pradesh	—	26.67	—	1.63	28.30	28.30	
22. West Bengal	4.39	3.04	0.94	1.91	5.89*	10.28	* Excludes Rs. 14.50 crore shown under land revenue
Total	39.10	110.59	30.93	36.32	163.34	202.44	

Source : Budget documents of State Governments.

not helped to arrest the decline of agricultural taxation in the country.

An important factor underlying the reluctance of the State Governments to embark on any major reform of the existing system of agricultural taxation seems to be the absence of a clearly superior and workable alternative to the traditional land revenue. While large-scale application of the agricultural income-tax is ruled out for practical as well as economic reasons—a land tax like the land revenue is more suited for taxing farmers, by tradition and also because it does not have the deleterious effects on production associated with the income tax[6]—suggestions for any radical reform of the existing land revenue system are looked upon with disfavour on the ground that any such reform would be beyond the administrative capacity of most states and that there might not be any tangible gain to revenue commensurate with the cost and effort that would be called for. Modification of the land revenue with suitably graded surcharges, etc, would, it is argued, achieve the same results at much less cost. The recommendations of the Raj Committee for an Agricultural Holdings Tax formulated after much deliberation were also put aside mainly on these grounds. This paper is an attempt to show that it is possible to modify the scheme of AHT as contemplated by the Raj Committee to make it administratively simpler while retaining its essential character and that even with moderately graduated rates its yield would in all probability be substantially higher than that of the land revenue including the cesses and surcharges.

II. AGRICULTURAL HOLDINGS TAX RECOMMENDED BY RAJ COMMITTEE

Taking note of the main deficiencies of the existing land revenue system, viz, wide variation in incidence as between regions and even between areas within the same region or districts (the settlements having been done 'under different systems and at different times, in different parts of the country'),[7] and lack of progressivity, and in order to introduce a system of land taxation which would provide an elastic and equitable source of revenue, the Raj Committee recommended the replacement of the land revenue by a graduated tax on agricultural holdings—the Agricultural Holdings Tax (AHT). The main features of the tax proposed by the Raj Committee are:

(i) Though described as a holding tax, the AHT would essentially be a tax on land leviable on the holdings of farm families according to the 'aggregate rateable value' (ARV) of each holding.

(ii) The ARV of a farm holding would be determined on the basis of the rateable value per hectare of land comprised in the holding. There would be a schedule of rateable value per hectare of land (hereafter RVL) for the principal crops grown in different areas derived from the norms of output, prices and costs which would be fixed on the basis of average of past years. For this purpose the country would be divided into a sufficiently large number of soil-climatically homogenous districts/tracts so that differences having any significant influence on productivity are taken into account. For each such district/tract norms of output of different crops per hectare would be worked out on the basis of the yield for the previous ten years, valued at the harvest prices of the preceding three years. From the value of gross output of the crops so arrived at allowance would be made for paid out costs of cultivation (other than expenses of irrigation) ordinarily at the rate of 40 per cent. Some allowance would be made separately for costs of irrigation. The schedule of RVL per hectare for each crop/crop group so derived would be revised every year.

(iii) There would be an annual assessment of rateable value/each holding on the basis of the schedule RVL per hectare for different crops grown in the area in which the lands of the holding in question are located, after taking into account the crops actually grown (harvested) on the land coming within the holding.

(iv) The tax would be levied on the aggregate operational holdings of each family, i.e., excluding lands leased out but including those leased in). There would be a prescribed schedule of tax rate and the tax payable by each family would be determined by applying the rates to the aggregate rateable value of the holding of the family so computed.

(v) Holdings would be aggregated on a family basis (the family consisting of husband, wife and minor children).

(vi) Relief would be allowed for crop failures in part or parts of a district/tract after inquiry by an appropriate authority).

(vii) An opportunity would be given to the taxpayer to challenge the correctness of the computation of the tax liability, including the cropping pattern recorded for his holding but not the schedule of RVL determined for his district/tract.

(viii) To ensure uniform and objective application of the AHT all over the country there would be a standing All-India Committee on AHT, consisting of experts and experienced administrators.

(ix) Trusts having operational holdings would also be liable to pay AHT. Trusts whose beneficiaries are unknown would be required to pay tax at the rate of 20 per cent of the ARV of their holdings. Companies too would have to pay AHT at a minimum rate of 20 per cent.

(x) The AHT would be implemented in two stages; first, by applying it to all operational holdings with ARV of the 5,000 or more. In the next phase, holdings with ARV of less than Rs. 5,000 would be covered. However, for holdings with ARV of less than Rs. 2,500 a census of land utilization and recording of crop averages would be carried out once in five years. All agricultural holdings with ARV of less than Rs. 600 would be required to pay AHT at a flat rate of Re. 1.

A tax on the above lines, the Committee felt, would remove the defects of the existing land revenue, as it would (i) provide for a built-in link with changes in productivity and prices; (ii) take into account the differences in the productivity of land all over the country on the basis of certain objective criteria and uniform procedures, and (iii) introduce a degree of progression comparable with that obtaining in other sectors of the economy.[8] At the same time, it would retain the merits of land revenue, being essentially in the nature of a land tax. It was estimated that the tax would yield an additional revenue of Rs. 150 crores per annum if applied to holdings with ARV of Rs. 5,000 or above. The yield, it was thought, would be larger, possibly around Rs. 200 crores, if it was extended to all holdings with ARV of Rs. 2,500 or more.[9]

Although the states were apparently consulted by the Raj Committee while formulating its recommendations practically none of the state has so far adopted the AHT. A progressive land holdings tax was introduced by Haryana in 1973,[10] but as will be shown presently, that is not quite what the Raj Committee had in mind. The principal consideration which dissuaded the states from implementing the AHT seems to be that it would involve insuperable administrative problems.

What may have made the AHT look so formidable are the requirements of:

(i) Drawing up of schedules of rateable value of land for each homogeneous district/tract separately for each major crop/crop-group grown in each such district/tract;

(ii) Revision of the schedules of rateable value of land every years;

(iii) Annual assessment of each holding on the basis of crops grown every year; and

(iv) Aggregation of holdings on operational basis.

It is however, possible to think of a tax on agricultural holdings which would avoid the difficulties mentioned above without sacrificing the merits of the AHT suggested by the Raj Committee. Lines on which such a holdings tax can be designed are indicated in the next section.

III. A MODIFIED SCHEME OF AGRICULTURAL HOLDINGS TAX

Although revision of assessments at regular intervals of not very long periods is required if the yield of the AHT is to be elastic, there is no reason to suppose that the merits of the tax would suffer unless the assessment of the holdings was carried out afresh every year and the schedules of RVL were revised annually. On the other hand considerable simplification could be achieved, if (i) the assessments once made were kept fixed for a period of say five years, and (ii) the schedules of RVL too were revised once every five years or so. It would also be simpler, and possibly some economic advantages derived, if the rateable value of land in a given area/tract was determined on the basis of the gross value of output of only one or two crops—the crops commonly grown in that area—and no attempt was made to lay down norms of net return per hectare separately for each crop which would be required under the Raj Committee's scheme. Also, the norms may be fixed on the basis of averages of output for five years, instead of ten as suggested by the Raj Committee, while prices may be taken at the average of three or four years. A further measure of simplification would be to assess the tax on the basis of ownership holdings, i.e., lands owned by a family instead of levying it on the operational holding basis. The broad features of the scheme of AHT modified on the lines suggested here and its points of departure from the AHT suggested by the Raj Committee are briefly considered below.

Periodical Assessment

While annual assessment of each holding as suggested by the Raj

Committee might not be altogether infeasible—it would call for collecting information regarding only the crops grown by the taxpayers from year to year and such information is collected now for purpose of annual assessment of crop cesses in states like Maharashtra—it is neither necessary nor desirable to require the AHT assessments to be made afresh every year. Certainty and fixity of liability is an important attribute of the land tax and fresh assessment every year would detract from this merit. To leave the assessments unchanged for long periods would however affect the elasticity of the yield of the tax as has happened now. A system of periodical assessment every five years would reconcile the conflicting aims of certainty and simplicity on the one hand and revenue elasticity on the other. Quinquennial assessments would also reduce the risks of avoidance and multiplication of disputes which a system of annual assessment on the basis of record of crops grown each year would inevitably give rise to.

With five-yearly revision of assessments, there would be no need for revising the schedules of rateable value of land every year. The period for revision of RVL schedules should ordinarily synchronise with the period fixed for revision of assessment of holdings.

Simpler Schedules of Rateable Value

A distinguishing feature of the land tax is that the tax to be paid in a year is usually determined on the basis of the average value of output over a few years in the area/tract where the land in question is located and not with reference to the actual annual income of individual holdings. This serves to provide an incentive for greater production as farmers who raise their productivity over the average are not subjected to any additional burden, while those whose productivity is below the average are penalised. This effect would be strengthened if the revision of RVL was carried out once in a few years instead of every year as marginal increases in productivity from year to year would not attract any additional tax burden.

The incentive effect of the said tax would be further reinforced if the rateable value of land in a particular area/tract is determined with reference to the average value of output of one or two crops commonly grown in the area instead of for each crop individually, leaving a choice to the farmers to go in for more remunerative crops without any fear of increased taxation. Such a system of taxation, it may be objected, would encourage the cultivation of cash crops in preference to foodgrains especially the coarse grains which would be socially undesirable. While there may be some merit in this contention it should be borne in mind that not all social objectives can be achieved with only one instrument. The land tax can be used in improving farm productivity and an element of graduation too can be built into it to serve the objective of vertical equity. It would be undesirable to load it on with the further objective of promoting the production of only selected crops regarded as socially desirable in preference to the farmers. If the production of foodgrains is to be encouraged even when they are less remunerative than cash crops the proper course would be to use some other instrument such as differential pricing of inputs like water for different crops.

Another direction in which the preparation and revision of the RVLs could be simplified with some justification on economic grounds is to adopt the gross value of output as the norm or basis and give up the attempt to assess the costs. Determination of costs in agriculture is notoriously difficult. Hence, while providing for deduction for costs, the Raj Committee laid down a ceiling to 40 per cent of the value of gross output excluding costs of irrigation (in exceptional cases 45 per cent to 50 per cent). A little reflection would show that there is no difference of substance between tax based on gross output and one based on net output where the 'net' is arrived at after allowing for costs as a fixed proportion of gross output. Identical yield and incidence can be obtained under the two systems if only the rates of the tax are adjusted suitably. The justification for adopting the gross output basis would be the same as advanced by the Raj Committee in support of provision for costs stated as percentages of gross value, viz, that it could serve "not only to simplify administration of the tax and avoid harassment but to give some inducement to farmers to reduce costs relatively to the value of their output and thus improve efficiency of cultivation while gaining at the same time some tax advantage for themselves".[12]

Levying a tax on farmers on the basis of gross output might give rise to objections on grounds of equity and the legality of such a tax might be challenged. Question regarding the legality of the tax and provisions needed to ensure its acceptance by judicial authorities are considered in some detail in Section V. Suffice it to note here that under the simplified AHT scheme even though the RV of land located in a given area would be fixed with reference to the gross value of output of the main crop or crops grown in that area, each state would still have to be divided into soil-climatically homogenous areas/tracts and the lands located in a given area/tract would be classified further 'under-irrigated' and 'unirrigated' and thus the major sources of difference in the productivity of land as between areas and even with a given area would be taken care of to the extent practicable and considered desirable, keeping in view the aims of promoting intensive and efficient agriculture in the country.

Under the Raj Committee's scheme it was envisaged that the AHT would be levied on ownership holdings, i.e., on the aggregate of lands owned by a family minus any land leased out but including lands leased in. Arguments put forward by the Committee for adopting the operational holding basis are

> .... the assessment of the AHT requires information on the crops grown and harvested on the agricultural holding. Moreover, the responsibility and decision to grow one crop or another, and how intensively to cultivate the land, are that of the farm operator—whether or not be is the owner. For the same reason, information on the crops grown and harvested could be legitimately and more conveniently obtained from the farm operator, whether or not he is the owner.[13]

Another consideration which seems to have weighed with the Raj Committee in recommending the levy of the AHT on the operational basis is that it would serve to bring out concealed leases since landowners would be liable to pay the tax in respect of leased out lands unless the lease was duly registered.[14]

Under the scheme proposed here the ARV of holdings would be determined on the basis of RVL per hectare without any reference to the crops actually grown on the land coming within the respective holdings. The consideration that collection of information regarding the crops grown would be easier with operational holdings would lose reliance under the modified AHT scheme as it would not involve collection of any information regarding crops grown on individual holdings. The argument that the levy of the tax on operational holding basis would bring out concealed or unregistered leases has some force but administratively a tax on operational holdings would be difficult to operate as it would involve keeping correct records of leading and call for cross checking as to who is leasing land to/from whom. Besides, such a system would leave a gap in the tax scheme in that there would be no liability on the landowners in respect of leased out lands although they could continue to derive some benefit from such lands in the form of a share of the produce.[15] Arrangements can no doubt be devised whereby that part of the rateable value which could be attributed to the owner's share would be included in the ARV of the owner or the share could be brought within the tax net through some other device[16] but that would be administratively burdensome. Under the system suggested here there would be no particular advantage in going in for operational holdings and the tax may be levied on the basis of ownership as under the existing land revenue system.

Since, under the modified AHT scheme, the tax liability for a given holding would be determined on the basis of the RVL per acre of the lands comprised in the holding and not on the RV per acre for each crop separately, there would be no need for the tax payers to furnish the particulars of crops grown by them. They would have to indicate only the size of the holdings and their location, and the particulars of the area where the land is located, whether irrigated and unirrigated and such other particulars as may be required for their classification under the prescribed categories. The ARV of a holding would be determined by aggregating the RVL of the land comprised in the holding by referring to the relevant schedules of RVLs.

To take a concrete example, suppose a farming family owns ten acres of land made up of six acres of irrigated land and four acres of unirrigated lands. Assuming that the irrigated land is located in an area/tract where mainly paddy is grown and that the gross value of the paddy output of irrigated land in a year that area is Rs. 1,000 per acre and of unirrigated land Rs. 600 per acre (average of five years) the RV of irrigated land in that area would be fixed at Rs. 1,000 per acre and that of unirrigated land at Rs. 600 per acre. The ARV of the holding of the family in question would then work out as follows:

(i)	Irrigated land 6 acres @ Rs. 1,000 per acre	Rs. 6000
(ii)	Unirrigated land 4 acres @ Rs. 600 per acre	Rs. 2400
	ARV of the holding	Rs. 8400

The tax would be levied on the ARV worked out in this way (i.e., Rs. 8,400) on the basis of a progressive rate schedule.

Under the scheme envisaged here all farm families liable to pay the AHT would be required to file a return once in five years. The return, as indicated, would only show the total area and particulars of location of land and their classification (irrigated, unirrigated, etc). After verification—which may be done on a sample basis—a demand notice with a brief order of assessment will be served by the competent revenue authorities. There should however be an appellate machinery and the tax-payer should have a right to appeal against the assessment. In order that the scheme does not get bogged down in interminable litigation, the scope of appellate review may be restricted to the determination of the size of the holdings, and the classification of the lands comprised therein and in computation of the tax. Schedules of RVL of different homogeneous areas/tracts should not be open to challenge by individual tax-payers. To ensure that the RVL schedules are drawn up in a fair manner, a statutory authority should be set up in each state with experts and administrators to undertake the task of preparing the schedules. The law should lay down the guidelines to be followed for determining the RVL. In order to allay any misgivings on the part of the tax-payers, the statutory authority may be required to publish the schedules in a draft from and invite comments from the public. Since different RVL schedules would be laid down for different areas and for different categories of land in a given area, the tax liability of a farm family would depend crucially on the classification of his land and since this classification would be open to appellate review, there can be no valid ground for apprehending arbitrariness or discrimination. Questions which may arise about the judicial acceptability of such a scheme are discussed in Section V.

Revision of Assessments and RVL Schedules

Under the modified scheme both the tax assessments of the holdings as well as the schedules of RVL would remain in force for a period of five years or so. Provision may however be made for revision of ARV of individual holdings within the said period under two circumstances, viz., (i) when the character or productivity of the land belonging to the holding undergoes a radical change, e.g., as a result of extension of public irrigation facility or of any special development programme leading to a change in the use of the land or (ii) when there is a change in the size of the holding because of transfers outside the family, acquisition by government, etc. Revision of RVL schedules (with consequent revision of tax assessments) may also be permitted under special circumstances such as when prices undergo a change by more than a specified percentage. Thus, when the price level of agricultural products goes up by say more than 10 per cent in a year for two or three consecutive years the RVLs may be revised upwards—not to the full extent of the price rise but by a specified fraction thereof (such as fifty per cent). There would however be no need for disturbing the schedules of RVL prematurely if the period laid down for revision is kept at five years. Such partial revision of RVL schedules might be carried out if for any reason the normal period of revision is fixed at more than five years or the revision is delayed beyond the normal period of five years.

Although under the proposed system it would be possible to vary the tax payable simply by changing the rates even while keeping the ARVs unchanged, in the interest of certainty, there should ordinarily be no change in the tax liability from year to year. Provision should however be made for granting full or partial remission, as may be required, in exigencies like floods and drought.

While it is possible to construct a tax schedule to achieve a high degree of progression, considering that the tax would be levied on the gross value of output, and that enforcement will be difficult if the progression is made too steep as there may be widespread evasion through splitting of holdings, etc, it would be advisable to keep the degree of progression somewhat mild. To afford relief to small holders it may be provided that no tax will be payable unless the size of the holding is more than one hectare, that is, 2.5 acres (5 acres if the land is unirrigated) and the progressive rate schedule will apply only to holdings whose ARV exceeds Rs. 5,000. For holdings whose area is more than 2.5 acres, but ARV is less than Rs. 5,000 the tax may be charged at a flat rate of say Rs. 7 per acre of unirrigated land and Rs. 10 for irrigated land. On all holdings with RV exceeding Rs. 5,000 the tax may be levied at the following rates:

Aggregate Rateable Value of Holdings	Rate of Tax
On the first Rs. 5000 of ARV	2.5%
On the next Rs. 5000 of ARV	5.0%
On the next Rs. 5000 of ARV	8.0%
On the rest	10%

For holdings whose ARV exceeds Rs. 5,000 by a small margin suitable marginal relief should be allowed.

In order to counter attempts to defeat the progressivity of the tax by forming trusts and companies, in the case of companies and trusts whose beneficiaries are unknown or the shares of the beneficiaries are indeterminate the tax may be levied at a flat rate of 8 per cent without any exemption. Plantations may be left out and may continue to be liable to agricultural income tax.

The exemption limit of Rs. 5,000 for the AHT, it may be objected, would be too low and the rates high, as compared to those laid down for taxation of non-agricultural incomes, for which the exemption limit is as high as Rs. 10,000 in the case of individuals, not to mention the various reliefs available by way of standard deduction, savings etc. It may also be argued that since the ARVs will be computed on the basis of gross value of output with no deduction for costs, the rates in effect would be much higher than suggested by the schedule.

Although a degree of parity in the tax levels between the farm and the non-farm sectors is desirable, it should be recognized that there are pitfalls in any straightforward comparison of nominal rates of a land tax and those of a tax on incomes. For one thing, the AHT is a tax on land based on averages of output of the crop commonly grown in a given area over past years while income tax is a tax on actual current incomes. The actual income, of a farmer from year to year would in most cases be substantially higher than the RV of land that would be fixed for purposes of AHT. That apart, every farm family liable to AHT would presumably have some income from subsidiary activities such as animal husbandry which, as the farm management studies show, are not inconsiderable. Prices of essential commodities like foodgrains and vegetables are also generally lower in the rural areas than in towns and thus the real incomes of farmers are often higher than what the income in money terms would suggest. It may also be pointed out that in the urban areas income earners in the low income brackets, though not liable to income tax, are subjected to imposts in other forms like profession tax while urban property owners have to pay municipal taxes without any exemption limit.

As for rates, even assuming that paid-out costs constitute roughly 30 per cent of the gross value of output the proposed rate schedule would read as follows if recast in terms of net output norms:

On the first Rs. 3,500 of ARV	@ 2.5%
On the next Rs. 3,500 of ARV	@ 5.0%
On the next Rs. 3,500 of ARV	@ 8.0%
On the rest	@ 10.0%

It may be seen that a holding with ARV of more than Rs. 10,500 (on net produce basis) would be taxed at the marginal rate of 10 per cent while in the case of personal income tax the initial rate (applicable to incomes exceeding Rs. 10,000) is currently 17.25 per cent.

In Taiwan and Korea, the two countries where the land tax has been relatively successful, the rates have been similar to those suggested here. In Taiwan the rates applied to the tax base range from 3.5 per cent to 9.6 per cent of the estimated production of the land which is taken to be the "standard yield" as established by land surveys. In Korea, the tax is levied at the rate of 6 per cent on the value of "standard" output of lands used for gains while land used for other crops is taxed on the basis of net income at progressive rates ranging from 10 per cent to 20 per cent.[17]

Lastly, land being the most valuable scarce resource of the community, the case for taxation of land designed to induce its optimal use has a rationale of its own. The burden of AHT at the rates suggested here would not be heavier than that of land revenue so far as the small holders are concerned as they would be mostly exempt.[18] Even for large holdings the burden would be less than under the Raj Committee's proposals and even less than what it was earlier in the century or for centuries past.

IV. YIELD AND INCIDENCE

For an idea of the likely yield of a tax on agricultural holdings on the lines indicated, above, one has to determine (i) the productivity of land in different parts of the country for each homogeneous area/tract separately for irrigated and unirrigated land, and (ii) the ARV of the holdings of different sizes in each such area/tract. The tax likely to be collected from different areas/tracts can then be worked out provided information is available regarding area-wise distribution of land holdings. Apart from the problem of obtaining data in such detail, even districtwise computation of the likely yield of the AHT would involve an exercise on a much larger scale than could be undertaken for this paper. As an alternative, an attempt has been made to estimate the probable yield of the AHT under the modified scheme based on statewise figures of productivity and farm prices available from the Report of National Commission on Agriculture (1976) and the publications of the Ministry of Agriculture[19] and NSS data on distribution of land holdings (as of

1971-72).[20] The estimates covering 15 major states are presented in Table 5.2. Details of the computation state by state are set out in the Appendix.

TABLE 5.2 : Current Receipts from Land Revenue and Estimated Yield of Modified AHT

States	*Land Revenue (BE 1978-79)*	*Estimated Yield of AHT (Modified)*	*Additional Revenue from Modified AHT (Col 3 to Col 2)*	*Proportion of Addl. Revenue to Existing Collection (Col 4 to Col 2)*	*Proportion of Land Revenue Collection to AHT Potential (Col 2 as % to Col 3)*
1	2	3	4	5	6
Andhra Pradesh	27.97	57.14	29.17	104	48.95
Assam	4.00	5.14	1.14	29	77.82
Bihar	17.59	29.89	12.30	70	58.85
Gujarat	7.50	28.60	21.10	281	26.22
Haryana	5.66	23.04	17.38	307	24.57
Karnataka	6.00	35.61	29.61	494	16.85
Kerala	3.53	19.96	16.43	465	17.69
Madhya Pradesh	13.76	49.17	35.41	257	27.98
Maharashtra	21.24	47.12	25.88	122	45.08
Orissa	3.63	8.19	4.56	126	44.32
Punjab	2.30	34.46	32.16	1398	6.67
Rajasthan	11.50	28.76	17.76	150	39.99
Tamil Nadu	2.12	38.38	36.26	1710	5.52
Uttar Pradesh	28.30	39.95	11.65	41	70.84
West Bengal	5.89	22.19	16.30	277	26.54
Total	160.99	467.60	306.61	190	34.43

From the estimates it would appear that if based on averages of output of principal crops for 1971-74 and prices prevailing during 1970-71 to 1974-75, an AHT formulated on the lines suggested in this paper should yield an additional revenue of around Rs. 300 crore per annum (Rs. 468 crore as against Rs. 161 crore collected from land revenue at present). While for some states (e.g. Assam) the additional revenue might not be sizeable, in most states the revenue from AHT would, it would appear to be double the amount accruing now from land revenue (*vide* col. 5 of Table 5.2). In two states—Tamil Nadu and Punjab—the revenue should increase by as much as seventeen and fourteen times the present yield of land revenue respectively. This brings out not only the revenue potential for land taxation but also the extent of inter-state disparity in the level of agricultural taxation at present. If a tax effort comparison of the states in India is made on the basis of the proportion of revenue actually raised to their taxable capacity determined on the basis of uniform principles the relatively poor states of Assam, UP and Bihar would appear to be doing much better than others in the matter of agricultural taxation.[21] If the states are ranked in order of this ratio, these three states come at the top (col. 6 of Table 5.2).

The estimates presented here, it should be added, are based on statewise data and suffer from all the limitations of averages. Also, it may be argued that with the lowering of ceilings on agricultural land since 1971 concentration of land holdings may have decreased, and, so, estimates of a progressive holdings tax based on 1971-72 distribution would have an upward bias. Reliance on prices of agricultural products during 1970-71 to 1974-75 in estimating the value of gross output could be another source of such a bias, since there was an abnormal spurt in prices during the year 1974-75 and agricultural prices registered a decline in the next two years.

Index numbers of wholesale prices of agricultural commodities with 1970-71 as the base are given below:

	Index Number (Average of the year)
1970-71	100.0
1971-72	100.4
1972-73	110.3
1973-74	139.2
1974-75	169.9
1975-76	157.8
1976-77	158.5
1977-78	174.3

Source : Economic Adviser's Wholesale Price Index.

It will be seen that although the prices went down in 1975-76 and 1976-77 as compared to the level prevailing in 1974-75, the average of the prices over the years 1970-71 to 1974-75 would be much lower than that over the period 1973-74 to 1977-78. Thus it cannot be said that the use of the average prices of the years 1970-71 to 1974-75 would have imparted an upward bias to the estimates of the value of output. As for the possibility of a decline in the degree of concentration in land holdings after 1971-72, while such a possibility cannot be ruled out, it is to be noted that a comparison of the NSS data on distribution of agricultural land for

1971-72 with the position obtaining in 1960-61 shows that during the 1960s there was probably a trend towards greater concentration in the distribution of ownership holdings of agricultural land in the country instead of towards greater equality. Even allowing for the fact that the large holdings might have been split up later in the seventies and therefore estimates of a progressive AHT based on 1971-72 data on land holdings could be on the high side, it is worth noting that the concentration in agricultural holdings revealed by the all-India agricultural census is even more acute than that suggested by the NSS data. In fact, if the probable yield of the AHT is estimated on the basis of the agricultural census data, the additional revenue works out to more than Rs. 400 crore a year as against Rs. 300 crore estimated on the NSS data. One reason for this difference probably is that the agricultural census covered holdings of institutions like trusts, co-operatives and companies and there has presumably been a shift towards holding of land by institutions on a large scale in recent years.[22] This trend is unlikely to have been reversed. For these and further reasons mentioned below the estimates of the yield of AHT presented here would seem to be on the conservative side:

(i) While deriving the rateable values of land for purpose of the estimates (*vide* Tables A.IV.1 to A.IV.15 of the Appendix) only the value of the main product has been taken into account. As will be seen from the available studies on farm management and also, the reports of the Agricultural Prices Commission, the values of the bye-product are not insignificant. In fact, they constitute around 12 to 29 per cent of the value of the main product.[23] Thus rateable values would actually be higher than assumed for the estimates here; and

(ii) The rateable values of land have been estimated on the basis of averages of gross output for the year 1971-74. As is well known, there has been further improvement in the productivity of agriculture almost all over the country in the last few years[24] and with the extension of irrigation facilities in different areas it would not be unreasonable to suppose that the averages of gross value of output, if worked out for the years 1971-72 to 1976-77, would be higher than what has been taken for the estimates of the probable yield of the AHT worked out here.

Incidence

The incidence of the tax on holdings of different sizes as a percentage of their rateable value (i.e., average gross output) is indicated in col 6 of Appendix Table A-4. Incidence in terms of tax per acre is shown in col. 7 of the Table.

Since the maximum marginal rate proposed here is 10 per cent, the average incidence will not exceed 10 per cent of the rateable value for any class of holdings. In most states the average incidence in the highest size class would be less than 8 per cent. The incidence would exceed this level (but of course cannot exceed 10 per cent only where the rateable value of land per acre is as high as Rs. 2,000 as in Kerala. For holdings of not more than 10 acres the incidence would ordinarily be between 2 and 3 per cent of the rateable value except in Kerala. Needless to add, even within a state like Kerala the incidence will be less if the holding of the taxpayer is located in an area/tract where the rateable value of land is low.

The tax per acre would, in the majority of the states, vary between Rs. 8 and Rs. 50. For the majority of the holdings liable to the tax the rate would be less than Rs. 20 per acre, although for holdings of more than 60 acres the tax might go up to as high as Rs. 191 per acre in some states (e.g., Kerala). While the maximum rate will not reach this level even for large holdings, unless the productivity is as high as in Kerala—in all other states it will be below Rs. 100—it may be noted that for the high-value crops the cesses payable by farmers at present in some states are not less onerous. In Maharashtra, for example, the tax on grapes is Rs. 380 per hectare, i.e., about Rs. 150 per acre.[25] For large holdings (12.50 acres) the cess comes to more than Rs. 200 per acre. Under the AHT scheme proposed here such high rates will not be attracted even if a farmer cultivates a high value crop unless his holding is sufficiently large, whereas under the crop cesses everyone has to pay the tax at a high rate, irrespective of the size of his holding. It is also relevant to note that even at present the maximum rate of land revenue goes up to Rs. 18 per acre (e.g. in West Bengal) and in some areas (e.g. in some of the districts of Andhra Pradesh) it is as high as Rs. 50 per acre.

However, as noted in no case would the tax under the AHT scheme exceed 10 per cent of the average value of output. Even as a proportion of average net output (allowing for costs at the rate of 30 per cent of gross output), the tax would come to not more than 15 per cent or so, whereas for non-agricultural incomes the average rates goes beyond 40 or even 50 per cent in the upper ranges.

V. CONSTITUTIONAL QUESTIONS

The Indian Constitution enjoins that no tax shall be levied or collected except by authority of law[26] and it is now well established that the law must not only be within the legislative competence of the authority enacting it but it must also not be inconsistent with any provision of the Constitution. Experience shows that enactments designed to levy a new tax are often challenged before the courts and their legal validity contested on constitutional grounds. The grounds usually advanced are that the tax in question does not fall within the legislative field of the authority imposing it and/or that some of the fundamental rights guaranteed by the Constitution are infringed. Several enactments authorising the levy of a tax were struck down by the courts in the past, wholly or in parts, as violative of the Constitution and quite a few of these related to the levy of a land or buildings tax. Instances are, the Travancore-Cochin Land Tax Act, 1955 as amended in 1957,[27] the Kerala Buildings Tax Act, 1961,[28] Madras Urban Land Tax Act, 1963[29], Mysore Buildings Tax Act, 1963[30] and the Andhra Pradesh Land Revenue (Additional Assessment) and Cess Revision Act, 1962.[31] Therefore, in framing any new tax scheme care has to be taken to see that the provisions of the Constitution are not violated.

Legislative Competence

As the proposed AHT would replace the land revenue which, under the Constitution, comes within the Legislative field of the states, (Entry 45 in List 2 of the Seventh Schedule to the Constitution) there should be no difficulty for the states to levy it. Even if the AHT is viewed as a tax on land,[32] there can be no objection to the levy of any such tax by the states as tax on land too comes within the purview of the states' legislative powers (Entry 49 in List 2 of the Seventh Schedule). The fact that the tax under the AHT scheme would be levied with reference to productivity of land should make no difference. The land revenue is also based on productivity and the courts recognize that a tax on land may be based on its productivity just as a tax on building can be levied on its rating value.[33] Thus, no valid objection can be raised against grounds of lack of legislative jurisdiction. Whether there could be any cause for complaint on grounds of violation of fundamental rights, however, calls for careful examination.

Article 19

The constitutional provisions relating to fundamental rights which are usually invoked in attacking tax laws are Article 14 which guarantees equality before law and Article 19(1)(f) which assures right to acquire, hold, and dispose of property. Any law which infringes the right to equality of all citizens is liable to be struck down as violative of Article 14. Article 19(1)(f) is violated if any unreasonable restriction is imposed on the right to acquire, hold and dispose of property. There is nothing in the proposed AHT scheme which could be said to lead to any unreasonable restriction on the right to acquire, hold and dispose of property. The tax would be related to the productivity and size of holdings and rates proposed could in no case be regarded as confiscatory or extortionist which is the principal test of violation of Article 19 in the case of tax laws.

Article 14

As for Article 14, pronouncements on the question would show that the guarantee of right to equality before law has a two-fold implication for the legal framework of the country, *viz.* (i) while a statute may classify persons or things to whom it is intended to apply, the classification should be reasonable, and (ii) no law should confer arbitrary and uncontrolled power on Government to discriminate between persons or things similarly situated, since, with such uncontrolled power, discrimination may result. The tests of reasonableness laid down by the courts in the matter of classification are (a) it must be founded on intelligible differential, and (b) the differential must have a rational relation to the object sought to be achieved by the statute in question.[34] In the case of tax laws, the courts, in view of the inherent complexities of fiscal adjustments of diverse elements permit a larger discretion to the legislature in the matter of classification.[35] even so, in order to avoid any conflict with Article 14, of tax law should ensure that persons similarly situated are not treated unequally and at the same time persons in dissimilar circumstances should not be subjected to uniform treatment. Where objects, persons or transactions essentially dissimilar are treated by the imposition of a uniform tax, the courts have observed, discrimination may result for 'refusal to make a rational classification may itself in some cases operate as denial of equality.[36] The Travancore-Cochin Land Tax Act of 1955 (as amended in 1957) which sought to impose a uniform tax of Rs. 2 per acre was struck down on the ground that it did not differentiate between productive and unproductive land. The Supreme Court held the Act to be violative of Article 14 observing that "it is one of those cases where the lack of classification creates inequality"[37]. The Mysore Buildings Tax Act 1963 which provided for the levy of a tax on buildings on floorage basis was also held unconstitutional by the Mysore High Court on the ground that the basis was 'unscientific, arbitrary and mechanical and that under that basis, the incidence fell unevenly on things similar.[38] The Kerala Buildings Tax Act, 1961 was also held by the Kerala High Court to be unconstitutional because of lack of classification resulting in inequality as it imposed a tax on buildings on the basis of floor area.[39]

An instance of a tax legislation being held to be violative of Article 14 on grounds of unreasonable classification as well as conferment of arbitrary powers on Government officials is the Andhra Pradesh Land Revenue (Additional Assessment) and Cess Revision Act 1962.[40] The Act enabled the Government to levy an additional Assessment at 75 per cent of the standard dry rates and to raise the wet assessment from 50 to 100 per cent subject to certain maximum and minimum. It was held discriminatory on the ground that the whole imposition of assessment was left to the arbitrary discretion of the officers not named in the Act without giving any remedy to the assessee for questioning the correctness of any of the important stages in the matter of assessment.[41] The Madras Urban Land Tax Act, 1963 which was levied on the basis of the average market value of urban land in respective sub-zones of the city was also held to be violative of Article 14 on the ground that it was levied not on the actual market value but the average value of lands in a locality known as a sub-zone and also because the Act did not furnish proper criteria and serve as a definite guide in the matter of classification of zones and sub-zones and determination of average market value.[42]

The AHT cannot possibly be faulted for lack of classification. The question for consideration is whether the classification contemplated under the proposed scheme would satisfy the tests of reasonableness laid down by the courts. The points repeatedly emphasised by the courts while laying down the test of reasonableness in the matter of classification are, as noted earlier (i) that it should be based on an intelligible differentia which distinguishes persons or things grouped together from others left out of the group and (ii) the differential must have a rational relation to the object sought to be achieved by the statute in question. In the proposed AHT scheme, the differentia for classification of assessee would be the rateable value of holdings, which would be derived on the basis of the rateable value of land coming within a holding. The RVL again would be worked out after classifying the different areas of a state into soil-climatically homogeneous areas/tracts. For this purpose, districts and sub-divisions would be sub-divided into blocks and sub-blocks wherever necessary. There would be a further classification of land in each block/sub-block into 'irrigated' and 'unirrigated. This would ensure that the major factors underlying the differences in productivity of land are taken into account. To accommodate the requirement of reasonable classification further, one can think of classifying both irrigated and unirrigated land in each homogeneous block/sub-block into two or three more categories such as 'good' and 'medium' over and above the classification under 'irrigated' and 'unirrigated'. With such classification, the schedule of rateable value of land in each block/sub-block in a district would read as follows:

State......../Dt......../Block/Sub-Block

	Rateable Value per Acre in Rs.	
	Irrigated	*Unirrigated*
(1) Good	1,200	900
(2) Medium	1,000	700

Note : Figures are only illustrative.

A further category may be provided for lands which are totally barren and their rateable value taken as 'zero'.

The scheme may yet be attacked on the ground that the classification suggested above would not provide for determination of productivity of land under each holding individually, and that absence of any provision to allow for costs in the determination of rateable value would create inequality. These arguments can, however, be rebutted by pointing out that the scheme, while providing for classification of land to take account of major sources of differences in productivity, would not permit individual assessment of the productivity of each holding for several important reasons. First—and this is a major virtue of the land tax—a tax based on average productivity obtaining in an area is intended to provide an incentive for landowners to raise their level of output and exceed the average as unlike in income tax marginal increases in production would not attract any additional liability and at the same time by punishing farmers whose productivity is below average and thereby inducing them to improve their productivity, it would secure optimal use of agricultural land.[43] Another reason is that individual assessment of farm holding in a country like India would throw up impossible administrative burden and if assessments were to be made by measuring the productivity of land under each holding plot by plot, no reform of land taxation would ever be possible. Besides, with the advent of modern technology in agriculture, the quality of the soil has ceased to be the sole determinant of productivity. The classification of land envisaged under the proposed scheme should be adequate to take account of significant differences in the productivity of farm lands. Assessment of productivity of each individual plot of land is, therefore, not needed. The Raj Committee, it may be recalled, had observed, "we do not consider it either feasible or desirable or necessary to arrive at the actual output on individual holdings".[44] The classification proposed for the modified AHT is thus clearly related to the objective in view and therefore the test of reasonableness laid down by the courts would be satisfied.

While striking down the Mysore Buildings tax Act, 1963 as unconstitutional for absence of reasonable classification the court had observed:

> The classification adopted appears to be a mere mechanical formality devised in a haphazard manner as cover for raising money. It does not disclose differentiation pertinent to the burden being heavy in some cases and light in others—the object of the Act is not to limit the floorage of the buildings in towns but to raise public revenue. Therefore, the classification on the basis of the floorage has no just relationship with the object of the Act.[45]

Such observations would surely not be applicable to the proposed AHT scheme. For, to repeat, it would provide for classification of land to take account of all major sources of differences in productivity but would refrain from carrying the classification beyond a point with a clear objective in view.

Similarly, assessment of productivity on the basis of gross output can be defended by pointing out that with the classification of land in the suggested manner the main factors underlying differences in productivity would be taken care of and with such classification reliance on gross output as an index of productivity in different regions and separately for each category of land cannot act as a source of significant inequity. Besides, determination of costs in agriculture is extremely difficult for both conceptual as well as practical reasons. Farm management studies show that four different concepts of costs are used[46] and choice between one for the other is not easy to justify. Moreover, adoption of gross output value as the base would serve the objective of promoting efficiency as, under a system of taxation based on gross output anyone who succeeds in economising on costs would bear no additional liability, unlike in a graduated income tax. As noted earlier, Raj Committee also had efficiency considerations in mind while suggesting allowances for costs as a fixed proportion of output. It is also worth noting that the principle of taxation on gross receipt basis is not unknown to the Indian tax system. Under the Income-Tax Act 1961, applicable to non-agricultural incomes, certain receipts are taxed on a gross basis.[47] The land revenue also was based on gross produce in almost all states of India before more scientific methods of assessment were evolved. In Assam, even after Independence gross output formed the basis of assessment and the limit of assessment for any tract was fixed at 10 per cent of the value of the gross produce.[48]

It should also be appreciated that to insist that no tax on land can be levied except on a net income or output basis would have serious implications for the entire tax system of the country. As observed by Sarkar J. in his dissenting judgment in the Moopil Nair's case arising out of the uniform Kerala Land tax, if this argument was accepted, "this would apply to all taxing laws which would then have to be correlated to the income or profit which would be made and this would have very far-reaching consequences".[49] To quote Seervai, the noted authority on constitutional law, "it would lead to the conclusion that unproductive land cannot be taxed at all which would mean that the state's power to tax land would be inoperative in respect of unproductive land".[50] If the argument that no tax can be levied on the gross output basis is carried to its logical end, value added tax in India should be permanently ruled out and as Seevai points out one can even argue that sales tax and excise duties which take no account of differences in profitability of the producers offend Article 14.[51]

As the tax would be levied on the basis of ownership, it may be argued that there should be some differentiation in the treatment of tenanted and self-cultivated land, since, in respect of leased out land, the owner is entitled only to a fixed proportion of the produce. Ownership basis for land taxation cannot, however, be regarded as a fatal infirmity. The existing land revenue is levied on ownership basis and the proposed tax scheme does not seek to disturb this aspect of the land tax system of the country. Moreover, the manner in which the produce is divided between the owner and a share-cropper depends on many factors such as who pays for the inputs etc, and it may be presumed that, as at present, part of the tax, would be passed on by its owners to the tenants. If an owner of leased out land cannot, for any reason, pass on any portion of the tax to the tenant and finds the levy too heavy, it would be open to him to reduce the burden by reducing the size of his holding. To the extent this takes place, the objective of reducing concentration of land holdings would have been achieved. The burden on the landowners can however be alleviated—and thereby the risks of attack on constitutional grounds minimised—by providing that when the land is in the occupation of a registered tenant and appropriate part of the tax may be recovered from the tenant. (Tax laws usually permit such recovery in the event of default by the assessee. See, for instance, Section 17 of the Karnataka Urban Land Tax Act, 1976) The 'appropriate part' may be fixed as the amount that would be payable if the rate prescribed for the lowest slab of the schedule was applied to the RVL of the land in question. The fact that the higher incidence arising from the size of the holding would have to be borne entirely by the owner cannot be a valid ground for assailing the legality of the tax, especially after the Supreme Court's ruling upholding the levy of an additional sales tax in the case of large dealers that cannot be passed on to the buyers. (*S. Kodar vs. State of Kerala and other cases* (1974) 34 STC 73. The levy of additional sales tax on dealers having turnover of over Rs. 10 lakhs a year was challenged in three cases on grounds of violation of Articles 14, 19(1)(f) and 19(1)(g) of the Constitution.)

Although the schedules of RVL would not be open to question, the scheme cannot be challenged on grounds of conferment of arbitrary power to assessing authorities since there would be a right to appeal to every tax payers to contest the determination of his tax liability. The right to appeal would extend to the determination of the size of holdings and the classification of land within each holding as well as assessment of tax and penalty. Schedules of RVL in different areas would not however be open to question. This cannot be a source of hostile discrimination as the task of determining the RVLs would be entrusted to an expert body who would invite comments from the public before finalisng the schedules for different areas and give due consideration to objection put forward if any. Suitable guidelines may be laid down in the law itself for the statutory expert body to follow.

It may also be added that courts have made it clear on more than one occasion that absolute equality is not to be expected especially in tax matters and the legislature is permitted some latitude in the matter of classification for purposes of tax legislation. The modified AHT would provide for classification to an extent consistent with its objectives. There is, therefore, no reason to apprehend that the proposed scheme would be vulnerable from the constitutional angle.

VI. AHT Vs LAND REVENUE RATIONALISATION

The case for continuing with the existing system of land revenue is sometimes supported with the argument that it is respected by the judicial authorities and anything which disturbs the *inter se* relation of the burden of land revenue assessments runs the risk of being struck down by the courts. As the discussion in the preceding section would show there is nothing in the judicial rulings to suggest that a tax on agricultural holdings which provides for classification of land in a manner related to the objectives in view and also allows scope for appeal against such classification in individual cases would not be upheld by the courts. Moreover, it needs to be pointed out that the impression that land revenue assessments are fair and equitable is not well founded and that there is no way of rationalising the existing land revenue system without going through the exercises which would be required for a modified AHT of the kind suggested here.

As is well known to students of public finance in India, land revenue in this country had varied origin and there is no uniformity in the principle applied in the assessments in different parts of the country. The basis of assessment differs widely. In some areas it is 'net assets or economic rent', in some it is the 'net produce or annual value' while in some the basis is 'capital value', or 'gross produce' and in others 'empirical'. Under the empirical methods, while several factors are supposed to be taken into account, as the Taxation Enquiry Commission of 1953-54 (TEC) noted "the assessment is ultimately based on the subjective judgment of the Settlement Officers'.[52] Way back, in the twenties, uncertainty of the base and the rates was commented upon as the main defect of the Indian land revenue system by the Todhunker Committee on Indian taxation.[53]

In the post-independence period this disparity widened because of several factors. With the merger of former princely states vast areas with widely different systems of revenue—in many states there was no regular statute to govern the settlements and the whole thing was based on custom and usage—were integrated with the states of the Indian Union. In many areas so integrated, no survey had been done. In the eight princely states which formed the PEPSU, "the policy in respect of land revenue was different in different states and there was great disparity not only in rates but also in the standards of assessment".[54] Attempts were no doubt made to being about some uniformity after integration but these were essentially *ad hoc* in nature. Land reform and abolition of intermediaries in land was another factor that aggravated the disparities. In most areas the cultivators continued to pay as land revenue the rents which they were paying to their landlords. To quote the TEC again, "as the rents which were being paid by the cultivators in the pre-abolition period varied for different classes of tenants, this has had the effect of introducing varying rates of land revenue for the same class of land depending on whether it is at present held by the previous intermediary or a particular class of tenant"[55]. It also meant "that those whose ability to pay is the highest are taxed at the lowest rates".[56] The result of all this was, as Bird puts it, "substantial interprovincial, interpersonal, and interclass inequalities in tax burdens".[57] It needs no argument to show that with a system of taxation which evolved in this fashion any attempt to raise the yield by superimposing surcharges or cesses on the basic rates can only serve to compound the inequalities. This point required to be laboured at some length as it is often overlooked by those who advocate the continuance of the *status quo* in the present system of land taxation for 'pragmatic' reasons.

It should also be recognised that any scheme of rationalisation of the existing land revenue which is based primarily on productivity has to proceed on the basis of the trends in productivity and prices in recent years and since allowance has to be made for differences in productivity as between areas, classification of different areas into broadly homogeneous areas and some classification within each area with an indication of the average level of productivity for different categories of land in each such area is unavoidable.

This is what the AHT is essentially designed to achieve. In fact, the modified AHT scheme put forward here provides an expeditious and not too ex-acting method of land revenue revision. A full scale revision of land revenue in the manner it was done in some of the ryotwari areas in the past would call for much greater administrative effort than would be needed to implement the AHT. It is possible to reduce the administrative burden even further, e.g., by reducing the number of years for purposes of averaging etc. But, unless linked to productivity a land tax cannot be rounded on equitable principles.

Crop Cesses

Crop cesses are sometimes suggested as an alternative to a radical reform of the land revenue. The equity of a crop cess levied uniformly for a particular crop all over a state is, however, not beyond question. For such a levy ignores the fact that the yield of a crop may not be the same in all areas even within a state. In fact the levy of crop cesses in Andhra Pradesh was challenged on this ground but the cesses were withdrawn before the court had given its verdict.[58] Moreover, if the cess is levied on more than one crop, the ratio between these rates should reflect their relative yield in value terms per acre which in turn calls for some idea of productivity and prices of different crops. Suggestions for a presumptive income-tax on agricultural incomes put forward by some as an alternative also tend to ignore the fact that a tax on presumed income whether 'potential' or 'normal' is essentially only a variety of land tax based on average productivity.[59]

Short-Cuts to AHT: The Haryana Land Holdings Tax

The Land Holdings Tax of Haryana is designed to short-circuit the process of measuring productivity by relying on a broad classification of soil, Schedule I to the Haryana Act, shows the classification of land, district-wise (broken down to assessment circles), under five classes, I to V. Soils are described as 'Abi', 'Chahi', 'Nehri', 'Barani', 'Banjar', 'Kallar', 'Thur', 'Sem', etc. 'Nehri', that is irrigated land, in all districts is placed generally under Class I while Banjar, Kallar, Thur and Sem come under Class V. The rates of tax are graduated and differ from class to class as may be seen from Table 5.3.

TABLE 5.3 : Haryana Land Holding Tax—Rates of Tax

	Class I	*Class II*	*Class III*	*Class IV*	*Class V*
For the first one hectare	70 p	60 p	40 p	25 p	10 p
For the next hectare	Re. 1.00	90 p	50 p	40 p	15 p
For the rest	Rs. 1.35	Rs. 1.20	60 p	50 p	20 p

Note : Adapted from Schedule II of Haryana Land Holdings Tax Act, 1973.

While assessing the tax on a land holding where it comprises two or more classes of land, the land of various classes is placed in different slabs so that the land of the highest or the next below class or classes is placed in the first slab, and the land of the next below class or classes is placed in the second slab and so on.[60]

While the classification of land in the manner prescribed in the Haryana Act helps to achieve some measure of equity, for various reasons such a classification cannot be regarded as satisfactory. Apart from the fact that even for such a classification an estimate of productivity of land in different districts/tracts has to be made, even though implicitly, unless related directly to productivity in value terms it cannot secure the aims of uniformity or objectivity and thus equity, the lack of which is considered to be the main deficiency of the existing land revenue. Under the system adopted in Harayan, 'Nehri' land is put in Class I in almost all districts. This amounts to assuming that the cropping pattern and productivity of irrigated land throughout the state is the same, which is unlikely to be true. Secondly, with such a classification it is not possible to ensure inter-regional or inter-district uniformity in the incidence of taxation even approximately. Class I land in one state (say Rajasthan) may not rank as Class I in another (such as Punjab). It would be patently inequitous to prescribe the same rate of tax for Class I lands in all states and even for lands categorised as Class I within a given state unless one assumes perfect homogeneity in the productivity of lands coming under a particular category in all parts of the country or the state.

If uniformity and equity are to be achieved one has to go in far an index of productivity like the rateable value. It is only with the help of such an index or yardstick that the incidence of the tax as between different areas and holdings of different sizes can be ascertained with a reasonable degree of reliability.

Besides, unless the tax on agricultural land is based on something like the RV which can be revised periodically the rates of the tax have to be changed from time to time, if the yield is to be responsive to changes in productivity and prices. If the rates are to be raised without raising the real burden when there is a rise in productivity or prices an implicit calculation of the yield and value of output is unavoidable. It is relevant to note that the yield of Haryana Land Holdings Tax is not commensurate with the productivity of agriculture in the state. A tax of this kind cannot thus be regarded as a substitute for an agricultural holdings tax designed to achieve the aims of uniformity, objectivity, progressivity and revenue elasticity.

VII. SUMMING UP

In the interests of both revenue as well as equity, the system of taxation of agriculture in India needs to be overhauled. Superimposition of surcharges or cesses on the basic assessments would only aggravate the inequities which have resulted from disparate principles and practices followed in the matter of settlement of land in different parts of the country over the years. In any case revision of the land revenue assessments on the basis of more recent measures of productivity and prices is long overdue. Assessment of the productivity or return on land for each holding individually is neither practicable nor necessary. A tax based on an index that reflects the productivity of land on an average, after making reasonable allowance for differences in soil and climate, is needed.

The AHT proposed by the Raj Committee was intended to provide a model for such a tax. In the interests of simplification the scheme recommended by the Raj Committee can be modified to suit the administrative resources of the State Governments. Such a scheme would retain the chief merits of the AHT and help to secure additional revenue of the order of Rs. 300 crores in the immediate future. Agriculture being the largest sector of the Indian economy a rational system of agricultural taxation is urgently called for *ad hoc* reforms of land revenue serve no useful purpose.

Note : Reproduced from *Economic and Political Weekly*, September 23, 1978, pp 1631-1648

NOTES

[Views expressed in this paper are purely personal. The author is indebted to D. Bandopadhyay for ideas on simplification of the AHT, and S.I.H. Naqvi for many useful suggestions especially in preparing the Appendix. He is also grateful to I.S. Gulati for going over a preliminary draft and for suggestions and comments at various stages in the preparation of the paper. Thanks are also due to S.G. Kale for pointing out several slips and suggesting improvements. The author alone is however, responsible for the error and short-comings of the paper.]

1. Strictly speaking, the wealth tax on agricultural land should also be counted among the direct taxes from agriculture. The yield from this tax has however, been negligible so far—less than a crore. The yield from partial integration of agricultural and non-agricultural incomes introduced in 1973-74 in the case of those whose non-agricultural income exceeds the exemption limit, is also yet to pick up.

   The market rates or fees charged in some states (notably Haryana and Punjab) on the produce brought to the regulated markets also fetch substantial revenue. But their incidence is uncertain. Besides they cannot be used for progressive taxation. Hence, these rates cannot be treated as a direct tax.

   In some states, e.g., Gujarat, the irrigation rates contain an element of tax in that the capacity of the user of irrigation water to pay the water rate is one of the factors kept in view while fixing these rates. The proportion of the tax element in the total realisation from irrigation rates in the state is however, not known. it is generally known that the irrigation rates in most states are inadequate even for meeting the cost of maintenance of irrigation works and the irrigation losses in 1978-79 is of the order of Rs. 300 crores (Draft Five Year Plan 1978-83, Vol. II, para 3.32).

   Compulsory procurement of foodgrains at fixed prices may also be cited as a form of taxation. However, currently, procurement through compulsory levy is not in operation in most states.
2. According to quick estimates of CSO, in 1976-77 NDP from agriculture was Rs. 39,276 crores. In 1978-79 it is unlikely to be less than Rs. 40,000 crores.
3. *Vide* Report of the Committee on Taxation of Agricultural Wealth and Income, Govt. of India, hereafter Raj Committee Report, Government of India, 1972, Appendix Table A-1.
4. In West Bengal, for instance, the rates averaged Rs. 2.00 per acre. At the time these were fixed, i.e. the thirties, the average yield of paddy was no more than 18-20 maunds per acre and the average price of paddy was about 12 annas per maund. The gross return per acre then was thus about Rs. 15 out of which land revenue took away nearly 15 per cent.

5. In West Bengal the rates were frozen at the level prevailing in 1937 while in the Punjab since 1935 they were allowed to slide when the agricultural prices changed, but only downward.
6. *Cf:* "Most authorities argue that a properly constructed tax on agricultural land is probably the most desirable, way to mobilise and transfer any needed resources from the agricultural to the non-agricultural sectors of the economy since only this tax can perform this task without affecting agricultural production too adversely". ("Taxing Agricultural Land in Developing Countries", by R.M. Bird, Harvard University Press, 1974, p. 74.).
7. Raj Committee Report, para 3.1.
8. *Ibid,* para 3.5
9. *Ibid,* para 3.61.
10. The Haryana Land Holdings Tax Act, 1973.
11. *Ibid,* para 3.12
12. *Ibid,* para 3.13
13. *Ibid,* para 3.31
14. *Ibid,* para 3.32
15. For a discussion of the relative merits and drawbacks of taxation on the basis of operational holding as against ownership holding see C.H. Hanumantha Rao, "Agricultural Taxation: Raj Committee's Report", *Economic and Political Weekly,* Nov. 25, 1972 and Sotthiya, "Agricultural Taxation: A Comment", *E.P.W.*, January 13, 1973.
16. See for a suggestion to get over this problem Sotthiya, *op cit.*.
17. R.M. Bird, *op cit,* pp. 69-70.
18. The flat rate of Rs. 10 proposed for those having more than 2.5 acres of land (irrigated), but whose A.R.V. is less than Rs. 5,000 is also not higher than the fee (Rs. 10 per acre) now charged in West Bengal from cultivators for an annual licence to cultivate lands which vested in the Government because' of the land ceilings etc.
19. Report of the National Commission on Agriculture (1976) Part I and Farm (Harvest) Prices of Principal Crops in India 1970-71 to 1974-75 (Directorate of Economics and Statistics, Ministry of Agriculture, 1977).
20. National Sample Survey, Twenty-sixth Round.
21. A crucial factor underlying the land tax potential, it hardly needs pointing out, is the distribution of the holdings. The exercises carried out to estimate the likely yield of the AHT brings this out clearly.
22. The total area as revealed by the Agricultural Census is larger than the NSS estimate by 32 million hectares and this difference is largely accounted for by the top size groups (above 10 hectares and 4 to 10 hectares). It seems that between 1961-62 and 1971-72 the area of land held by institutions (like cooperatives and trusts) has increased significantly ["NSS Survey on Land Holdings (1971-72) and All India Agricultural Census (1970-71)", unpublished note by S.I.M. Naqvi].

23. For wheat, the income of the by-product as estimated for Madhya Pradesh and Punjab is as follows:

		M.P. (1975-76)	Punjab (1976-77)
(A)	Value of main product per hectare (Rs.)	1275.41	2525.13
(B)	Value of by-product per hectare (Rs)	157.23	306.28
(C)	B as % to A	12.33	12.13

Source: Agricultural Prices Commission, Report on Price Policy for Wheat for 1978-79 Season.

For Bajra, the value of the by product as estimated for Rajasthan during 1970-71 is:—

(A)	Gross Income from main product per hectare (Rs)	331.65
(B)	From by-product (Rs)	71.85
(C)	B as % to A	21.66

Source: Report on Cost of Production of Bajra in Rajasthan during the 1970-71. Crop Season (Directorate of Economics and Statistics, Ministry of Agriculture, 1973).

24. The Index Numbers of the yield of agricultural production for the relevant years are:

(yield per acre)

1970-71	106.8
1971-72	107.2
1972-73	106.8
1973-74	106.4
1974-75	111.5
1975-76	113.5

(1969-70 = 100)

Source: Economic Survey, 1977-78, Government of India.

25. The rates of special assessment on agricultural lands growing certain crops charges under Maharashtra Education Cess Act, 1962 are as follows

Agricultural Land on which Following Crops are Raised	*Rate of Special assessment Per Hectare (Rs)*
1. Sugarcane, grown land perennially irrigated	190
2. Sugarcane, grown on any other land	110
3. Irrigated cotton (except hybrid cotton seed)	40
4. Hybrid cotton seed	110
5. Hybrid jawar seed	40
6. Hybrid maize seed	40
7. Hybrid bajra seed	40
8. Irrigated groundnut	40
9. Betel leaves	190
10. Cirtrus Fruits	80
11. Bananas	110
12. Grapes	380
13. Chikus	80
14. Turmeric	80
15. Arecanut	300
16. Tobacco (irrigated)	130

Under the Maharashtra Increase in Land Revenue and Special Assessment Act, 1974, the cesses indicated above are further increased at the following rates:

Where Amount of Special Assessment Payable under the Maharashtra Education (Cess) Act, 1962	*Rate of Increase*
(a) does not exceed Rs. 200	Nil
(b) exceeds Rs. 200 but does not exceed Rs. 500	25 per centum of such amount as in excess of Rs. 200.
(c) exceeds Rs. 500 but does not exceed Rs. 1,000	Rs. 75 plus 50 per centum of such amount as is in excess of Rs. 500.
(d) exceeds Rs. 1,000 but does not exceed Rs. 2,000	Rs. 325 plus 150 per centum of such amount as is in excess of Rs. 2,000
(d) exceeds Rs. 2,000	Rs. 1,325 plus 150 per centum of such amount as is in excess of Rs. 2,000

26. Article 265 of the Constitution of India.
27. *K.T. Moopil Nair* v. *State of Kerala and another,* AIR 1961 SC 552.
28. *N. Kunhalji Haji* v. *State of Kerala,* AIR 1966 Ker. 14.
29. *Buckingham and Carnatic Co. Ltd.* v. *State of Madras,* (1966) 2 MLJ 1972.
30. *Bhuvaneswariah* v. *State of Mysore,* AIR 1965 Mys 1970.
31. *State of Andhra Pradesh and another* v. *Mulla Raja Reddy and other,* AIR 1967 SC 1438.
32. From the pronouncements of the Supreme Court it would appear that land revenue under entry 45 and a tax on land under entry 49 are indistinguishable. ("Constitutional Law of India" by H.M. Seervai, Second Edition, 1975 Volume 2, p. 1291).
33. Madras High Court in *Buckingham and Carnatic Co. Ltd.* v. *State of Madras,* (1966) 2 MLJ 1972.
34. These propositions follow from the judgment of the Supreme Court in *Ramkrishna Dalmia v. Justice Tendulkar,* AIR 1958 SC 538. See "Constitutional Law of India" by H.M. Seervai, Second Edition, Vol. I, pp. 210-215.
35. *Gopal Narain* v. *State of UP*, AIR 1964 SC 370.
36. AIR 1969 SC 378.
37. AIR 1961 SC 552.
38. AIR 1965 Mys. 170.
39. AIR 1966 Ker 14.
40. AIR 1967 SC 1458.
41. *Ibid*
42. (1966) 2 MLJ 172.
43. The progressive structure of the tax also serves to promote the same objective: of 'there is a strong case for a progressive taxation of the upper income-group in the agricultural sector. The case is strong not only from the point of view of equity, but also from the point of view of productivity. Highly progressive

taxation of land would force the big farmers either to use land more productively or else to sell it to small farmers who, as most farm studies in India and abroad have shown, tond to secure a higher yield per acre (I. S. Gulati, in "Public Finance", 1967, Volume XXII No. 3.

44. Para 3.49 of the Raj Committee Report.
45. AIR 1965 Mys 170, at p. 183.
46. See for instance, "Studies in the Economics of Farm Management in Hooghly District" (West Bengal) for 1970-71 to 1972-73, (Directorate of Economics and Statistics, Ministry of Agriculture, Govt. of India, 1977), Chapter V.
47. See 44D of the Income-Tax Act, 1961.
48. Report of the Indian Taxation Enquiry Commission, 1953-54, Volume III, p. 186.
49. AIR 1961 SC 552.
50. Seervai, *op cit,* p. 414.
51. *Ibid,* Seervai makes this point in support of the state's right to tax unproductive land. The argument would apply a fortriori to taxation of land leased on norms of gross output.
52. Report of the Indian Taxation Enquiry Commission 1953-54, Volume III, pp. 185-86.
53. Report of the Indian Taxation Enquiry Committee, 1924-25, (Govt. of India 1928), Volume I, pp. 81-82.
54. Report of the Indian Taxation Enquiry Commission, 1953-54, *op cit.*
55. *Ibid,* p. 194.
56. *Ibid,* p. 194.
57. Bird, *op cit* p. 132.
58. Writ Petition No. 2639 of 1975 before the Andhra Pradesh High Court. The cess was imposed in 1974 but withdrawn in 1976. It had been imposed first in 1955 and was challenged. But it was abolished in 1962 before the case was heard.
59. Bird, *op cit,* p. 66.
60. The Haryana Land Holdings Tax Act, 1973 Section 5(4).

APPENDIX

The estimate of the probable yield of the AHT put forward in this paper is based on (i) the rateable value of land derived on the basis of average value of output of principal crops in each state and (ii) the state-wise distribution of land holdings:

The RVL, for a state was worked out by taking the weighted average value of principal crops grown in the State. Computation of RVLs state by state, is set out in Appendix Table A-1. Col. I of the Tables lists the principal crops grown in the state. Col. 2 gives the gross area sown in hectare for each crop (average of 1970-71), Col. 3 the average output of 1971-74 in tonnes, col. 4 the price per tonne, Col. 5 gives the value of output as worked out by multiplying col 3 by Col. 4. The average value of output was then derived by dividing that total value of output of principal crops grown (total of col 5) by the net sown area under the crops in question. The net sown area was computed by applying the proportion of net area sown to gross sown area in each state to the total gross area sown under the principal crops. (i.e. total of col 2). The proportion of the net area sown to gross sown area in each state is given in Appendix Table II (Col. 4). Figures of production of principal crops as also gross area sown are based on data available from the Report of the National Commission on Agriculture (1976). For the prices of crops, the farm harvest prices published by the Ministry of Agriculture have been relied upon. The average value of the output of main crops grown in a state derived in this way was taken to be the RVL of agricultural land in the state. The RVLs for each state so worked out are set out in Appendix Table III.

It will be noticed that there is considerable variation in the RVLs as between states. Taking Punjab's RVL as 100, the index of productivity varies between 22.89 and 90.49 if Kerala is excluded (Col. 3 of Appendix Table III). For Kerala the index is 186,37 which is largely due to the relatively high price of rice in that state during the period taken for averaging. Such variations, it may be noted in passing, also mark the index of productivity worked out by Gulati and Krishnan sometime ago (see "Public Distribution and Procurement of Foodgrains: A Proposal" by I.S. Gulati and T. N. Krishnan, *EPW,* May 24, 1975).

Appendix Table IV gives the estimates of the probable yield of the

AHT for each state. Col 1 of the Table shows the mild-value of the holdings in size class from 2.5 acres upwards (shown in brackets). Col. 2 gives the number of holdings in each size class, Col. 3 the aggregate rateable value (ARV) of the holdings worked out by multiplying the RVL for the state as given in Appendix Table III with the size of the holding as given in col. 1. The tax payable on the ARV of a holding of the size indicated in col. 1 was then computed by applying the rate schedule suggested in Section IV. The tax determined in this way is shown in Col. 4. For holdings whose ARV is less than Rs. 5,000 the tax payable has been taken at an average rate of Rs. 8 per acre as, under the suggested scheme, such holdings are proposed to be taxed at the rate of Rs. 7 per acre of unirrigated land and Rs. 10 per acre of irrigated land.

The total amount of revenue likely to be collected from holdings of each size class was estimated by multiplying the tax payable by a holding in a given size class by the number of holding in that class (col 4 × col 2). Aggregate of the amount of tax payable so estimated (i.e. total of col 5) was taken to indicate the likely yield of the AHT in the state. Col 6 of the Tables shows the incidence of the tax as a proportion of ARV of the holding in each size class and col 7 shows the incidence in terms of tax payable per acre.

TABLE A-1 : Average Output and Value of Principal Crops

A-1.1 : Andhra Pradesh

Principal Crops	*Gross Area Sown (Avg of 1971-74 in '000 hectares)*	*Output (Avg. of 1971-74 in '000 tonnes)*	*Price per Tonne (Av. of 1970-71 to 1974-75) (Rs.)*	*Value of Output (Rs. '000) (Col. 3 × 4)*
1	2	3	4	5
Rice	3076	4775	1259.20	6012680.00
Jowar	2716	1231	939.30	1156278.30
Bajra	530	252	862.40	217324.80
Maize	277	319	820.70	261803.30
Groundnut	1428	1161	1841.90	2138445.90
Cotton	348	38	2871.00	109098.00
Sugarcane (Gur)	136	1210	1253.30	1516493.00
Total	8511 (A)	—	—	11412123.30
Net area sown (90% of A)*	7660			

Average value of output per hectare† = Rs. 1489.83 or Rs. 595.93 per acre

A-1.2 : Assam

1	2	3	4	5
Rice	2038	2051	1216.30	2494631.30
Jute	142	197	1068.10	210415.70
Total	2180 (A)	—	—	2705047.00
Net area sown (76.9% of A)*	1676			

Average value of output per hectare† = Rs. 1613.99 or Rs. 645.60 per acre

A-1.3 : Bihar

1	2	3	4	5
Rice	5082	4626	1414.70	6544402.20
Wheat	1828	2427	1159.20	2813378.40
Maize	885	625	1006.20	628875.00
Jute	135	118	1213.30	143157.60
Sugarcane (Gur)	141	479	1349.80	646554.20
Total	8071(A)	—	—	10776367.40
Net area sown (77.6% of A)*	6263			

Average value of output per hectare† = Rs. 1720.64 or Rs. 688.26 per acre

A-1.4 : Gujarat

1	2	3	4	5
Rice	442	378	2194.70	829596.60
Wheat	489	784	1255.00	983920.00
Jowar	1071	376	1152.50	433340.00
Bajra	1773	1234	1027.40	1267811.60
Maize	259	294	1024.30	301144.20
Groundnut	1700	1044	1820.50	1900602.00
Oilseeds	1907	1143	1820.50	2080831.50
Cotton	1833	317	3001.50	951475.50
Total	9474(A)	—	—	8748721.40
Net area sown (94.1% of A)*	8915			

Average value of output per hectare† = Rs. 981.35 or Rs. 392.54 per acre

A-1.5 : Haryana

1	2	3	4	5
Rice	291	513	1170.60	600517.80
Wheat	1208	2148	958.80	2059502.40
Bajra	921	593	800.80	528244.40
Cotton	253	79	2730.40	215701.60
Sugarcane	133	570	1579.50	900315.00
Total	2797(A)	—	—	4304280.60
Net area sown (68.5% of A)*	1916			

Average value of output per hectare † = Rs. 2446.49 or Rs. 898.60 per acre

A-1.6 : Karnataka

1	2	3	4	5
Rice	1091	1972	1477.20	2913038.40
Wheat	331	162	1526.40	247276.80
Jowar	2143	1547	1006.10	1556436.70
Bajra	410	156	923.20	144019.20
Groundnut	870	606	1750.10	1060803.00
Oilseeds	1023	656	1750.50	1148328.00
Cotton	988	97	2512.70	243731.90
Sugarcane (Gur)	107	880	115.20**	101376.00
Total	6963(A)	—	—	7415010.00
Net area sown (94.2% of A)*	6559			

Average value of output per hectare† = Rs. 1130.51 or Rs. 452.20 per acre

**Sugarcane price

A-1.7 : Kerala

1	2	3	4	5
Rice	876	1323	2377.00	3144771.00
Coconut	738	3876**	651.20††	2524051.20
Tapiocca	305	5405	278.20	1503671.00
Total	1919(A)	—	—	7172493.20
Net area sown (73.6% of A)*	1412			

Average value of output per hectare† = Rs. 5079.67 or Rs. 2031.87 per acre

**Million nuts

†† per thousand nuts

A-1.8 : Madhya Pradesh

1	2	3	4	5
Rice	4524	3441	1217.40	4189073.40
Wheat	3439	2612	1106.10	2889133.20
Jowar	2102	1382	960.30	1327134.60
Bajra	216	160	945.40	151264.00
Maize	614	506	805.20	407431.20
Groundnut	454	267	1777.20	474512.40
Oilseeds	1668	588	1951.50	1147482.00
Cotton	689	53	2527.60	133962.80
Sugarcane (Gur)	50	135	1608.10	217093.50
Total	13756(A)	—	—	10937087.10
Net area sown (89.2% of A)*	12270			

Average value of output per hectare† = Rs. 891.37 or Rs. 356.56 per acre

A-1.9 : Maharashtra

1	2	3	4	5
Rice	1325	1252	1662.30	2081199.60
Wheat	888	429	1490.90	639596.10
Jowar	5770	2016	1191.10	2401257.60
Bajra	1549	433	1154.90	500071.70
Groundnut	723	383	2110.40	808283.20
Oilseeds	1083	457	2110.40	964452.80
Cotton	2376	175	2305.20	403410.00
Sugarcane (Gur)	163	1332	1737.40	2314216.80
Total	13877(A)	—	—	10112487.80
Net area sown (94.3% of A)*	13086			

Average value of output per hectare† = Rs. 772.71 or Rs. 309.08 per acre.

A-1.10 : Punjab

1	2	3	4	5
Rice	478	1013	1161.70	1176802.10
Wheat	2375	5414	906.10	4905625.40
Bajra	136	136	897.40	122046.40
Maize	553	834	747.60	623498.40
Groundnut	162	161	1574.90	253558.90
Oilseeds	306	275	1574.90	433097.50
Cotton	493	177	2566.40	454252.80
Sugarcane (Gur)	106	490	1394.60	683354.00
Total	4609	—	—	8652235.5
Net area sown (68.9% of A)*	3176			

Average value of output per hectare† = Rs. 2724.26 or Rs. 1089.70 per acre.

A-1.11 : Orissa

1	2	3	4	5
Rice	4625 (A)	4014	884.00	3548376.00
Net area sown (81.1% of A)*	3751			

Average value of output per hectare† = Rs. 946.00 or Rs. 378.40 per acre

A-1.12 : Rajasthan

1	2	3	4	5
Rice	134	119	1176.20	211367.80
Wheat	1528	1811	1160.00	2100760.00
Jowar	1006	316	949.40	300010.40
Bajra	5349	1470	943.40	1386798.00
Maize	801	585	987.80	577863.00
Groundnut	246	142	1780.80	252783.60
Oilseeds	1198	372	2994.10	1113805.20
Cotton	326	58	2598.70	150724.60
Total	10588 (A)	—	—	
Net area sown (92.3% of A)*	9773			

Average value of output per hectare† = Rs. 623.58 or Rs. 249.43 per acre.

A-1-13 : Tamil Nadu

1	2	3	4	5
Rice	2764	5489	1257.10	6900221.90
Jowar	689	538	965.10	519223.80
Bajra	450	295	860.70	253906.50
Groundnut	1095	1180	1796.80	2120224.00
Oilseeds	1237	1226	1796.80	220876.80
Cotton	306	63	2215.70	139589.10
Sugarcane (Gur)	137	1199	1163.40	1394916.60
Total	66781(A)	—	—	1350958.70
Net area sown (82.2% of A)*	5489			

Average value of output per hectare† = Rs. 2465.10 or Rs. 986.04 per acre.

A-1-14 : Uttar Pradesh

1	2	3	4	5
Rice	4522	3630	1106.80	4017684.00
Wheat	6079	7026	1029.60	7233969.60
Jowar	685	412	751.20	309494.40
Bajra	1033	686	849.80	582968.20
Maize	1482	1100	794.50	873950.00
Groundnut	332	252	1388.60	349927.20
Oilseeds	3690	1481	2111.40	3126983.40
Jute	10	16	953.80	15260.80
Sugarcane (Gur)	1352	5574	102.70††	572449.80
Total	19175 (A)	—	—	17082687.40
Net area sown (75% of A)*	14381			

Average value of output per hectare† = Rs. 1187.86 or Rs. 475.14 per acre.

†† Price of sugarcane

A-1-15 : West Bengal

1	2	3	4	5
Rice	5071	6008	1699.60	10211196.80
Wheat	373	746	1183.20	882667.20
Jute	416	591	1477.10	872966.10
Total	5860 (A)	—	—	11966830.10
Net area sown (80.3% of A)*	4706			

Average value of output per hectare† = Rs. 2542.89 or Rs. 1017.16 per acre.

* For proportion of net area to gross area sown, vide Table A-2.

† Total of Col 5 net ÷ area sown.

Note : Production figures of rice relate to clean rice.
Price of rice has been derived from that of paddy, when rice price is not available, by using the conversion factor 1.6.

Sources :

1. For principal crops and gross area sown : ''Report of the National Commission on Agriculture'' (1976) Part I, Appendix 3.13.
2. For price per tonne : "Farm (Harvest) Prices of Principal Crops in India 1970-71 to 1974-75" (Directorate of Economics and Statistics, Ministry of Agriculture) Table 22.1.

TABLE A-2 : Gross Cropped Area and Net Area Sown Statewise

('000 Hectares)

State	*Total Cropped Area*	*Net Area Sown*	*Percentage Col 3 to Col 2*
1	2	3	4
Andhra Pradesh	12314	11078	90.0
Assam	3035	2335	76.9
Bihar	10383	8053	77.6
Gujarat	9628	9062	94.1
Haryana	5188	3555	68.5
Himachal Pradesh	929	559	60.2
Jammu and Kashmir	891	688	77.2
Karnataka	10410	9808	94.2
Kerala	2987	2197	73.6
Madhya Pradesh	20742	18497	89.2
Maharashtra	17372	16382	94.3
Manipur	162	140	86.4
Meghalaya	182	164	90.1
Nagaland	103	103	100.0
Orissa	6936	5622	81.1
Punjab	5932	4086	68.9
Rajasthan	16097	14858	92.3
Tamil Nadu	7699	6332	82.2
Tripura	356	236	66.3
Uttar Pradesh	22927	17194	75.0
West Bengal	7091	5691	80.3
All-India	161903	137099	84.7

Source : Stastistical Abstract (India) 1975.

TABLE A-3 : Rateable Value of Land per acre Statewise

State	*Value of Output of Main Crops or Rateable Value per Acre (Rs.)*	*Index of RVL**
1	2	3
Andhra Pradesh	595.93	54.69
Assam	645.60	59.25
Bihar	688.26	63.16
Gujarat	392.54	36.02
Haryana	898.60	82.46
Karnataka	452.20	41.50
Kerala	2031.87	186.37
Madhya Pradesh	356.55	32.72
Maharashtra	309.08	28.36
Orissa	378.40	34.73
Punjab	1089.70	100.00
Rajasthan	249.43	22.89
Tamil Nadu	986.04	90.49
Uttar Pradesh	475.14	43.60
West Bengal	1017.16	93.34

* RVL of Punjab = 100

Note : Details of computation of Col 2 are given in Appendix Table I.

TABLE A-4. Estimate of Yield and Incidence of Modified Agricultural Holding Tax

A-4.1 : Andhra Pradesh

Size of Holdings (Acres)	*No. of Holdings ('000)*	*Rateable Value of Holdings (Col. 1 × Rs. 596* (Rs.)*	*Tax on Col. (3) (Rs.)*	*Total Revenue (Col. 4 × (Col. 2) (Rs. '000)*	*Tax as % to Rateable Value (Col. 4 as % of Col. 3)*	*Tax per Acre (Col. 4 ÷ Col. 1) (Rs.)*
1	2	3	4	5	6	7
3.75 (2.5-5.0)	1018.3	2235	30	30549	1.34	8.00
6.25 (5.0-7.5)	545.0	3725	50	27250	1.34	8.00
8.75 (7.5-10.0)	277.3	5215	136	37713	2.61	15.54
11.25 (10.0-12.5)	227.6	6705	210	47796	3.13	18.67
13.75 (12.5-15.0)	104.7	8195	285	29840	3.48	20.73
17.50 (15.0-20.0)	145.3	10430	409	59428	3.92	23.37
22.50 (20.0-25.0)	70.5	13410	648	45684	4.83	28.80
27.50 (25.0-30.0)	53.7	16390	914	49082	5.58	33.24
40.00 (30.0-50.0)	83.0	23840	1659	137697	6.96	41.48
60.00	37.3	35760	2851	106342	7.97	47.52
(Above 50.0)	2562.7			571381		

* Rateable value of land per acre for the State.

A-4.2 : Assam

1	2	3	4	5	6	7
3.75 (2.5-5.0)	392.0	2423	30	11760	1.24	8.00
6.25 (5.0-7.5)	138.8	4038	50	6940	1.24	8.00
8.75 (7.5-10.0)	70.6	5653	158	11155	2.79	18.06
11.25 (10.0-12.5)	23.6	7268	238	5617	3.27	21.16
13.75 (12.5-15.0)	10.0	8883	319	3190	3.59	23.20
17.50 (15.0-20.0)	11.9	11305	479	5700	4.24	27.37
22.50 (20.0-25.0)	5.0	14535	738	3690	5.08	32.80
27.50 (25.0-30.0)	1.1	17765	1052	1157	5.92	38.25
40.00 (30.0-50.0)	1.2	25840	1859	2231	7.19	46.48
60.00	—	—	—	—	—	—
(Above 50.0)	654.2			51440		

A-4.3 : Bihar

1	2	3	4	5	6	7
3.75 (2.5-5.0)	1275.3	2580	30	38259	1.16	8.00
6.25 (5.0-7.5)	534.4	4300	50	26720	1.16	8.00
8.75 (7.5-10.0)	244.2	6020	176	42979	2.92	20.11
11.25 (10.0-12.5)	138.0	7740	262	36156	3.39	23.29
13.75 (12.5-15.0)	61.0	9460	348	21228	3.68	25.31
17.50 (15.0-20.0)	79.7	12040	538	42879	4.47	30.74
22.50 (20.0-25.0)	35.6	15480	823	29299	5.32	36.58
27.50 (25.0-30.0)	11.2	18920	1167	13070	6.17	42.44
40.00 (30.0-50.0)	13.4	27520	2027	27162	7.37	50.68
60.00	6.2	41280	3403	21099	8.24	56.72
(Above 50.0)	2399.0			298851		

A-4.4 : Gujarat

1	2	3	4	5	6	7
3.75 (2.5-5.0)	564.8	1474	30	16944	2.04	8.00
6.25 (5.0-7.5)	337.6	2456	50	16880	2.04	8.00
8.75 (7.5-10.0)	182.0	3439	70	12740	2.04	8.00
11.25 (10.0-12.5)	198.5	4421	90	17865	2.04	8.00
13.75 (12.5-15.0)	94.2	5404	145	13659	2.68	10.55
17.50 (15.0-20.0)	144.0	6878	219	31536	3.18	12.51
22.50 (20.0-25.0)	89.4	8843	317	28340	3.58	14.09
27.50 (25.0-30.0)	63.8	10808	440	28072	4.07	16.00
40.00 (30.0-50.0)	102.7	15720	847	86987	5.39	21.18
60.00	20.2	23580	1633	32987	6.93	27.22
(Above 50.0)	17972			286010		

A-4.5 : Haryana

1	2	3	4	5	6	7
3.75 (2.5-5.0)	108.8	3371	30	3264	0.89	8.00
6.25 (5.0-7.5)	106.0	5619	156	16536	2.78	24.96
8.75 (7.5-10.0)	37.7	7866	268	10104	3.41	30.63
11.25 (10.0-12.5)	54.9	10114	384	21082	3.80	34.13
13.75 (12.5-15.0)	29.3	12361	564	16525	4.56	41.02
17.50 (15.0-20.0)	46.6	15733	848	39517	5.39	48.46
22.50 (20.0-25.0)	25.7	20228	1298	33359	6.42	57.69
27.50 (25.0-30.0)	7.9	24723	1747	13801	7.07	63.53
40.00 (30.0-50.0)	19.4	35960	2871	55697	7.98	71.78
60.00	4.4	53940	4669	20544	8.66	77.82
(Above 50.0)	440.7			230429		

A 4-6 : Karnataka

1	2	3	4	5	6	7
3.75 (2.5-5.0)	738.5	1695	30	22155	1.77	8.00
6.25 (5.0-7.5)	503.6	2825	50	25180	1.77	8.00
8.75 (7.5-10.0)	321.8	3955	70	22526	1.77	8.00
11.25 (10.0-12.5)	204.3	5085	129	26355	2.54	11.47
13.75 (12.5-15.0)	93.7	6215	186	17428	2.99	13.53
17.50 (15.0-20.0)	143.6	7910	271	38916	3.43	15.49
22.50 (20.0-25.0)	109.6	10170	389	42634	3.82	17.29
27.50 (25.0-30.0)	38.8	12430	569	22077	4.58	20.69
40.00 (30.0-50.0)	68.0	18080	1083	73644	5.99	27.08
60.00	32.8	27120	1987	65174	7.33	33.12
(Above 50.0)	2254.7			356089		

A-4.7 : Kerala

1	2	3	4	5	6	7
3.75 (2.5-5.0)	203.8	7620	256	52173	3.36	68.27
6.25 (5.0-7.5)	58.2	12700	591	52126	4.65	94.56
8.75 (7.5-10.0)	28.5	17780	1053	30011	5.92	120.34
11.25 (10.0-12.5)	9.6	22860	1561	14986	6.83	138.76
13.75 (12.5-15.0)	7.5	27940	2069	15518	7.41	150.47
17.50 (15.0-20.0)	6.0	35560	2831	16986	7.96	161.77
22.50 (20.0-25.0)	1.1	45720	3847	4232	8.41	170.98
27.50 (25.0-30.0)	1.1	55880	4863	5349	8.70	176.84
40.00 (30.0-50.0)	0.8	81280	7403	5922	9.11	185.08
60.00	0.2	121920	11467	2293	9.41	191.12
(Above 50.0)	316.8			199596		

A-4.8 : Madhya Pradesh

1	2	3	4	5	6	7
3.75 (2.5-5.0)	1063.7	1339	30	31911	2.24	8.00
6.25 (5.0-7.5)	811.0	2231	50	40550	2.24	8.00
8.75 (7.5-10.0)	485.6	3124	70	33992	2.24	8.00
11.25 (10.0-12.5)	382.0	4016	90	34380	2.24	8.00
13.75 (12.5-15.0)	237.6	4909	110	26136	2.24	8.00
17.50 (15.0-20.0)	294.8	6248	187	55128	2.99	10.69
22.50 (20.0-25.0)	163.0	8033	277	45151	3.45	12.31
27.50 (25.0-30.0)	86.0	9818	366	31476	3.73	13.31
40.00 (30.0-50.0)	160.8	14280	717	115294	5.02	17.93
60.00	54.8	21420	1417	7752	6.62	23.62
(Above 50.0)	3739.3			491670		

A-4.9 : Maharashtra

1	2	3	4	5	6	7
3.75 (2.5-5.0)	1099.9	1159	30	32997	2.59	8.00
6.25 (5.0-7.5)	753.6	1931	50	37680	2.59	8.00
8.75 (7.5-10.0)	442.4	2704	70	30968	2.59	8.00
11.25 (10.0-12.5)	408.6	3476	90	36774	2.59	8.00
17.50 (15.0-20.0)	328.4	5408	145	47618	2.68	8.29
22.50 (20.0-25.0)	151.6	6953	225	34110	3.24	10.00
27.50 (25.0-30.0)	133.1	8498	300	39930	3.53	10.91
40.00 (30.0-50.0)	197.6	12360	564	111446	4.56	14.10
60.00	68.4	18540	1129	77224	6.09	18.82
(Above 50.0)	3787.4			471165		

A-4.10 : Orissa

1	2	3	4	5	6	7
3.75 (2.5-5.0)	758.3	1418	30	22749	2.12	8.00
6.25 (5.0-7.5)	270.8	2363	50	13540	2.12	8.00
8.75 (7.5-10.0)	119.5	3308	70	8365	2.12	8.00
11.25 (10.0-12.5)	74.2	4253	90	6678	2.12	8.00
13.75 (12.5-15.0)	29.1	5198	135	3929	2.60	9.82
17.50 (15.0-20.0)	29.3	6615	206	6036	3.11	11.77
22.50 (20.0-25.0)	18.9	8505	300	5670	3.53	13.33
27.50 (25.0-30.0)	8.8	10395	407	3582	3.92	14.80
40.00 (30.0-50.0)	8.4	15120	787	6611	5.21	19.68
60.00	3.1	22680	1543	4783	6.80	25.72
(Above 50.0)	1320.4			81943		

A-4.11 : Punjab

1	2	3	4	5	6	7
3.75 (2.5-5.0)	153.2	4088	30	4596	0.73	8.00
6.25 (5.0-7.5)	138.6	6813	216	29938	3.17	34.56
8.75 (7.5-10.0)	80.1	9538	352	28195	3.69	40.23
11.25 (10.0-12.5)	59.1	12263	556	32860	4.53	49.42
13.75 (12.5-15.0)	42.0	14988	774	32508	5.16	56.29
17.50 (15.0-20.0)	30.2	19075	1183	35727	6.20	67.60
22.50 (20.0-25.0)	26.3	24525	1728	45446	7.05	76.80
27.50 (25.0-30.0)	18.1	29975	2273	41141	7.58	82.65
40.00 (30.0-50.0)	17.6	43600	3635	63976	8.34	90.88
60.00	5.2	65400	5815	30238	8.89	96.92
(Above 50.0)	570.4			344625		

A-4.12 : Rajasthan

1	2	3	4	5	6	7
3.75 (2.5-5.0)	641.8	934	30	19254	3.21	8.00
6.25 (5.0-7.5)	454.0	1556	50	22700	3.21	8.00
8.75 (7.5-10.0)	195.1	2179	70	13657	3.21	8.00
11.25 (10.0-12.5)	233.8	2801	90	21042	3.21	8.00
13.75 (12.5-15.0)	138.6	3424	110	15246	3.21	8.00
17.50 (15.0-20.0)	183.7	4358	140	25718	3.21	8.00
22.50 (20.0-25.0)	161.3	5603	155	25002	2.77	6.89
27.50 (25.0-30.0)	83.0	6848	217	18011	3.17	7.89
40.00 (30.0-50.0)	127.4	9960	373	47520	3.74	9.33
60.00	103.2	14940	770	79464	5.15	12.83
(Above 50.0)	2321.9			287614		

A-4.13 : Tamil Nadu

1	2	3	4	5	6	7
3.75 (2.5-5.0)	750.5	3698	30	22515	0.81	8.00
6.25 (5.0-7.5)	319.4	6163	183	58450	2.97	29.28
8.75 (7.5-10.0)	130.2	8628	306	39841	3.55	34.97
11.25 (10.0-12.5)	86.0	11093	462	39732	4.16	41.07
13.75 (12.5-15.0)	39.5	13558	660	26070	4.87	48.00
17.50 (15.0-20.0)	47.9	17225	1001	47947	5.80	57.20
22.50 (20.0-25.0)	16.0	22185	1494	23904	6.73	66.40
27.50 (25.0-30.0)	9.0	27115	1987	17883	7.33	72.25
40.00 (30.0-50.0)	18.4	39440	3219	59230	8.16	80.48
60.00	9.3	59160	5191	48276	8.77	86.52
(Above 50.0)	1426.2			383849		

A-4.14 : Uttar Pradesh

1	2	3	4	5	6	7
3.75 (2.5-5.0)	2789.9	1781	30	83697	1.68	8.00
6.25 (5.0-7.5)	1089.9	2968	50	54495	1.68	8.00
8.75 (7.5-10.0)	535.1	4156	70	37457	1.68	8.00
11.25 (10.0-12.5)	278.5	5344	142	39547	2.66	12.62
13.75 (12.5-15.0)	165.2	6531	202	33370	3.09	14.69
17.50 (15.0-20.0)	156.2	8313	291	45454	3.50	16.63
22.50 (20.0-25.0)	74.6	10688	430	32078	4.02	19.11
27.50 (25.0-30.0)	28.3	13063	620	17546	4.75	22.55
40.00 (30.0-50.0)	40.1	19000	1175	47118	6.18	29.38
60.00	4.1	28500	2125	8713	7.46	35.42
(Above 50.0)	5161.9			399475		

A-4-15 : West Bengal

1	2	3	4	5	6	7
3.75 (2.5-5.0)	722.2	3814	30	21666	0.79	8.00
6.25 (5.0-7.5)	309.2	6356	193	59676	3.04	30.88
8.75 (7.5-10.0)	110.4	8899	320	35328	3.60	36.57
11.25 (10.0-12.5)	74.0	11441	490	36260	4.28	43.56
13.75 (12.5-15.0)	25.5	13984	694	17697	4.96	50.47
17.50 (15.0-20.0)	27.7	17798	1055	29224	5.93	60.29
22.50 (20.0-25.0)	10.8	22883	1563	16880	6.83	69.47
27.50 (25.0-30.0)	2.5	27968	2072	5180	7.41	75.35
40.00 (30.0-50.0)	—	—	—	—	—	—
60.00	—	—	—	—	—	—
(Above 50.0)	1282.3			221911		

Note : Figures in brackets indicates size-class of holding

6

Edible Oil Crisis in India

Some Policy Options

S.S. JOHL

Since 1976-77 India has spent some Rs. 8000 crore on the import of edible oils. In 1976-77 edible oils worth about Rs. 100 crore only were imported. These imports in 1987-88 are expected to be of the value of over Rs. 1000 crore. Quantity wise imports of edible oils during the period jumped from 0.169 million tonnes in 1976-77 to over 2.0 million tonnes (expected) in 1987-88 (Table 6.1.). These imports have put a heavy drain on foreign exchange resources of the country, which is mounting every year. The problem is of serious concern to the nation; hence considerable amount of financial resources are being allocated for increasing the production of edible oils. For the Seventh Plan period about Rs. 399 crore have been provided for the development of oilseeds crop production technology and its dissemination.

The year 1976-77 was a crucial year when the country started importing edible oils. That year per capita domestic production of edible oils dropped to 3.7 kg per annum from a level of 4.6 kg per annum in the previous year (Table 6.1). However, per capita domestic production during the triennium ending 1978-79 increased to 4.23 kg per annum from a level of 4.20 kg during the triennium ending 1976-77. Over time the per capita domestic production increased to 4.27 kg per year by the triennium ending 1986-87. Thus, whereas per capita domestic production of edible oils over time has not decreased, rather has increased marginally, the total per capita availability has been increased considerably from 4.3 kg per annum during the triennium ending 1976-77 to 6.0 kg per annum during the triennium ending 1986-87.

TABLE 6.1 : Production and Availability of Edible Oils in India 1971-72 through 1986-87

Year	*Population (millions)*	*Domestic production of edible oil (m. tonnes)*	*Per capita availability (kg/year) (col 3/ col 2)*	*Imports of edible oils*		*Total availability of edible oils (m. tonnes)*	*Per capita availability of edible oils (kg/ year) col 7/ col 2*	*Economic survey per capita availability (kg/year)*
				Quantity (m. tonnes)	*Value (m. Rs.)*			
1	2	3	4	5	6	7	8	9
1971-72	551.3	2.512	4.6	—	—	2.512	4.6	3.0
1972-73	563.9	2.077	3.6	—	—	2.077	3.6	2.4
1973-74	576.8	2.541	4.4	—	—	2.541	4.4	3.4
1974-75	590.0	2.558	4.3	—	—	2.558	4.3	3.3
1975-76	603.5	2.788	4.6	—	—	2.788	4.6	3.5
1976-77	617.2	2.270	3.7	0.169	999.1	2.439	4.0	3.2
1977-78	631.3	2.635	4.7	1.288	7115.4	3.923	6.2	3.8
1978-79	645.7	2.799	4.3	0.821	4798.3	3.810	5.6	3.5
1979-80	660.3	2.411	3.7	1.149	6171.8	3.560	5.4	3.7
1980-81	675.2	2.500	3.8	1.074	5160.0	3.634	5.4	3.8
1981-82	690.1	3.219	4.7	0.996	4496.9	4.217	6.1	5.1
1982-83	706.2	2.728	3.9	1.150	5070.0	3.878	5.9	4.5
1983-84	720.4	3.282	4.6	1.634	13190.0	4.916	6.8	5.8
1984-85	735.6	3.446	4.7	1.368	11221.3	4.814	6.5	5.5
1985-86	750.9	3.028	4.0	1.180	4890.0	4.208	5.6	5.0
1986-87	766.1	3.100	4.1	1.477	6676.7	4.597	6.0	5.9

* Includes Groundnut oil, Rapeseed and Mustard oil, Sesamum oil, Nigerseed oil, Sunflower oil, Soyabean oil, Sunflower seed oil and excludes Oil used for manufacture of Vanaspati.

Source : From data reported by (i) Directorate of Economics and Statistics. Ministry of Agriculture, (ii) Department of Civil Supplies and (iii) Economic Survey 1987-88, Ministry of Finance, Government of India

Overtime butter oil production (*Ghee*) in the country has also increased substantially apart from some free gifts of butter oil from the western countries. *Under this situation when per capita domestic production of edible oils has been steadily increasing over time, though marginally, it is questionable whether country should have imported edible oils and should continue importing in such huge quantities inducing a hike in consumption by some fifty per cent.* Surprising aspect of the situation is that imports bear little relationship with the level of domestic production. Coefficient of determination (R^2) between the annual per capita availability of domestically produced oil and imports is 0.2125 which is illogical, because it indicates that imports were increased as domestic production increased. With a standard error of 0.2374 the coefficient R^2 is of course, not significant.

Price Distortions

While average per capita domestic production of edible oils in India has not decreased and per capita total availability has been increased by about 50 per cent during the last decade, edible oil prices have increased by 170.8 per cent over this period (Table 6.2). However, deflated with general wholesale price index, prices of edible oils in real terms have increased only by 27.7 per cent over this period. Further except for 1986-87, prices of edible oils in real terms during the decade remained lower than the general price level based on all commodities price index. This happened because imports of edible oils and their distribution have been effected at prices much below the level of domestic prices. Thus price distortions in the market have been introduced which have helped the oil consumers, of course at a huge cost of foreign exchange resources. Yet, this situation has gone against the oilseed producers by rendering the general terms of trade unfavourable to them. In the process, the government realized considerable rupee resources at the cost of foreign exchange by importing the edible oils at low prices and supplying it to the consumers at higher prices. This margin has been fluctuating between Rs. 7.00 to Rs. 13.00 per kg of the oil imported and distributed/sold in the market.

Over this period, a large proportion, between 30 to over 60 per cent of the imported edible oil, has been allocated to the vanaspati industry. The release of imported oil to the vanaspati industry has been meeting between 40 per cent to 95 per cent of their requirements in different periods of time. These supplies to the vanaspati industry have been throughout effected at concessional rates, which has distorted the normal price structure of the refined edible oil vis-a-vis vanaspati. While vanaspati was priced at more than 30 per cent higher than the groundnut oil in 1976-77, this price differential came down to only 2.7 per cent in 1986-87. *If this imported oil could be directly released into the market, even at a marginally higher price than that at which supplies have been made to the Vanaspati Industry, it could have had better overall disciplining effect on oil prices in the market.* After all the conversion of otherwise edible oils to vanaspati does not add to the

TABLE 6.2 : Price Behaviour of edible oils in India, 1976-77 through 1986-87

	Oil Prices (Rs. per the tonne)@			*Oil price index Nov.-Oct. 1970-71 = 100*	*General price index Nov.-Oct. 1970-71 =100*	*Oil price in real terms 1970-71 = 100*	*Average unit value of imported oil (Rs. per tonnne)***
	Ground-nut (Rajkot)	*Rapeseed and Mustard (Kanpur)*	*Soya-bean (Devas)*				
1976-77	6855	10058	—	176.6	184.1	95.9	5912
1977-78	7444	9475	—	162.6	184.7	88.0	5521
1978-79	8575	9275	—	173.8	200.2	86.8	5844
1979-80	9771	11938	—	212.0	242.5	87.4	5371
1980-81	13553	13083	—	256.0	274.6	93.2	4804
1981-82	13198	12175	—	256.9	283.7	90.6	4505
1982-83	15248	12883	—	283.7	303.6	93.4	4408
1983-84	16535	15217	—	324.3	330.6	96.5	8072
1984-85	15926	12046	—	294.1	350.5	83.9	8202
1985-86	19243	14250	14350	333.1	368.5	90.4	4144
1986-87	22713	21000	22400	478.2	390.4	122.5	4520

@ Average month and wholesale prices.

** Cif value plus customs duty and other charges such as loading/unloading, letter of credit etc.

Customs duty on an average has been 50 per cent of cif value. Other charges work out to about 2 per cent of cif value.

Source :

(i) Ministry of Commerce, Government of India.

(ii) Directorate of Economics and Statistics, Ministry of Agriculture, Government of India.

(iii) Ministry of Food and Civil Supplies, Government of India.

quantity, rather the process reduces the quantity marginally.

Quality Considerations

On quality, hydrogenation has an adverse effect on the edible oil. For example soybean oil has 85 per cent unsaturated fatty acids and 51 per cent essential fatty acid (linoleic). Similarly, sunflower has some 90 per cent unsaturated fatty acids and 53 per cent linoleic acid, cottonseed has 71 per cent and 50 per cent of these oils respectively (Table 6.3). After hydrogenation the unsaturated fatty-acids are reduced to less than 2 per cent and the essential fattyacid (linoleic) is reduced to zero in vanaspati. Thus from health point of view the quality of edible oil deteriorates totally when it is processed into vanaspati. One can understand it, if otherwise non-edible oils are processed into vanaspati

TABLE 6.3 : Per cent of Unsaturated fatty acids and essential fatty acids (linoleic) in different oils (on oil basis)

	Protein	*Unsaturated fatty acid**	*Essential fatty acid linolic*
Groundnut	27	81	25
Cotton seeds	18	71	50
Rapeseed/Mustard	24	90	19
Copra	9	9	2
Sesamum	20	87	42
Sunflower [EC-684]	14	90	53
Safflower	15	95	62
Soyabean	40	85	51
Nigerseed	16	88	53
Rice Bran	12	78	38
Vanaspati		2	0
Desi ghee		20	2

*Unsaturated fatty acids = Oleic, linoleic, linolenic, Arachidonic acids.

Source : Records of CFTRI, Mysore (Courtesy Director CFTRI)

thereby increasing the total availability of edible oils. Such non-edible oils are generally priced low in the market, which should make it possible for the Vanaspati Industry to offer hydrogenated oils at competitive rates vis-a-vis refined oils. But one fails to comprehend the logic when even otherwise edible oils are supplied to the industry and that too at concessional price and these oils after having been rendered inferior in quality are put in the market at a higher price in the form of vanaspati ghee. If these imported edible oils were sold through the public distribution system and/or were offered in the open market at going prices or were auctioned in bulk, it could perhaps have had a more sobering effect in the edible oils market and also could have saved this edible oil from getting totally damaged in quality from health point of view.

Domestic Production Possibilities

Imports of edible oils to supplement domestic supplies in a shortage situation is quite understandable. Yet, ultimate answer to this national problem of shortages lies in increasing domestic production of edible oils to meet the effective demand. Area under oilseed crops (nine oilseeds) in the country during the triennium ending 1986-87 was 18.88 million hectares, which was a 2.31 million hectares increase over an area of 16.57 million hectares reported during the triennium ending 1972-73. Maximum area under these oilseeds so far has been 19.02

million hectares in 1985-86. Productivity average has increased from 520 kg. per hectare during the triennium ending 1972-73 to 622 kg. per hectare during the triennium ending 1986-87. The maximum average productivity of nine oilseeds so far achieved has been 684 kg. per hectare in 1984-85 (Table 6.4). With this increase in area and productivity over time average total production of nine oilseeds has increased from 8.62 million tonnes during the triennium ending 1972-73 to 11.74 million tonnes during the triennium ending 1986-87. The maximum production achieved so far (nine oilseeds) has been 12.95 million tonnes in 1984-85. In the immediate future, in the absence of any major breakthrough in production technology, based on the maximum area covered and the maximum level of productivity achieved so far, total production of nine oilseeds would not exceed 13 million tonnes. Even if we superimpose a growth rate of 2.24 per cent on this level of production, which has been achieved between the triennium ending 1972-73 and 1986-87, total production would work out to be not more than 13.9 million tonnes by 1989-90. With this level of production, per capita production of edible oils in the country will increase to 5.05 kg per annum. This is an increase of only 0.3 kg over the maximum of 4.7 kg achieved during 1984-85 and 0.78 kg over 4.27 kg per annum during the triennium ending 1986-87. If it is intended to maintain the per capita availability of 6.0 kg (average of triennium ending 1986-87), we have to continue to take a heavy recourse to the imports of edible oils. This is the scenario that unfolds with the existing level of production technology and land resources that can be diverted to the oilseeds production under the assumption of existing cropping patterns and under excellent weather conditions. If, however, through some deliberate concerted efforts considerable land area is diverted from cereal crops and put under oilseeds and if production technology is improved perceptibly and adopted quickly, that would be a different situation. Yet, given the pressing need for increasing foodgrains production to around 175 million tonnes within next two years, there is no great scope or even advantage in diverting the area from under cereal crops to oilseeds to the extent that edible oil shortage is wiped out even in the long run. Thus under these circumstances, even if unconventional sources of edible oils such as oil yielding trees, bushes and other seed are exploited and solvent extraction is expanded, oil shortage is not expected to be covered from domestic resources in the near future.

Above that more than 83 per cent of the area under oilseeds is

TABLE 6.4 : Oilseeds Area, Production and Yield All India

Area : 000 Hectares
Production : 000 tonnes
Yield : Kgs/hectare

	1970-71	*1971-72*	*1972-73*	*Trienium ending 1972-73*	*1984-85*	*1985-86*	*1986-87*	*Triennium ending 1986-87*
Total 9 Oilseeds								
Area	16644	17274	15790	16569	18924	19020	18689	18878
Production	9630	9083	7137	8617	12946	10833	11453	11744
Yield	579	526	452	520	684	570	613	622
Total 7 Oilseeds (Excluding Linseed and Castor)								
Area	10368	9961	9314	9881	16585	17159	16884	16967
Production	6872	6053	3807	5577	12087	10749	10873	11036
Yield	663	608	409	564	717	591	644	650

unirrigated and is heavily dependent upon the extent and spread of rainfall that determines the level of soil-moisture at the planting time and availability of water during the growth period. It is not only the scarcity of rainfall and soil moisture but also the excess of it too that would adversely affect the area under oilseed crops and their productivity. In these crops even price incentives would not work that good in the absence of favourable weather and soil moisture conditions. Thus production of oilseeds would remain highly unstable so long as such a high proportion of the cropped area under these crops remains unirrigated. Even within the recent period of 1980-81 through 1986-87, area under nine oilseeds has fluctuated by over 1.4 million hectares, between 17.60 and 19.02 million hectares, and production fluctuated between 10 million tonnes and 12.95 million tonnes over this period.

On irrigated area also the production of oilseeds is not as stable as that of cereal crops. With a little excessive rainfall or prolonged cloudy weather, oilseed crops in irrigated areas get adversely affected and damaged by insect pests and diseases. Thus right from the seeding stage of the crop and its germination through growth and maturity oilseeds are a highly uncertain crop both on unirrigated as well as irrigated areas. Therefore, whatever be the claims, with the existing level of production technology India is not expected to achieve self-sufficiency in edible oils through the production of oilseed crops in the near future. Under the situation if gross availability of edible oils is to be maintained at the existing levels, there is no other alternative but to heavily depend upon imports.

A Viable Alternative

The question that begs answer, therefore, is what is the alternative. Answer, as it looks, does not rest with the oilseed crops. In fact a little reflection would reveal that the countries which are surplus in edible oils today are in such happier position not because of the production of edible vegetable oils *per se.* In the western countries, especially USA, major oilseed crop is soybean. But, for them soybean is not an oil crop. Soybean is produced for soy-meal which is so rich in proteins (at 40 per cent) but yields only 16 per cent oil. The oil is in fact a bye-product for them which is in excess of their requirements and can be exported at any price. This is the reason India is able to import soybean oil at such a low price while total per hectare economic costs of production of soybean in USA is more than three times that in India. On per tonne basis the cost is more than double of that in India (Table 6.5). Unfortunately in India soybean crop is getting a major thrust as an oil crop and soybean meal, much needed for improving our diets and animal feed, is exported to earn foreign exchange.

TABLE 6.5 : Comparative Economics of Production of Groundnut and Soyabean in India and USA for 1981-82 and 1982-83

Cost categories	*Cost in Rupees*			
	1981-82		*1982-83*	
	India	*USA*	*India*	*USA*
1. Variable expenses per acre:				
Groundnut	680.00	3331.00	675.00	3323.00
Soybeans	293.00	707.00	330.00	1117.00
2. Total economic costs:				
2.1 Per acre:	1147.00	6177.00	1099.00	6110.00
2.1.1 Goundnut	670.00	2114.00	754.00	2262.00
2.2.2 Soybeans				
3. Per quintal				
Groundnut	258.00	503.00	311.00	556.00
Soybeans	157.00	248.00	158.00	323.00

1. For USA, calculations are based on data available in Economic Indicators of Farm Sector (Cost of Production, USDA) Economic Research Service, 1984.
2. US Dollars are converted into Rupees @ Rs. 12.00 per USS.
3. Figures for India are based on average of Karnataka and Tamil Nadu in case of Groundnut and for Madhya Pradesh and Uttar Pradesh in case of Soyabeans. These figures are based on the data available from the Directorate of Economics and Statistics, Ministry of Agriculture, Government of India, New Delhi.

Alternatively, the answer to the edible oil crisis in India lies in the development of dairy industry well organized on scientific lines. It may sound strange, yet a careful reflection would unfold the fact that *if western world today is not facing any cooking media shortage or is rather surplus in edible oils, it is precisely because of the highly developed dairy industry in those countries.* Surplus of butter oil along with other milk products and milk, leaves only a marginal need for edible vegetable oils, except on grounds of health consciousness which makes the people avoid consuming animal fats. Most of the oilseeds and edible vegetable oil they produce are thus available for export. If these countries did not have such a developed dairy industry, they, many of them, would have perhaps themselves landed into the problem of edible oil shortages. *It is an encouraging fact that in India also solution of the edible oil problem through development of dairy industry is technically feasible and economically justifiable, provided the industry is developed in an organised manner and on scientific lines.* The dairy industry can be easily developed through improved milch cows (cross-bred and/or pure bred) yielding milk not less than 5000 liters per lactation of ten months (13 months total including dry period and advanced pregnancy). This in fact is a level quite low by western standards, where animals yielding below 6000 liters are often culled out. In India eight such cows with their followers can easily be maintained for green fodder on one hectare of land with assured irrigation. A better level of management can maintain even ten milch cows with their followers. With eight cows the average yield of milk per hectare would amount to 40,000 liters per lactation of thirteen months. This milk will contain 3.5 to 4.00 per cent fat. If only 3 per cent fat is extracted from this milk, it will yield 1.20 tonnes of pure butter oil per hectare. Rest of the 0.5 to 1.0 per cent fat can be left in the milk. This milk can be standardized at 0.75 per cent level of fat and easily sold in the market at the same price (or even higher price) at which it is procured from the milk producers, *thus yielding butter oil at only the handling and processing costs involved or even less.* At a level of 5000 liters per lactation, dairy would pay the farmer better than any competing crop on two crops irrigated land even if the milk is priced at Rs. 2.00 per liter. As shown in the normative enterprise budgets prepared in the Punjab Agricultural University, cost of production of milk with cows yielding 5000 liters per lactation comes to Rs. 1.63 per liter (Appendix I). If milk is priced at Rs. 2.00 per liter for the producer, the farmer's returns would be about Rs. 14,800 per hectare. Reduced to twelve months, it works out to be over

Rs. 13,700 per hectare. In comparison the best of wheat-rice rotation gives returns over variable costs not more than Rs. 9600 per hectare and sugarcane does not yield returns over variable costs more than Rs. 8250 per hectare (Appendices III and IV). Further while estimating the returns over variable costs from crops fixed costs have not been accounted for whereas for milk production all fixed costs, except land rent, have been accounted for. Thus real difference in returns would be still wider giving added advantage to the dairy enterprise over general crop cultivation. Thus there is a tremendous economic advantage for the farmer in diverting irrigated area to the dairy enterprise at the margin. If only one million hectares of assured irrigated land is diverted to growing of forages for milk production through improved dairy cows this can yield some 1.2 million tonnes of butter oil per lactation period of 13 months or a little over 1.11 million tonnes per annum. Thus where meeting the edible oil requirements of the country on a sustained basis through growing of oilseed crops appears to be a remote possibility or atleast a very difficult proposition, development of dairy industry offers quite a feasible alternative. Even a single state, like Punjab, which even otherwise needs to diversify its agricultural production pattern by shifting some 20 to 25 per cent of its cropped area from under wheat and rice can meet the total shortage of edible oil in the country through developing organized dairy industry on scientific lines. Here point of emphasis is that well organized and scientifically managed dairy enterprise based on improved cows yielding an average of 5000 liters per lactation can produce about 1.11 tonnes of butter oil per hectare per year. Even when the area required for foodgrains is added to the areas required for forages, the butter oil production per hectare of land comes to 825 kg per annum through the organized dairy enterprise. No single oilseed crop yields even half of this quantity of edible oil. A good irrigated rapeseed/mustard crop yields about 0.35 tonnes of oil per hectare. A good groundnut crop also does not yield more than 0.35 tonnes of oil per hectare. Double crop (Rapeseed/Mustard followed by Groundnut) will yield only about 0.7 tonnes of oil per hectare per year. Above all, even this much production of vegetable oil from oilseed crops is totally uncertain because of high degree sensitivity of the crops to the changes in weather. Butter oil production is comparatively much more certain.

Cost of Production of Butter Oil

This pure butter oil can be obtained at a residual cost equal to or even less than handling and processing costs of milk only. After removing 3 per cent fat, the milk standardized at 0.75 per cent fat can be easily sold at Rs. 2.00 per liter. At this price and even at Rs. 2.50 per litre or so, there will be no shortage of consumer demand. If it comes to that milk becomes surplus (which is a remote possibility at a price of Rs. 2.00 or Rs. 2.50 per litre), the surplus milk can be processed into milk powder, for mid-day lunch programmes for school children and also for health ameliorating programmes for the vulnerable sections of the society. This way the milk industry in India which has priced itself out limiting consumer demand to a very low level, would easily get out of the vicious circle of high cost of production and high consumer prices resulting in low demand.

Cost of transportation, handling and processing of milk even at present level of efficiency and volume of business ranges between Rs. 1.00 to 1.50 per litre, which includes collection, transportation, cold storing, processing, distribution, back delivery of unsold milk to be reprocessed, over heads, staff and all other costs. This cost can be easily reduced to even below Re. 0.25 per litre through increasing the volume of business and improvement in the efficiency on transport, handling, processing and distribution of milk. As may be seen in Appendix V, depending upon the volume of milk collected and distance of collection centres from milk plants, transport costs have varied from Re. 0.10 per litre to Re. 0.52 per litre in the month of April 1987. Detailed study of the data has shown that these costs get reduced to even Re. 0.03 per litre in flush season at several centres. If volume of milk handled increases as envisaged, this cost can be reduced to even below Re. 0.01 per litre. Similarly, other processing, handling and distribution costs can be reduced very considerably. Information in Appendix VI indicates that in flush season cost of processing of milk comes down to only 3.6 paise per litre. Thus even at present level of efficiency the collection and processing costs in flush season come to 6.6 paise per litre. These costs can be easily reduced to not more than 3 or 4 paise per litre if milk production is developed through organized dairies and efficiency of the processing units is improved. Thus, even if residual costs of transport, handling and processing are charged to the butter-oil so separated from milk, these costs would come to below Rs. 4.00 per three kg of butteroil or about Rs. 1.33 per kg. If these processing and handling and even distribution costs are also passed on to the consumers of milk by pricing milk at Rs. 2.00 per litre for the farmer and Rs. 2.25 per litre for the consumer, the butter oil can be obtained absolutely free of cost. No vegetable oil can be produced that cheap. This system will not work with buffaloes. The cost of production of buffalo milk even with improved animals yielding 2000 liter per lactation comes to Rs. 4.13 per litre (Appendix II). At this price consumer demand will not pick up and if milk is standardized at one per cent fat, it cannot sell in the market at this price.

Given the availability of oilcakes and other feed materials used in India, it would require less than 3 million tonnes of foodgrains for producing 1.2 million tonnes of butter oil per million hectare of land put under forages on a sustained basis. By shifting marginal lands to coarse grains it should not be impossible to grow this much quantity of feedgrains in the country. And also, rain spoiled and damaged foodgrains, which is a common phenomenon in India, can be used as feedgrains for dairy cows. As an alternative, even if 3 million tonnes of feedgrains are imported these will not cost more than Rs. 350 crore whereas imports of 1.2 million tonnes of edible oil will cost about Rs. 600 crore. Also advantage of 39 million litres of milk is lost in this trade off in favour of importing oil instead of feedgrains. Even if area required for green fodder as well as that for feed is considered, butter oil production comes to 825 kg per hectare per annum (Appendix I). Still it is higher than the quantity of edible vegetable oil any two oilseed crops in rotation would yield on irrigated lands assuming high degree certainty of the oilseed crops. Thus even if 1.4 to 1.5 million hectares of two crop irrigated land is diverted to produce forages and feed for dairy enterprise, this will produce some 1.2 million tonnes of butter oil which will wipe out the shortage of edible oils at present level of consumption and in addition would produce 39 million tonnes of standardized milk to make it an operation flood in the real sense.

Precautions and Prerequisites

It is as simple as that! Yet there are some prerequisites that must be ensured and precautions that must be observed in order to make the whole system work for expected results.

1. Emphasis must shift from buffaloes to cows of improved breeds. No country in the world has ever developed its dairy industry around buffaloes. Best possible germ plasm of buffaloes available the world over is in India. Here if a buffalo yields 1500 litres per lactation the

animals is categorized as an improved one. Buffaloes with 2000 litre yields are very few and animals yielding beyond that area rare. On the other hand cows in the western countries yield even beyond 7000 litres per lactation and the animals yielding less than 6000 litres are often culled out. Semen of progeny tested bulls of even 10000 litres capacity is available in these countries. Thus scope of improvement in cows is quite vast, and buffaloes present a poor, rather no alternative. Thus so long we keep sticking to buffaloes as milch animals, there is little scope of the dairy industry providing this country with milk and butter oil enough to resolve the edible oil crisis.

2. The countries that started late in dairy development, improved their animals from the level already available in the developed countries. Israel for example started with pure-bred Holstein-Freisen cows available in Europe and USA and today that country has the best animals of their own. In India even for cows we have so far concentrated on cross-breeding with our own breeds, which are dismally low yielders of milk. Even now in the era of fast scientific developments, we are hesitant to increase the exotic blood level of our animals beyond 75 per cent. This conservative approach has got to be shed and bold actions taken by importing good quality pure-bred dairy cows in a substantial number to provide a reasonable base for improving our dairy cows. From that base we need to improve our animals with imported semen of high quality from progeny tested bulls of standing from USA and Europe. We must set our standards not to accept a cow yielding less than 5000 liters per lactation and our goal should be to improve production beyond this level.

3. In this system of improvement of milch animals and dairy development, *dairies should be organised in units of economic size not less than 25 milking cows.* This minimum size unit can be easily maintained on a 4 hectare holding using 75 per cent of the area for forages and rest 25 per cent for other crops. This system has to be operated on commercial basis and should in no case be mixed up with poverty alleviation programmes and other programmes involving the provision of one or two milch animals to the poor, as is being done today. This programme should also not get mixed up with the existing programmes of dairy development, followed on Anand pattern or otherwise which are based on squeezing out the milk from villages. Most of the milk collected at present is a distress sale by the producers for cash income. These sales are only a marketed surplus if some

respectable terminology is to be used for it. In no way these are a marketable surplus. On the other hand the suggested system is envisioned to produce real milk surpluses on commercial basis, which would yield good incomes to the producer farmers and would make milk and butter oil available to the consumer at low costs.

4. Health and sanitation cover for dairy animals, is an essential component of this programme. Cross-bred animals and/or pure bred animals imported are more sensitive and prone to diseases. Trained veterinarians have to be appointed for specific number of dairies/ animals. They have to be made responsible for mandatory examination of every dairy and every animal once a week. For violating health and sanitation provisions, punitive measures have to be legislated and executed. For adequate professional and clinical support to these veterinarians, policlinics have to be set up at every district and even tehsil level, fully equipped with laboratories and specialists in various branches of animal health. For meeting these health service costs, dairy farmers can be charged a monthly fee of Rs. 5 to Rs. 10 per animal. The farmer will not mind paying these costs, if the dairy enterprise turns out to be an economic proposition for him. The system will thus be able to meet its own costs through its normal course of growth and development.

5. In order to put the system on a sound footing, farmers have to be trained in technical and management aspects of the dairy enterprise. The organized dairy enterprise with improved milch animals requires a high degree skill and competence in management. The existing level of management skills with the dairy farmers is no doubt an encouraging sign of the possible improvements that can be brought about. Yet it is not adequate enough to handle organized dairy enterprises on an optimum level.

6. Last but not the least the process of development of dairy industry must start from the marketing end with adequate and efficient facilities for prompt collection of milk, transport, handling and processing of milk into butter oil, standardized milk and other milk products, storing and distribution of the same. The processing plants should have adequate forward linkages with the distribution and sale system and backward tie-ups with the collecting centres and milk producers. Handling of milk, its processing and marketing is the most crucial factor that would make the system success or render it a failure.

Thus dairy enterprise and dairy industry as a whole based on high yielding cross-bred and pure bred cows hold great promise for solving the problem of edible oil shortages in India and for providing adequate milk to the rural people and the urban centres. It should be kept in mind that the system has to develop on efficient and surplus production of milk in the farm sector and in no case on squeezing the trickles of milk from the villages as distress sales by the producers, as is the case at present. The existing dairy development programmes are in fact a misnomer and most of the time these programmes exploit the poverty of the milk producers.

Dairy development also has good scope for expansion of semiskilled, skilled and professional employment and also for enhancing income levels in the rural areas. In turn, the need for skilled and professional personnel will require expansion of facilities for training of a large number of veterinarians, health specialists, animal scientists and skilled technical workers. In consequence, an upward spiral of income and enhancement of employment would take place, leaving a very healthy influence on the socio-economic fabric of the society.

Thus answer to the edible oils problem in India in the long-run lies in the development of dairy industry and not so much in expansion of area under oilseed crops, provided the dairy industry is developed in an organized manner on the strength of cross-bred and pure bred cows yielding not less than 5000 litres of milk per lactation. With this approach India is capable of not only meeting its internal demand for edible oils on a sustained basis, but also the country can be in a position to export edible oils and milk products. The approach needs some bold policy decisions in an integrated manner. Yet, if edible oil crisis is to be resolved permanently, there is no escape from going this way.

APPENDIX-I

ECONOMICS OF DAIRY FARMING WITH 25 CROSSBRED COWS

A. *Livestock Number*

1. Milch cattle = 25
2. Total adult units with followers = 30

B. *Capital Investment (Rs.)*		With Average Milk Yield/Cow/Lactation (4000 Lit)	(5000 Lit)
1. Value of cows	Rs.	1,50,000	1,75,000
2. Housing cost (cattle shed, calf pen and feed store etc.)	Rs.	90,000	90,000
3. Equipment (Chaff cutter, utensils, milking machine and power etc.)	Rs.	60,000	60,000
Total Capital Investment	Rs.	3,00,000	3,25,000
C. *Fixed Cost (On capital investment)*			
1. Interest on capital investment @ 12.5%	Rs.	37,500	40,625
2. Cattle insurance @ 3%	Rs.	4,500	5,250
3. Depreciation and repair @ 4% of the value of shed	Rs.	3,600	3,600
4. Depreciation of equipment @ 10%	Rs.	6,000	6,000
Total fixed cost Annual	Rs.	51,600	55,475
Total fixed cost per lactation	Rs.	55,900	60,098

D. *Variable Cost*

1. Green fodder (Legume, non-legume)

(i) Lactation period 325 × 50 × 25	4,062.5 qtls
(ii) Advanced pregnancy 70 × 50 × 25	875.0 qtls
(iii) Young Stock 270 × 10 × 10	270.0 qtls
Total	5,207.5 qtls

Approximately 5,210 qtls @ Rs. 10/- per qtl (Farm gate costs)	Rs.	52 ,100	52,100
2. Dry fodder Dry fodder for adult stock for 395 days (13 months) 395 × 5 × 25 = 493.75 qtl Approximately 494 quintals @ Rs. 25/- per quintal (Farm gate value)	Rs.	12,350	12,350

3. Concentrates			
(i) Lactation period			
395 × 2.5/3 × 25 =	247 qtls	296 qtls	
(ii) Young stock			
240 ×,0.5 × 10 =	12 qtls	12 qtls	
	259 qtls	308 qtls	
Approximately 260 and 310 quintals of concentrate mixture @ Rs. 150 per quintal	Rs.	39,600	46,500
4. Labour cost – three man units @ Rs. 20 per day i.e. 395 × 20 × 3	Rs.	23,700	23,700
5. Veterinary expenses	Rs.	4,000	4,000
6. Electricity charges	Rs.	2,600	2,600
7. Misc. Expenses	Rs.	2,500	2,500
Total Variable Cost	Rs.	1,36,250	1,43,750
Total Cost (Fixed + Variable)	Rs.	1,92,150	2,03,848
Cost of production per litre milk		1.92	1.63
Area requirements for green fodder		3.0 ha	3.0 ha
Area requirements for grain		1.0 ha	1.2 ha
Total area required		4.0 ha	4.2 ha
Total milk production : per year		92,308 lit	1,15,384 lit
Fat @ 3%		2,769 lit	3,461 kg
Fat per hectare		689 kg	825 kg

Note: Depreciation on cow has not ben accounted for because appreciation of some animals and followers compensate for the depreciation of others throughout.

Source : Calculated from Normative Enterprise Budgets prepared in the Department of Economics and Sociology, PAU, Ludhiana. (*Courtesy* : Professor Joginder Singh)

APPENDIX-II

ECONOMICS OF DAIRY FARMING WITH 25 BUFFALOES

A. *Livestock Number*

1. Milch cattle = 25
2. Total adult units with followers = 30

B. *Capital Investment (Rs.)* — *With Average Milk Yield/of 2000 lit/lactation*

25 buffaloes @ Rs. 5500 each	1,37,500
Housing cost	90,000

C. *Fixed Cost (on capital investment)*

Interest on capital investment @ 12,5%	35,938
Cattle Insurance @ 3%	4,125
Depreciation and repairs @ 4% on the value of shed	4,600
Depreciation on equipment @ 10%	6,000
Total fixed cost (Annual)	49,663
Total fixed cost (per lactation)	66,217

D. *Variable Cost*

1. Green fodder		
Lactation period 300 × 55 × 25	4,125.0 qtls	
Dry period 120 × 30 × 25	900.0 qtls	
Advanced pregnancy 60 × 55 × 25	420.0 qtls	
Total	6,270.0 qtls	
6270 quintals @ Rs. 10 per quintal		Rs. 62,700
2. Dry fodder		
Lactation period 300 × 5 × 25	375.0 qtls	
Dry period 120 × 6 × 25	180.0 qtls	
Advanced pregnancy 60 × 5 × 23	75.0 qtls	
Young stock 210 × 1 × 210	21.0 qtls	
Total	651.0 qtls	
651 quintals @ Rs. 25 per quintal		Rs. 16,275
3. Concentrates		
(i) Lactation period 300 × 1/5 × 25	112.5 qtls	
(ii) Advanced pregnancy 60 × 2 × 25	30.0 qtls	
(iii) Young stock 210 × 1/2 × 10	10.5 qtls	
Total	153.0 qtls	
153 quintals @ Rs. 150 per quintal		Rs. 22,950
4. Labour 3 man unit @ Rs. 20 = Rs. 480 × 20 × 3		Rs. 28,800

5.	Veterinary expenses	Rs. 4,000
6.	Electricity charges	Rs. 3,200
7.	Misc. Expenses	Rs. 2,500
	Total Variable Cost	Rs. 1,40,425
	Total Cost (Fixed + Variable)	Rs. 2,06,642
	Cost of milk production	Rs. 4.13 per litre

Requirements (Qtls.)

1. Per lactation :	
Green fodder	6,270 qtls
Dry fodder	651 qtls
Grain (30% of concentrate)	46 qtls
2. Per Year :	
Green fodder	4702 qtls
Dry fodder	488 qtls
Grain	35 qtls
Area requirements for green fodder	3.00 ha
Area requirement for grians	0.60 ha
Total area required	3.60 ha
Total milk production per year	37,500 lit
Fat @ 6%	2,250 kg
Fat per hectare	625 kg

Source : Calculated from Normative Enterprise Budgets prepared in the Department of Economics and Sociology, PAU, Ludhiana. (*Courtesy* : Professor Joginder Singh)

APPENDIX-III

ENTERPRISE BUDGETS FOR RABI CROPS 1986-87

(*Rs./Acre*)

Sl. No.	*Crop*	*Variety*	*Yield (Qtls)*	*Gross returns*	*Variable cost*	*Returns over variable costs*
1.	Wheat	WL 711/HD 2009	20.00	3920.00	1946.68	1973.32
		HD 2329/WL 5023				
		WL 1562	17.00	3422.00	1883.55	1538.45
2.	Barley	PL 56 (Unirrigated)	12.00	1660.00	1049.80	610.20
		DL 70 (Irrigated)	16.00	2100.00	1172.97	927.03
3.	Winter Maize	Partap 1	22.00	4260.00	2586.38	1673.62
4.	Gram	GL 769/C 214 (UI)	5.00	1918.00	1102.95	815.65
		G 130 (Irrigated)	6.00	2298.00	1167.12	1130.88
5.	Lentil	LL 56	5.15	1947.50	920.12	1026.88
6.	Rapeseed & Mustard	Gohbi Sarson	6.00	3112.00	1439.17	1672.83
		Toria TL 15	8.90	2001.00	1327.39	673.61
		Raya RLM 14/	7.00	3738.00	1547.84	2190.16
7.	Linseed	K2/LC 54	5.00	2765.00	1279.10	1485.90
8.	Field pea	PG/3	7.00	2450.00	1442.69	1007.31
9.	Minor crops	Celery	4.00	2238.00	1706.51	501.49
		Fennel	4.00	2010.00	845.85	1164.15
		Corriander	3.00	1506.00	925.24	580.76
		Isabgol	4.00	1608.00	515.43	1092.57

Source : Department of Economics and Sociology, Punjab Agriculture University, Ludhiana. (*Courtesy*: Professor Joginder Singh)

APPENDIX-IV

ENTERPRISE BUDGETS FOR KHARIF CROPS, 1988

(Based on PAU Recommendations)

(Rs./Acre)

Sl. No.	*Crop*	*Variety*	*Yield (Qtls)*	*Gross returns*	*Total variable*	*Returns over variable costs*
1.	Paddy	Basmati 370	10.00	3800.00	2050.17	1749.83
		Jaya/IR-8	30.00	4500.00	2851.52	1648.48
		PR-106/PR103/PR 108	30.00	4740.00	2853.59	1886.41
2.	Maize	Ageti 76	15.60	3672.00	2165.84	1506.16
		Partap	16.00	3770.00	2344.21	1425.79
3.	Bajra	PHB 10	13.00	2782.00	1409.31	1372.69
		PHB 47	13.60	2902.00	1411.38	1490.62
4.	Pulses	Moong G-65/SML-32	4.00	2168.00	1197.74	1048.26
		Moong ML-5/ML-131	4.50	2436.00	1120.25	1315.75
		Mash 1-1	3.00	1760.00	770.94	989.06
		Arhar AL-15	5.00	2850.00	1041.04	1808.96
5.	Cotton	F-28/LH 900	7.00	4910.00	2151.52	2758.48
		LD-230	6.00	3990.00	2063.54	1926.46
6.	Sugarcane	COj-64	300	10050.00	7159.90	2990.10
		COj-67	350	10550.00	7236.95	3313.05
		CO-1148	350	10550.00	7338.06	3211.94
	Miscellaneous					
	Soybean SL-4		6.00	1956.00	1088.56	867.64
	Sesamum Pb. Til No. 1		2.00	1650.00	1217.44	432.56
	Sunflower (Remsun Record)		6.00	2442.00	1454.60	987.40
	Mentha		45 lit.	6300.00	5019.18	1280.82

Source : Department of Economics and Sociology, Punjab Agriculture University, Ludhiana. (*Courtesy*: Professor Joginder Singh)

APPENDIX-V

COMPARATIVE STATEMENT ON AVERAGE TRANSPORTATION COST PER KG MILK IN PUNJAB FOR THE MONTH OF APRIL 1987

	Jalan - dhar	*Hoshiar pur*	*Sangrur*	*Bath- inda*	*Farid- kot*	*Feroze- pur*	*Ropar*	*Ludhi- ana*	*Patiala*	*Gurdas- pur*	*Amrit- sar*	*Total*
Average daily milk production (kg)	96900	30067	11267	11900	14633	23068	35267	89367	31833	34567	52268	431137
Average milk collected per day per society (kg)	160	78	48	46	71	76	88	189	91	109	114	106
Average price paid to milk producer per kg of milk	2.76	2.90	2.45	3.09	2.61	2.84	3.13	3.11	2.87	2.79	2.78	2.85
Average transportation cost per kg paise	13	20	42	46	52	31	20	10	25	19	18	20

Source : Courtesy of Milk Plant, Verka

7

Restructuring of Agricultural Planning and Policies for Alleviating Rural Poverty

N.S. RANDHAWA

I. POVERTY

Poverty is a perpetual human tragedy. The poor often becomes vulnerable and insecure which lead to loss of human dignity and optimism and human waste. Poverty existed in all times in various forms and in different intensities. Long ago it used to be tackled by communities and was of a local concern. With improvements in communications, distance barriers were broken. Poverty became a national issue and flow of knowledge played a significant role in bringing about international focus on poverty. (Table 7.1)

Poverty is predominantly a rural phenomenon. Agriculture is still, directly or indirectly, the major source of income of the rural poor. The paper, therefore, focuses on the impact on poverty of the overall development strategies and of the agricultural development strategies in particular, followed by the developing countries, traces their inadequacies to deal with poverty sufficiently; and suggests restructuring of agricultural development strategy and planning to combat poverty more effectively.

Nearly one half of the extremely poor and of the total poor were concentrated in South Asia, predominantly in India. The incidence of the poverty was also largest in South Asia. The poor in South Asia and East Asia combined comprises 71.4 per cent of the total poor.

TABLE 7.1 : Poverty Situations in 1985

	Extremely Poor		*Poor Including Extremely Poor*	
Region	*No. Million*	*Head Count Index %*	*No. Million*	*Head Count Index %*
East Asia	120	9	280	20
China	80	8	210	20
South Asia	300	29	520	51
India	250	33	420	55
Sub-Saharan Africa	120	30	180	47
All Developing Countries	633	18	1,116	33

Source : World Development Report, 1990.

The incidence of rural poverty is greater than that of urban poverty as indicated in Table 7.2 for the selected Asian countries.

TABLE 7.2 : Rural and Urban Poverty in the 1980s

Region and Country	*Rural Population as Percentage of Total Population*	*Rural Poor as Percentage of Total Poor*
ASIA		
India	77	79
Indonesia	73	91
Malaysia	62	67
Philippines	60	67
Thailand	70	80

Source : World Development Report, 1990 (abridged).

Small and Marginal Farmers and Agricultural Labourers and Poverty

The poor in agriculture comprise small and marginal farmers drawing major portions of their income from self employment and agricultural labourers who are entirely dependent on wage earnings. The World Development Report (1990) quotes studies of 1981 of Thailand that 75 per cent of the poor rural households were self employed in agriculture; of Indonesia that in 1987 self-employed households in farming accounted for 58 per cent of the poor and 41 per cent of the total population and wage earners formed 14 per cent of the poor and 9 per cent of the population of India that self-employed in agriculture accounted for 35 per cent of the poor rural households, in 1977-78 and for 46 per cent of all households, while households engaged in agricultural labour accounted for a further 44 per cent of poor rural households but for only 30 per cent of all households; of Pakistan that suggested that among the poor there were fewer farm labourers than own-account cultivators; and of Bangladesh that the opposite was true.

Self-employed poor in agriculture are small and marginal farmers.[19] Some of their characteristics are given in Table 7.3. As broad guidelines, the proportion of small farmers can be taken as 60-70 percent of total farmers, operating about 25 to 30 per cent of total area.[19]

Table 7.4 gives the changes in poverty incidence in 1980 for selected Asian Countries.

The reduction in the number of poor and in the incidence of poverty is not striking in India and Malaysia though some gains have been experienced. The number of poor has in fact increased in China, Pakistan and Thailand, so has the incidence of poverty in China and Thailand. Indonesia is the only country where impressive achievements have been made in the alleviation of poverty. There does not seem to be close relation between reduction in poverty incidence and the performance index in some countries discussed in Section II. For example, India classified as a below average performer reduced poverty incidence while in Thailand (better performer) poverty incidence has increased. The net impact on poverty reduction depends upon several measures which influence it positively, negatively or neutrally. The performance index and its component indices are the results of several such policy measures, programmes and projects meshed into planning. Therefore, attacking poverty requires a broader poverty-reducing strategy and planning. Therein lies its comprehensive solution as has been discussed in Sections IV, V and VI. It may, however, be noted that poverty incidence based on single year is considerably influenced by fluctuations in agricultural production caused by weather and other natural factors.

It is evident that development strategies, planning methodologies and policies followed by the developing countries could at best only contain the poverty situation. But they have not shown the potential to eradicate it or reduce it substantially in the foreseeable future. Therefore, a critical examination of the development strategies is needed to assess their relevance and adequacy to make a significant dent on poverty and to formulate an appropriate restructuring of strategy and planning to combat poverty.

TABLE 7.3 : Some Characteristics of Small Farms, 1971-82

	No. of small frams (SF)	*% of SF to total no. of farms*	*% of pop. living in SF households*	*% of total land area cultivated by SF*	*% of SF contribution to Total agri. Production*	*Average farm size*
	(millions)	%	%	%	%	%
	—	74.6% (1980-81)	50% (rural households & rural pop.)	26.3% (1980-81)	20-25%	1.82 ha
	—	66.9% (1982)	66.6% (rural pop.)	35.1% (1982)	32.8%[a]	0.5-1.6 acres
	9.54	—	—	—	—	—
	1.438[b]	61% (1971)	80.5% (total farm pop.)	24.1% (1971)	73.7% (total agri. crop production, 1981)	
	3.4	80% (1978)	62.1%	21.5% (1978)	64.1% (Paddy production (1983-84)	Central 40 rai[c] North 30" South 20" N/E 50"
		75% (1980)		40% (1980)	—	—

a. Paddy production.,

b. Farms less than 3 ha.,

c. 1 rai = 0.16 ha.

Country case studies presented at the FAO Regional Workshop on "Impact of Policies and Programmes for Increased Agricultural Production Particularly Food by the Small Farmers", 11-15 November 1986.

Source : Compiled in reference [19].

TABLE 7.4 : Changes in Poverty in the 1980s

Country and Period (Years)	*Length of Period*	*Head Count Index*		*No. of Poor (Million)*		*Average Income Shortfall (Per cent)*	
		First Year	*Last Year*	*First Year*	*Last Year*	*First Year*	*Last Year*
China (1985-88)	3	10	14	79.2	101.3	25	24
India (1977-83)	6	50	43	324.9	315.0	29	28
Indonesia (1984-87)	3	28	17	45.4	30.0	24	17
Malaysia (1984-87)	3	15	14	2.3	2.2	26	24
Pakistan (1979-84)	5	21	20	17.1	18.7	19	19
Thailand (1981-86)	5	20	26	9.5	13.6	27	35

Source : World Development Report ,1990 (abridged).

II. PERFORMANCE OF COUNTRIES

Poverty incidence is one of the important indicators of performance of countries for human welfare. It is linked with growth, social and other distribution indicators. The overall performance of countries for human welfare has, therefore, been attempted by constructing a composite index using the ten indicators subdivided under three groups[18] (i) Growth indicators, including GDP growth rate (1961-83) GDP per capita (1983US$) and inverse of population growth rate (1961-83); (ii) Distribution indicators based on income equality (1 minus Gini Coefficient of income inequality, latest information available), poverty alleviation (100 minus incidence of poverty in percentage terms, latest available information) and degree of equality in landholding distribution (1 minus Gini coefficient of landholding distribution, latest available information); (iii) Social indicators comprising nutrition measured by average daily calorie intake per person or Dietary Energy Supply (DES) in 1983, life expectancy at birth in 1981, survival rate of infant (1000 minus infant mortality per 1000) in 1981 and adult literacy in 1981 as a proxy for level of education. It should be noted that in the case of population growth, income inequality, incidence of poverty, distribution of landholding and infant mortality increasing indicator values correspond to deteriorating performance. Therefore, adjustments have to be made for these indicators as stated above to make the adjusted indicators to show improvement in performance by their increasing values.

These 10 indicators subdivided into three groups were given different weights to make a composite index of overall performance. For a standard solution the weights are for growth subgroup, GDP 20, GDP per capita 20 and population 20; for distribution indicator index based on income equality, poverty alleviation and land holding distribution equality 20 (in fact each of the three gets about 6.67) and for social indicator index comprising nutrition, life expectancy, infant survival rate and adult literacy 20 (each of the four components gets 5). four weighting systems were used; the results of the standard weighting system are given in Table 7.5.

The magnitude of difference in the composite index between the countries reflects the quantitative difference in their performance.

— The Philippines (99) and Indonesia (98) could be considered average performers, with their index numbers nearly 100;
— Best performers were Korea, Rep. (160) and Malaysia (145);
— Thailand (112) and China (109) were better performers;
— Sri Lanka (91), Pakistan (85) and India (78) were below average performers; and
— Bangladesh (62) and Nepal (61) were poor performers.

The countries of South Asia, except Sri Lanka, where poverty incidence is greater are either poor or below average performers. The South East Asian Countries and China, where poverty incidence is low, with the exception of the Philippines, are best, better or average performers. All low-income countries except Indonesia are either poor or below average performers.

The poor performers were predominantly agricultural countries, as their share of agriculture in total GDP was highest for them: Nepal (59%) and Bangladesh (47%). On the other hand, the share was lowest for the best performers; Korea, Rep. (14%) and Malaysia (21%), both comparatively more industrialized. For the remaining countries there was no consistent pattern. The share of agriculture in total GDP was as follows:

— Below average performance countries: Sri Lanka (27%); Pakistan (27%); India (36%)
— Average performance countries: the Philippines (22%); Indonesia (26%).
— Better performance countries: Thailand (23%); China (37%).

Comparing component indexes with the composite index of performance showed a strong relationship between GDP growth rate and

TABLE 7.5 : Composite Index Numbers of Performance and Ten Selected Indicators

Country	*Composite index (standard solution)*	*Index of poverty alleviation*	*Income equality*	*Land-holding equality*	*Inverse of population growth rate*	*CDP per caput 1983*	*Life expectancy*	*Infant survial rate*	*Daily per caput DES* [a]	*Adult literacy rate*	*GDP growth rate*
Korea, Rep.	160.24	131.19	108.34	135.98	116.75	275.15	113.09	105.47	119.17	161.12	159.42
Malaysia	144.74	108.31	90.58	94.14	92.85	284.11	111.37	105.80	110.68	103.95	141.11
Thailand	11.64	115.94	94.13	115.06	87.67	112.33	107.95	103.29	104.00	148.99	133.78
China	109.30	135.77	138.54	140.16	114.49	56.91	114.80	101.32	107.76	119.54	126.11
Philippines	99.13	94.58	83.48	98.32	87.35	109.04	107.95	103.29	98.80	129.93	97.15
Indonesia	97.83	111.36	87.03	79.49	102.54	77.67	92.52	97.62	105.81	107.41	115.47
Sri Lanka	91.49	117.46	104.79	75.31	121.57	46.00	118.23	104.38	99.81	147.26	73.26
Pakistan	84.90	106.78	117.22	96.23	85.14	54.37	85.67	95.65	93.72	41.58	99.07
India	77.80	96.11	108.34	79.49	106.23	38.83	89.10	95.87	91.65	62.37	64.53
Bangladesh	62.27	21.35	97.69	110.87	85.45	25.54	82.24	94.34	79.26	45.04	48.49
Nepal	61.11	61.02	71.04	75.31	101.65	20.01	77.10	92.93	89.28	32.91	41.68

a. Dietary energy supplies.

Sources : Tables 3.1, 4.1 5.1.

Note : Adopted from reference [18].

performance (correlation coefficient *r*=0.92 and level of significance *p*=0.0001) and per caput GDP and performance (*r*=0.93, *p*=0.0001). The rank of countries in composite index and GNP per caput in US$ in 1988 (Table 7.6) is the same except China and Nepal despite difference in the time periods. However, such an association between performance and population was weak (*r*=0.27, *p*=0.41); between performance and distribution was moderate (*r*=0.61, *p*=0.05); and between performance and social indicators was high (*r*=0.80, *p*=0.003). It means that, with few exceptions (Pakistan and the Philippines), the higher the GDP growth rate, the better the performance index, and the same was true for GDP per capita, with an exception of Pakistan and China. These two indicators thus had a very dominant influence on the performance level of a country. Stepping up GDP growth rate, particularly for poor and below-average performers, is therefore, very crucial to break the impasse of low-growth equilibrium status (a poverty trap). Distribution policies alone for such low income and low GNP growth countries cannot do much to improve human well-being. A sustained economic growth in fact is needed in almost all the countries under discussion to finance meso policies and social development over the long run.

For instance, without sufficient growth, required domestic investment cannot be undertaken. But substantial poverty/and or inequality coexisted with high growth in some countries in the study. The "trickle-down" effect of growth on poverty alleviation and inequality reduction is also determined by other concomitant measures. Low-performance countries are still faced with the dilemma of formulating "balanced" strategies for accelerating their growth rates as well as reducing poverty and income disparities. It is important for them to assess also whether their slow growth is related to lack of resources and/or their maldistribution and patterns of growth.

The relationship between performance and level of GDP per caput (1983) was also very strong. The GDP per caput embodies the cumulative effect of past growth and also helps to estimate the rate of growth in the future. The level of available domestic funds depends on it and its influence on population control is also significant higher levels of GDP per capita encourage a low population growth rate. The countries with high GDP per caput can thus afford to support more and better social measures.

The growth and its pattern has direct effect on social indicators through the Government's experience on social services e.g. education and health. The poor needs not only access to them but capacity to bear associate costs to utilize these services even though they may be provided free and to divert time from their income earning opportunities. The analysis shows only the impact of these services on social indicators on the whole. In fact distributions of these indicators according to income will be more revealing.

The Republic of Korea, as a developing market economy, achieved the highest degree of performance, while China as a centrally planned economy, despite starting from a lower level of development, was nevertheless able to attain fourth rank in performance. Both had radical land reforms in the very early stages of their development (Korea, Rep. during 1947-50 and China 1949-52). While industrialization is centrally planned in China, it is also strongly guided by the Government in the Republic of Korea. Exports provide a major source of development income in the Republic of Korea, while import substitution plays a dominant role in China. During the period of study, 1961-83, poverty in both cases was low, but income inequality was somewhat higher in the Republic of Korea than in China. However, the incidence of poverty decreased in the Republic of Korea, particularly by keeping a parity between average monthly farm incomes to average monthly urban-wage incomes (This ratio was about 65 per cent in the 1960s and reached almost 100 per cent in 1984; it even surpassed 100 per cent in some intermittent years). Strong determination to fulfil the commitments, particularly for more even redistribution of resources, in the short term, is the common thread in both theses economies despite different ideologies.

In several other countries, there was a wide gap between declared objectives relating to redistribution of wealth and income and their achievement because of a lack of a corresponding commitment. Changes needed in redistribution of resources should be done in a manner so as not to retard growth by creating uncertainties that curb incentives for development. The countries intending or pursuing such changes will be better off if these are accomplished as rapidly as possible.

Composite Index and Human Development Index

The composite index indicating performance of the countries is based on 10 indicators (Randhawa, Antiporta and Syed, 1990). While the human development index (UNDP 1990) on 3 indicators only, purchasing power adjusted GDP per capita, life expectancy and adult literacy, also included in the composite index. The base of the composite index is thus much broader than the HDI. There are also the differences in the time periods covered and weighing systems of the indicators. In the HDI two social indicators, life expectancy and adult literacy, are given much higher weights than in the composite index. Also adjusted GDP per capita is used in the HDI converting it into logarithm. Despite these differences in the two indices the ranking of countries for them is the same except that Sri Lanka rank has improved to 9 in the HDI from 5 in the composite index and of India to 4 in the HDI from 3 in the composite index (Table 7.6). Although the two indices of human development give similar ranking of countries, indices vary substantially in terms of percentage difference in their values of corresponding pairs of contiguous countries. Therefore, the ability of each index to discriminate between ranks would seem to differ widely.

TABLE 7.6 : Composite Index and Human Development Index

Country/ Value	*GNP Per capita*** US 1988*	*Composite Index* [18]*		*Human Development Index** HDI UNDP Report 1990 [28]*		*HDI 1992*	
		Value	*Rank*	*Value*	*Rank*	*Value*	*Rank*
Korea Rep.	3600	160.24	11	0.903	11	0.859	11
Malaysia	1940	144.74	10	0.800	10	0.794	10
Thailand	1000	111.64	9	0.783	8	0.798	9
China	330	109.30	8	0.716	7	0.644	8
Philippines	630	99.13	7	0.714	6	0.621	7
Indonesia	440	97.83	6	0.591	5	0.586	5
Sri Lanka	420	91.49	5	0.789	9	0.665	6
Pakistan	350	84.90	4	0.423	3	0.397	4
India	340	77.80	3	0.439	4	0.382	3
Bangladesh	170	62.27	2	0.318	2	0.304	2
Nepal	180	61.11	1	0.273	1	0.289	1

Note : The ranks for HDI for 1992 are the same as those of the composite Index with the exception that Sri Lanka has become 6 and Indonesia 5.

* Composite Index is a weighted average of 10 component indicators, taken as hundred.

** HDI is a simple average of three indicators [adult literacy, life expectancy, (adjusted) GDP per capita]. It varies 0 to 1, the higher value shows better performance.

*** Low-income economies with GNP per capita of US$ 545 (or less; lower-middle income economies with GNP per capita of more than US$ 545) upto $2,200; and upper-middle income economies with GNP per capita of more than US$ 2,200 but less than $ 6,000.

III. PATTERNS OF HUMAN DEVELOPMENT

Table 7.5 provides useful insights into the pattern of development which emerged in the countries covered in the study.

Best Performers

Korea Republic (upper-middle-income country) achieved highest degree of performance in human development by rapid and equitable growth and balanced social development facilities through appropriate economic management policies.

Malaysia (lower-middle-income country) has less than average income equality and land holding equality distributions and relatively high population growth rate. But with high GDP growth rate and high GDP per caput, it achieved a good record on poverty alleviation, nutrition, life expectancy and infant survival rate. Its human development performance can, therefore, be enhanced by improving income and land distribution and adult literacy. The country was also able to make effective adjustment in the early 1980s with the fall in prices of oil without adversely affecting poverty incidence by making swift changes in macro economic policies.

Two best performers (Korea Republic and Malaysia) had significantly different patterns of development. The former indicates balanced and the latter comparatively unbalanced patterns. Both are middle income economies.

Better Performers

Thailand (lower-middle-income country) attained 9th rank in human development performance with relatively balanced development in respect of GDP growth rate, poverty alleviation, land holding equality distribution and social development particularly high adult literacy despite comparatively high population growth rate and greater overall income inequality.

China is a low-income country but it ranks 8th in human development because of its better performance in social and distribution indicators indicating less income inequalities and more social development facilities. But the social development has recently been disrupted to some extent in the transition period of economic liberalization. The Human Development Report 1990 appropriately stressed (UNDP 1990). "There is no rationale for neglecting social development in a period of accelerated economic growth".

Economic, social and political freedom is also vital to develop human potential and utilize it fully but the 10 indicators do not include the impact of human freedom for which there is yet no satisfactory quantitative indicator.

Thailand is a middle income country while China is a low income economy. They attained the status of better performers by a comparatively balanced pattern of growth, particularly China.

Average Performers

In the Philippines (lower-middle-income country) significant income inequalities and substantial poverty coexist with higher social development particularly high adult literacy, above average life expectancy and infant survival rate, and nearly average nutrition level. The GDP growth rate was nearly average. The poor was adversely affected in the recent adjustment period. "The recession of 1983-85 in the Philippines illustrates the costs to the poor of declining labour demand, lack of real exchange rate movement inflation..... The country's approach to stabilization was unfavourable to the poor in this period (World Bank, 1990).

Among the ASEAN, Indonesia is low-income country. The incidence of poverty in Indonesia is lower than in the Philippines and Malaysia. Poverty reduction in numbers and incidence in Indonesia has been outstanding in the recent past despite relatively high income and land holding inequalities. It achieved moderate social development accompanied by above average GDP growth rate. The country made effective adjustment with the fall in oil prices in early 1980s.

The two average performers in human development have in their pattern a common element of relatively high income inequality. Indonesia with its superior economic management than the Philippines strode over the adjustment more successfully.

Below Average Performers

Sri Lanka, Pakistan and India are below average performers in human progress. All three are low-income countries.

Sri Lanka is a distinct example of attaining high adult literacy rate and infant survival rate, highest life expectancy, lowest population growth rate, low income inequality and low incidence of poverty with low GDP per capita and medium GDP growth rate of 4.20 per cent per annum between 1961-1983. However, land holding distribution is significantly unequal. The country relatively devoted more resources to the above stated aspects of human progress. The human capital thus created needs to be fully utilized as an engine of growth by implementing an appropriate mix of policies accelerating growth with equity and sustaining human progress. Lack of opportunities for use of human capital already created especially among poor, leads to wastage of an important source and may cause dissatisfaction and unrest because of disillusionment regarding their well-being not being met to their expectations generated by more education, better health and improved skills; Life expectancy, infant survival rate and adult literacy the three important elements contributing to social development, are low in Pakistan compared to its performance in GDP growth rate and GDP per capita. Income equality and poverty alleviation are above average but land holding distribution is substantially unequal and population growth is also high. Military expenditure was disproportionately high, about three times the expenditure on education and health. Internal political conditions and forces and external conflicts must have contributed to this pattern of expenditure. The Human Development Report 1990 very appropriately categorizes Pakistan a country of missed opportunities for human development. The progress in democratization may enable the country to design appropriate policies to utilize its economic potential for improving human development particularly by increasing social services and improving their access to the poor.

The record of India on social services as indicated by life expectancy, infant survival rate and adult literacy, is better relative to its low GDP per caput and its steady GDP growth rate of 3.70 per cent per annum (1961-83). Its population growth rate is also comparatively low but poverty alleviation and land holding equality is less than average. It has, however, attained income equality slightly above average. Accelerating growth, rendering its distribution more equal, poverty alleviation and keeping up the progress on social development seem to be an appropriate mix of priorities.

Below average performance of these countries resulted from different routes, namely high and low social development and comparatively balanced development.

Poor Performers

Bangladesh and Nepal (low-income countries) could achieve only GDP growth rate of 2.78 and 2.62 per cent per annum respectively between 1963-83, the lowest among the countries studied. Bangladesh followed by Nepal have all round performance in growth, distribution of income, poverty alleviation and social indicators (adult literacy, infant survival rate and life expectancy). The rate of population growth was also quite high in Bangladesh and land holding inequality in Nepal. The major emphasis of these countries should on strategies and policies to spur growth, and enhance the participation of the poor in it.

To sum up, various degrees of performance have been attained by the countries under review with widely divergent development patterns. Balanced as well as relatively unbalanced development has led to the best performance. Comparatively balanced growth also culminated in better performance. Average performance coexisted with relatively high income inequality. Below average performance has been achieved with greater emphasis given to social progress as well as low priority given to it and with a comparatively balanced development. All round poor performance resulted in poor performers status. Therefore, there cannot be a uniform system to enhance human development for the countries with widely diverse socio-economic conditions and resource structures and at different stages of development. Furthermore, the horoscope, indicating values of 10 indicators in Table 7.5, shows the expected and unexpected results of strategies, policies, programmes and projects judiciously and deliberately or inadvertently pursued by the countries. In depth evaluation of strategies and policies followed by each country is, therefore, needed to arrive at a rational strategy and judicious mix of policies to make a steady and sustained improvements in human development. Hence, each country must review its own performance vis-a-vis that of other countries and determine its own course of action for accelerating growth with equity, reducing poverty and increasing social services. However, it is pointed out in the following sections that the small farmer development is a common thread which can greatly enhance human progress, directly or indirectly, in various aspects.

IV. AGRICULTURAL DEVELOPMENT AND RURAL POVERTY[19]

A brief review of agricultural development strategy pursued in the past is discussed in this section to assist in identifying necessary adjustments, and to bring an appropriate focus on small farmer development.

In the economic literature, one discovers several strands of views on agricultural development philosophy, strategy and practice. Particularly in developing countries, the evolutionary history of economic thought has often been so closely meshed with policy action that it has been difficult to isolate the influence of one from the other and determine the cause effect relationship between the two with analytical precision.

The traditional and most enduring view that held sway until about the middle of the present century had been to treat agriculture as a sole provider of surpluses for development. The "passive" role that agriculture had to play in economic development was further propagated by Lewis 'Economic Development with Unlimited Supplies of Labour' (Lewis 1954). The 'secular decline' hypothesis had a special appeal (warning!) to those large numbers of developing countries that exported primary products and imported manufactures in as much as the empirical evidence appeared to demonstrate a secular tendency for unfavourable terms of trade (Prebisch 1959). Then, with the emergence of the concept of 'forward and backward linkages' as an important tool for determining the allocation of investment in the late 1950s, the 'industry first' strategy gained more credibility. It was pleaded that industry had greater linkage and multiplier effects than agriculture and that a larger share of investment allocated to it would generate a more rapid and broad-based economic growth than otherwise (Hirshman, 1958, Enke 1962).

The inherent weakness of the 'back-seat-agriculture-and-front-seat-industry' strategy surfaced in many developing countries shortly after they launched their programmes of economic development. It was realized that the relative neglect of agriculture, *inter alia*, leading to food shortages could choke off growth in the non-farm sector (Enke 1962a).

In the early 1960s, interdependence between agriculture and industrial growth was beginning to be felt. Five positive contributions of agriculture to economic development started getting attention. It could provide labour, capital, foreign exchange and food to a growing industrial sector besides an assured market for domestically produced industrial goods (Johnston and Mellor, 1961). Some investment in agriculture to accelerate growth of surpluses was advocated to avoid falling into a low equilibrium trap in the early stages of development.

Yet another plea was that agriculture was to act as a holding sector until the rate of population growth started declining on the one hand, and, industry could absorb the entire surplus labour from agriculture on the other. In most development experiences, these shifts took quite a long time and agriculture could not be left to stagnate in the meantime. (Ohkawa and Rosovsky, 1960).

A common belief that agriculture was pursued by the "tradition bound peasant" also contributed to the low priority given the sector. Schultz's "Transforming traditional agriculture" was a major step in breaking this myth. It argued that farmers and herders in developing countries, far from being irrational and fatalistic, were calculating economic agents who carefully weighed the marginal costs and benefits associated with different agricultural techniques, and despite low levels of per caput output, traditional agriculture was characterized by allocative efficiency known as the "efficient-but-poor hypothesis". To lift agriculture from a low-level equilibrium, a major shift from agricultural extension towards investment in agricultural research and human capital was advocated. (Schultz 1964). Equally comforting is the conclusion that planted acreage, output and market supply in traditional agriculture do respond, in varying forms and content, to price incentives (Krishna 1962).

The understanding of the process of agricultural growth improved overtime, and by the mid-1960s the role of agriculture in economic development was appreciated more fully. Indeed, it was beginning to be viewed as a potential positive force in development. This turning point, and some of the agricultural development experiences of the 1950s 1960s led to a growing interest in empirical analysis.

The empirical evidence showed that industry-led growth in the 1950s and 1960s did not meet expectations. The core assumption that large surpluses of labour in the agricultural sector could be gainfully absorbed in industry by its rapid development did not work in reality. The "trickle-down effect" of growth was found weak, as demonstrated by the devastating poverty and hunger, particularly in rural areas in developing countries. The evidence of increasing poverty and income inequalities further triggered the critical view of industry-led growth philosophy. Economic development thought therefore took an important turn, and growth with equity" became a major issue in the 1970s. This change in orientation of development economics implied much greater role for agriculture and particular emphasis was focused on small farmers who form the bulk of rural poor. A greater recognition of the role of inter-sectoral links in development also took place (Staatz and Eicher, 1984). The Green Revolution in Asia in the 1970s also contributed to increasing priority to agriculture. Since agricultural production needed more supplies from other sectors, the inter-sectoral linkages were further strengthened.

The economic recession of the early 1980s, the heavy debt burden of developing countries, and the growing imbalances in developing economies brought to the fore the server problem of structural adjustment in economic development. These adjustments aimed to re-establish an equilibrium in external accounts and to control inflation, while creating conditions for resumed economic growth. The emphasis in the early 1980s was on "growth-oriented structural adjustment" and later on "adjustment with a human face" embodying, among other things, poverty equity considerations (Cornia, Jolly and Stewart 1987). The agricultural sector in general and small farmers, in particular, bore the initial brunt of growth-oriented structural adjustment.

The passive role given to agriculture in economic development literature is confirmed by developing countries' low investment allocation to agriculture in the past, compared to its contributions to total GDP. Planning and other policy documents of these countries reflect the low priority the agricultural sector held in the 1950s and 1960s. "Developing countries commonly follow strategies that foster dualism, concentrating capital in large-scale, capital-intensive industries. The intellectual base for India's Second Five Year Plan provides a notable case of one approach..... It was understood that improved welfare for the labouring classes was to be postponed in favour of large increases in the future. In the short run, agriculture and the cottage industries were to provide employment and consumer goods without drawing on the scarce capital of other sectors. Thus, the approach was explicitly highly dualistic in capital intensity (Mellor and Johnston, 1987). This approach was corrected, to some extent, in India's later development plans. While this general bias against investment in agriculture has continued into the 1970s and the 1980s, there has been significant improvement in the share of agriculture in total investment in this period. India, Indonesia and Thailand provide good examples in this regard (Edgran and Muqtada, 1989).

The low priority given to agriculture in development, and consequently the low investment allocated to it from public sector, discouraged its development, including that of small farmers. It affected small farmers more adversely than large farmers in certain areas, as will be indicated later.

Summing up, the role of agriculture in development strategy has improved over time in economic development literature. Agriculture has been and continues to be a basic sector in developing countries. The difficulties in understanding the events adequately and appropriately by those who tackled them at the time are often underestimated in hindsight. Nevertheless, the agricultural sector's evolutionary role in development literature gives the impression that comprehension of its rightful place in development could have been less difficult and more rapid. Its sheer size alone should have been argument enough in judging its appropriate weight. A start could have been made by according it a positive and active role. One wonders why it has taken so long to decrease the gap between economic thought and reality.

But still there is a missing link in understanding the role of agriculture in alleviating poverty, one of the major goals of development. Poverty still persists despite the fact there is strong evidence that agricultural growth helps the rest of the economy. The World Development Report 1986 (World Bank, 1990) pointed out countries with rapid agricultural growth have also had rapid industrial growth.

Defining high, medium and low annual growth rates of GDP as 7-9 per cent, 3-6 per cent and less than 3 per cent respectively and high, medium and low annual growth rates of AGDP as 4 to less than 6 per cent 2 to less than 4 per cent and less than 2 per cent correspondingly, it was found that Nepal had a low GDP as well as AGDP growth rate and Bangladesh had low GDP and medium AGDP growth rates between 1961-83. Pakistan, Sri Lanka and India experienced medium AGDP and GDP growth rates. In the remaining six countries (Table 29.6) high and medium AGDP growth rate along with high or medium GDP growth rates were achieved (Randhawa, Antiporta and Syed, 1986).

The World Development Report of 1990 (World Bank, 1990) based on growth path of some countries also invalidates the strongly held belief that the conflict between growth and income equality reduction is a must in development planning. It is recognized that growth is, generally, essential for a large scale poverty alleviation but it is not sufficient. It is the pattern of growth which is also an important determinant of changes in poverty. The pattern of growth itself influences the growth. Some patterns may be neutral, some helpful and others harmful to growth. There is, thus, a circular interdependency between growth and pattern of growth. The same is true of inequality; the income inequality may, therefore, decrease, remain the same or even increase while poverty reduction takes place.

The relationship appears to be that rapid agricultural development accelerate industrial growth under appropriate stimuli. Faster industrial growth than that of agricultural growth leads to declining agricultural share in total GDP, a positive interpretation of the "secular decline" hypothesis. The patterns of growth in agriculture and industry determine their impact on poverty and equity and also they have a strong bearing on growth itself. In large countries expansion of domestic demand is more relevant while in small countries export is more important. They need more an export-led growth.

Small farm development, which reduces poverty and would be helpful to growth, is the crux of the paper.

Some Adjustments in Development Strategy

A review of performance of some major adjustments advocated by international agencies in development strategies of developing countries elicits some important lessons that might help eliminate repetition of past errors.

With the introduction in 1970 of the "growth-with-equity" strategy, several economists spoke of lowering the dominant influence of GNP as the target and indicator of development, and an employment-oriented strategy was brought forward as a strong and appropriate replacement (Seers, 1970). The International Labour Organization (ILO) took the lead with the launching of several employment strategy missions during the first half of the 1970s. It was realized then that absolute unemployment was not the main cause, as the rural poor were generally employed in some way or another, though most of them were severely underemployed. The creation of jobs was not the real issue for eliminating poverty and hunger. Increasing the productivity and purchasing power of a large number of small farmers and other rural poor were the main problems. This initiative, because of its weak foundation, withered away by the mid 1970s. But it did stimulate research in several aspects of labour economics.

In the mid 1970s, the employment-oriented strategy gave way to a "basic needs approach", which was also spearheaded by the ILO, and subsequently supported by the World Bank and adopted as a criteria in its support of projects. Needless to say, investment in social services and amenities greatly contributes to the welfare of the poor and improves the rate of economic growth. But a prerequisite to such large scale investment is a sound economic foundation of a country. The additional cost involved in the "basic needs approach" discouraged its wider adoption, and by the early 1980s, it disappeared into oblivion. Ironically, the growth-oriented strategy was reborn with a new fervor in the early 1980s under growth-oriented adjustment. At present, "adjustment with a human face" is considered a realistic proposition, as it incorporates, among other things, major elements of growth, distribution, employment and basic needs strategies.

At the programme and project levels, the integrated rural development projects of the 1970s, when they became too complex because of an all-inclusive approach, lost balance between social and economic objectives, as was the case with the community development approach. They became expensive as well. It hampered their replicability and validity. Therefore, several such previous attempts, though with a sound logical base failed to take off beyond experimental and pilot demonstration stage, being too expensive and complicated, and because they did not adequately grasp the real issues to be addressed. Consequently, they did not succeed to serve their initial purpose of bringing about with equity and the alleviation of poverty.

There World Bank's evaluation of 112 rural development projects completed recently (including integrated rural development projects) which were started during McNamara's years, bears also testimony to the above conclusions; (World Bank, 1990).

— Many integrated projects were too ambitious and complex and they often placed impossible demands on local leadership and institutions.
— In many cases, the benefits flowing from projects were not sustained beyond project completion. One reason was the Bank's tendency to create independent project management units.
— Moreover, the projects themselves often proved ineffective levers for influencing overall national policies for agricultural development.
— Many tended to be successful "enclaves" within national systems that were still largely inefficient and inattentive to the needs of poor farmers.

There were also other attempts directed towards the improvement of some other aspects of development strategy, which were sooner or later abandoned or diluted. Therefore, if past experience is a guide, it is essential that the adjustments address the precisely identified realities of the situation; be focused on the improvement of the economic base around which social and welfare aspects are properly interwoven; be readily made within economic, social, political and administrative structures of the countries, at least in the short and medium run; and pass the basic and crucial test of widespread replicability and sustainability to ensure their validity beyond an experimental and demonstration stage.

Following these guides, the paper attempts to formulate in Section 6 restructuring of agricultural planning and policies for alleviation of poverty which would be also helpful to growth.

Small farmers and landless labourers (with small farmers predominating) form the bulk of the rural poor. Small farmers are a diversified group; their main characteristics are determined largely by the socio-economic and political structures of the country concerned and its stage of development. By and large, they are engaged in subsistence agriculture, though there are exceptions as regards commercialized small farmers, for example. Generally, they eke out an inadequate living from farming and their basic concern is survival. Agriculture is a way of life to most of them, rather than just an economic activity. They work mostly in harsh and risky environments and face more economic, social or political barriers than do large farmers. Natural calamities namely famine and pestilence, disproportionately affect them. They are frequently accused of being irrational, fatalistic and conservative in their attitudes to farming. These factors have led to several misapprehensions about this group, and have kept them out of the mainstream of agricultural development or they have received in general much smaller benefits compared to large farmers. Their development has been considered more on "humanitarian grounds" than as a true economic necessity (Randhawa and Sundaram, 1990).

The number of small farmers is substantial (60-70 per cent of total farmers).They operate on a considerable area (25-30 per cent of total area) and make significant contributions to agricultural production (30-35 per cent of total production). A rational agricultural development strategy should not permit them to be by passed by the development process. Their welfare should be an integral part of development strategy, as they are underprivileged in respect to resources and political power and are thus vulnerable. Small farmers can be scattered among large farmers and may be located in well-developed areas. They can be location- specific, for example, in remote areas that are comparatively less developed and have poor infrastructure (location inaccessibility). Pursuance of strategies giving high priority to developed areas with a greater immediate potential to solving pressing agricultural problems, have widened inter-regional inequality in many countries. The concentration of poor farmers has, therefore, increased in such cases (World Bank, 1990).

V. GREEN REVOLUTION, FARM SIZE, PRODUCTIVITY, INCOME DISTRIBUTION AND EMPLOYMENT AND SMALL FARMERS

The interrelationship among Green Revolution, Farm Size, Productivity, Income Distribution and Employment have been critically examined in detail in reference (World Bank 1990).

The main findings are quoted below:

(i) A widespread belief that economies of scale were important in agriculture led to more emphasis on large farmer development .It was contradicted by the inverse relationship between farm size and output per unit of land in the pre-Green Revolution period, and now it has been seriously challenged by the experience of the Green Revolution;

(ii) Early adopters in the Green Revolution were primarily large, better-off farmers, having controlled production condition, suitable for HYV. Small farmers, having a low-risk bearing capacity, rapidly followed when the success of new technologies was proven by large farmers who took initial risk, while the demonstrative impact of large farmers on small farmers needs hardly to be stressed. The inverse relationship between farm size and output per unit of land, which was witnessed in the pre-Green Revolution period, passed through the following stages because of the above sequence of adoption of HYV by large and small farmers. In the "transitory" period, immediately following the start of the Green Revolution, the inverse relationship weakened, disappeared or was even reversed because of higher capital intensity and risk-bearing capacity of large farmers, which compensated for greater labour intensity and better land quality of small farmers. As small farmers caught up rapidly, the inverse relationship where it was reversed, disappeared, and in some cases, the inverse relationship in favour of small farmers started appearing again depending upon the institutional support available to them and the general agrarian situation under which they operate;

(iii) Given the appropriate environment and the technology specifically suitable for yield income enhancement, small farmers have been found to adopt new technologies as quickly and, in some cases, even earlier and more intensively than large farmers.

(iv) The Green Revolution primarily benefitted producers of wheat and rice, large or small, who had farmed under a relatively controlled environment (e.g. irrigation). Improved technologies suited to less favourable areas(e.g. dry land),even for these two crops and for several other crops growing in these areas and for livestock enterprises, are extremely limited, though emphasis on research for them has increased in recent years. Small farmer development with suitable technologies will diversify agriculture rendering it less risky;

(v) Hindsight reveals that the time lag between large and small farmers in the adoption of new technologies can be substantially decreased by proper policies and by improving access to credit and input supplies;

(vi) Mechanization triggered by the Green Revolution has not always reduced employment. Where mechanization for land preparation and harvesting increased the intensity of land use by double-cropping and multiple cropping, employment has increased. Also, significant losses and damages to output have been avoided through mechanized harvesting in short periods (e.g., before rains for wheat in the Punjab, India);

(vii) The poverty-reduction contribution due to adoption of HYV by small farmers is important. Since the gains from modern technologies were closely associated with farm size, farm income inequality could, however, increase. But this negative impact on income distribution could be corrected by evolving a widespread adoption of modern technologies for crop and livestock enterprises predominantly pursued by small farmers. Similarly, the widening of regional income inequality could be reduced or checked through appropriate technologies suitable for less favourable regions. Therefore, the most serious criticism that can be levelled at the Green Revolution is that it has not become sufficiently pervasive (Ruttan, 1977).

Small farmer development will create employment and income-earning opportunities for landless labourers directly on farms and indirectly through the stimulus of overall development. They themselves will also benefit by augmentation of their income from off-farm activities; and

(viii) As regards political and social implications, the contribution of the new seed-fertilizer technology for good grain production has weakened the potential for revolutionary change in political and economic institutions in rural areas in many countries in Asia and in other parts of the developing world. Despite widening income differentials, the gains from productivity growth, in those areas where the new seed-fertilizer technology has been effective, have been sufficiently diffused to reinforce interests of most classes in an evolutionary, rather than revolutionary pattern of rural development (Ruttan, 1977).

It may also be added:

(i) "Most horticultural products in development countries are produced on small farms and in labour intensive ways, though there is wide variation among crops and countries. With appropriate policies and technology, horticultural production could significantly supplement the income of small farmers and provide additional employment in labour abundant developing countries. Also, given the urgent need for increased export earnings and the not-too-bright prospects for traditional agricultural exports, the horticultural sector could be an important source of additional foreign exchange." (Islam, 1990). The constraints to overcome are development of infrastructure and marketing to benefit from economies of scale in post harvest activities, evolution and dissemination of improved technologies and improvements in market intelligence and information on world horticultural trade.

(ii) Poverty and environmental damage in developing countries are generally interrelated. Poverty on account of lack of resources and opportunities, leads to degradation, desertification, poor sanitation, and polluted and unsafe water. Immediate survival is the major concern of the poor. The environmental damage in turn reinforces poverty. The agricultural development in general and of small farmers in particular can reduce and check the environment degradation.

Indirect effects of small farmer development are also very important and manifold (World Bank, 1990).

(i) Since small farmers constitute the bulk of rural poor, the poverty reduction impact of small farmer development will therefore be widespread and large;
(ii) The majority of small farmers can grow their own food and can thus contribute substantially in solving the problems of hunger and undernourishment. It appears a less costly option because collection, storage distribution of food for them will not be required. Increases in their incomes will further strengthen the effect on food security by enabling them to buy the additional food they will need;
(iii) Improvement in small farmer incomes will increase the demand for goods and services produced, particularly in rural areas, and will thus give a stimulus to economic growth. It will also alleviate the limitation caused by constrained demand for agricultural commodities, particularly food, on agricultural development.
(iv) The rural labour force engaged in non-farm activities constituted 49.3 per cent in Malaysia (1980), 45.8 per cent in Sri Lanka (1981), 37.9 per cent in Java, Indonesia (1980), 32.3 per cent in Pakistan (1982-83), 31.9 per cent in the Philippines (1982) and 19.0 per cent in India (1981) (Riswanul, 1987). Emphasis on small farmer development will strengthen links between agricultural households and non-farm economies, particularly in rural areas. It will encourage and expand many economic activities. These links arise from expenditures on local goods and services and agricultural inputs, and from local processing of farm produce (Longhurst, 1987). In the Philippines, for example, it has been noted that a 1 percent increase in agricultural output is associated with a 1 percent increase in non-agricultural rural employment, with ratio increasing the greater the concentration of the output increase on

small farmers. (Ranis and Stewart, 1987). It will, therefore, alleviate the problem of rural unemployment and under employment. Its spread effect on rural poverty and malnutrition will be substantial;

(v) The urban poor will also benefit in several ways, for instance, out-migration to urban areas will slow down and consequently reduce urban poverty; and

(vi) The urban poor, in several cases, have ties with rural areas, and improvement in income of small farmers will provide a fallback position to urban relatives after opportunities cease to exist there. Thus they provide flexibility for rural-to-urban migration and return-migration, depending on comparative opportunities "a very effective low-cost special security system" (Longhurst, 1987).

In brief, small farmers are not fatalistic or conservative, but are active entrepreneurs and respond rationally to economic incentives and opportunities within their reach. Their contribution to the alleviation of poverty, equity, food security, reducing food imports, decreasing rural-urban migration, creating more markets for non-agricultural goods and services, and providing a stimulus to overall growth, clearly justifies their rapid development. The overall national welfare will substantially improve because the marginal utility of each increment in income is much higher for the poor than it is for rich. Rapid small farmer development will therefore accelerate growth, enhance efficiency and promote equity. There is thus, an urgent need to incorporate fully small farmers into national development strategies, rather than treat them as an adjunct to it.

Some countries have made serious attempts to help small farmers and promote their growth through additive special programmes. Such programmes, for instance, have been implemented in India, Bangladesh and Indonesia. The very introduction of such programmes is proof that overall development strategy has been inadequate to deal with their problems. The coverage of such programmes has been relatively limited, and they have not succeeded in "meshing" with general development strategy. They have remained essentially piecemeal, isolated interventions and have failed to make a discernible impact. This is another reason small farmer development should be an integral part of national planning and policy.

VI. NEED FOR RESTRUCTURING AGRICULTURAL DEVELOPMENT PLANNING METHODOLOGY AND PRACTICE[19]

Growth-oriented agricultural development has not been evenly distributed. It generally either accentuated interpersonal and inter-regional (within a country) income inequalities or did not lessen them. Special programmes, undertaken as corrective measures by some countries to improve the economic position of small farmers, have not succeeded in arresting these trends. It is, therefore, imperative that appropriate adjustments in agricultural development strategies be made in order to distribute more evenly the benefits of growth among the various categories of farmers.

The growth-oriented strategy emphasized increasing agricultural production predominantly of superior food grains, wheat and rice, to attain or improve self-sufficiency. Planning was primarily concerned with growth, that is, additional output that could be achieved in a planning period. Therefore, the implicit weights accorded to various categories of farmers in this strategy were based on the perceived contributions they could make to meet production targets in terms of additional output.

Small farmers, constituting 60 to 70 per cent of total farmers, contribute only 30 to 35 per cent of total output, while large (and medium) farmers, constituting 30 to 40 per cent of total farmers, contribute 60 to 70 per cent to total production; almost the reverse share in terms of number and contribution to total output. These facts indicate that the role of small farmers in meeting additional output targets is less than their share in total output at a particular time. They, more often than not, are severely constrained by the non-availability of appropriate technology. Activities primarily pursued by them get lower priority in production targets (e.g. inferior coarse grains, root and tuber crops, fruits and vegetables and livestock production). Above all, they lack adequate access to credit farm inputs, services and markets. These barriers might have pushed down their contribution to additional production in the planning period, to one half, 15 to 20 per cent. This is planning dilemma of growth-oriented selected commodity based strategy.

With the same level of production, the marketable surplus for food can vary considerably, depending on the relative share of small and large farmers in total production. The high share of small farmers in total production will reduce the marketable surplus, as the larger proportion of increased production is likely to be consumed by themselves rather than sold in the market or *vice-versa.*

Urban dwellers are more vocal and politically more influential than rural people. The growth-oriented strategy seems to have been influenced also by an overriding concern of developing countries to generate marketable surplus of food to provide cheap and suitable food to urban areas, through market mechanisms or the public distribution system. Considerations of foreign exchange earnings by raising production of certain commodities for export appear to have also supported growth oriented strategy.

This strategy, though suited to resolve the food insecurity problems in urban areas, might have actually helped to increase the number of poor and undernourished in rural areas and therefore augmented food security problem in these areas. The allocation of disproportionately high implicit weight to large farmers and small weight to small farmers is reflected in objectives of growth, quantitative targets of production of commodities, allocation of investment, supporting infrastructure and agricultural policies of development plans. These weights have operated at almost every stage and in each component of planning and policies. For instance, the regions with a greater potential of increasing agricultural production in the short and medium term got precedence over less developed and remote areas, though they might have greater potential in the long term. Inter-regional inequality (within a country) was accentuated in this process. Within region, development programmes and projects and institutional support were mostly determined by these weights. Consequently, small farmers received less emphasis in them than their contribution to total output. Agricultural policies designed to implement growth-oriented strategy, plans, programmes and projects followed suit, and small farmers were deprived of an appropriate degree of benefits from them. Research priorities were also similarly determined.

Welfare and equity objectives have been mentioned in several development plans. They remained more as pious declarations, however, than concrete actions plans. The hidden force of these implicit weights has always been overwhelming in causing a bias against small farmers in implementation.

In some countries, when small farmer problems became more

accentuated or were better understood, these problems tended to be dealt with through special programmes. However, the sheer number of small farmers, their contributions to agricultural production, prevalence of poverty and the intensity of food insecurity should have made their development the central consideration in agricultural development strategy.

Awareness about the need for small farmer development is not lacking. Concern about small farmers has been growing since the 1970s. National governments and international agencies, particularly the Food and Agriculture Organization of the United Nations (FAO), the International Fund for Agricultural Development (IFAD), the World Bank and UNICEF have all stressed small farmer development through various measures.

The direct and indirect effects of small farmer development discussed earlier have demonstrated clearly that their development is viable, feasible and necessary. Besides social justification and economic rationale, it is politically compulsive to launch small farmer development programmes. In any democratic setup, they are a big force to reckon with.

Successful experience in East Asia countries were based on a "unimodal" pattern of development characterized by a widespread and gradual increase in productivity and expansion in the use of purchased inputs by small farmers, which was also consistent with the inverse relationship between farm size and productivity. Considering, the dominant position of agriculture as a source of income and employment, the central importance of a broad based "unimodal" pattern of agricultural development was emphasized in contrast with the dualistic or "bimodal" pattern, which was based on rapid modernization of subsector of large capital-intensive farm units, along with capital-intensive industrialization, a pattern that many developing countries pursued. (Mellor and Johnston, 1987).

A necessary prelude for successful "unimodal" development are appropriate economic, technical and social conditions, which are often lacking in the desired quanta in most developing countries.

Since the bulk of output is still produced on large farms, and political, economic and social structure are still highly skewed in favour of large farms, unimodal development can, at this stage, be considered a long term proposition. Furthermore, existing development systems are bimodal in a very broad sense, as the various regions, farm sizes and activities receive implicit weights. Therefore, in the short and medium term, particularly within the prevailing economic, political and social structures, it seems less problematic to change the weights by according higher priority to small farmer development.

In addition to according an appropriate priority to agriculture in development plan as stated earlier, the weight of small farmers in development strategy should be increased to accelerate their development. An appropriate change in small farmer weight in development strategy is the remedy. This change is termed as the restructuring of agricultural development planning methodology and practice (Randhawa, 1988). It is debatable issue by how much the weight should be increased. Several considerations will determine the degree of increment, particularly its effect on large farmers, production targets, export earnings, etc. The maximum limit may correspond to the proportion of small farmers in the total number of farmers. The minimum limit is what they are allocated now. Minimum and maximum limits can be set approximately at 20 to 70 per cent, respectively.

The aim should be to raise the small farmer weight as rapidly as possible to match their potential contributions to production targets and which remains consistent with equity and poverty considerations. It seems appropriate to raise their weight to 30 to 40 per cent in the medium term, in may countries. Introducing explicitly the weighing system changes the focus from commodity planning to planning for farmers, large and small according to their appropriate priorities assigned to them in development strategy incorporating growth, poverty, and income inequality objectives. Each country must however determine the existing weights of small and large farmers, and the degree to which small farmer weight can be increased and, correspondingly, large farmer weight, decreased. This change should also be consistent with declared growth equity poverty objectives.

This restructuring does not claim philosophical "elegance", but its significance lies in its practical validity. It provides a pragmatic framework to intergrate investment, development policies and measures in an appropriate mix to achieve and sustain rapid agricultural growth and combat poverty. While it can be perceived easily, its implementation has far reaching implications which must be threshed out thoroughly before its adoption. Therefore, the crucial issue is to evaluate the restructuring by developing countries and the international agencies concerned with agricultural development for further assessing its relevance and suitability as a main guide for achieving growth with equity, reducing poverty and attaining sustained agricultural development in general and food security in particular. The restructuring is likely to face less resistance now because most of the developing countries in Asia have already made great strides in solving their immediate food problems by increasing rice and wheat production.

The implications of the restructuring suggested in agricultural development strategy are manifold and extensive. The intention is not to explore these implication comprehensively, as the foremost task is acceptance in principle of the necessity for the restructuring. Nonetheless, some implications are examined as illustration to enhance this understanding further. These implications may be divided into two categories—substance and logistics. The most important changes of this restructuring in substance will be needed in the planning process and methodology, agricultural policies, technological advancements and access to land.

Agricultural Planning and Policies

Agriculture is pursued by a significant number of large and small family production units. Therefore, it is the response of a large number of decision makers to the economic, social, physical and technical environment that *inter alia* determines agricultural growth and its distribution. Small farmers are a heterogeneous group in respect to their resources, output and location. Therefore, their importance may differ widely among agricultural regions with in a country. Consequently, the weight given them in national development strategy should preferably be formulated on the basis of the appropriate weight accorded them in well defined agricultural or administrative regions.

Restructuring entails a greater emphasis on multilevel (decentralized) planning. It may be considered a major departure from overall national planning. Recognizing the necessity of decentralization in decision making for planning, over the years emphasis on multilevel planning has increased in several countries, but a lot of remains to be achieved (FAO, 1985). Since the more visible opportunities of agricultural development have become steadily scarcer and more costly to exploit, correspondingly, pressure on planning has been increasing to improve resource utilization. A necessary prelude to achieve this objective is the knowledge of resource potential and capabilities at sub-national and local levels. It will involve meaningful participation of people in making the planning process harmonious, interdependent and participatory, and as well as relevant, authentic and implementable. Two significant tendencies will emerge in this process. Investment through domestic resource mobilization in most countries is much greater than investment from external sources, and the efficiency of domestic investment will increase. Relative emphasis will increase further for small scale programmes and projects compared with large scale prestigious projects, although small farmers too have benefitted substantially from large scale projects (e.g. canal irrigation projects in India).

Small farmers organized into politically active, local and regional associations and fora, can transmit their felt needs forcefully; (i) for technology to research stations; (ii) for suitable organizational and administrative structures to administrators; and (iii) for incentives to policy-makers. Crops and livestock activities particularly pursued by small farmers remain mostly outside the orbit of positive price policies. A more rapid progress on multilevel planning (decentralization) will therefore be conducive and complementary to effective implementation of the restructuring suggested. This restructuring will have an important bearing also on the planning and development of other sectors. For example, complementary rural industrialization may become more prominent to further enhance the income of small farmers. A comprehensive methodology is needed for restructuring the agricultural planning system. But in the beginning, pragmatism may be the best guide. Practical procedures and guidelines for this purpose should be formulated. These guidelines should be pursued at various stages of planning. With time, a more formal methodology and procedures can be further improved upon and refined. In short production structure evolved by restructuring will be different than that for major commodities oriented planning biased in favour of large farmers.

Technology

As mentioned, there is a severe dearth of appropriate technologies suited to small farmers operating under harsh and risky environment. The rationale for small farmer development has shown that given the appropriate opportunity and environment, they will adopt new technologies willingly and enthusiastically, provided its profitability is demonstrated to them. Appropriate technologies within their reach are the main vehicle for their development. To achieve this objective, several changes will have to be made in the allocation of research funds and in research priorities.

It is well known that agricultural research in developing countries is severely under funded. "On average, developing countries spend only one third as much of their agricultural GDP on agricultural research as do developed countries, although agriculture is more than six times as important as a proportion of their overall economies and investments in agricultural research has the highest rate of return among investments in the agricultural sector. (Donovan, 1987).

Agricultural technology consists mainly of; (i) mechanical technology (labour saving) and (ii) biological and chemical technology (land saving). Land being the major constraint of small farmers, biological technology, which is yield and income-increasing and also often cost-reducing per unit of output, is more relevant to them. Biological technology is urgently required by small farmers, though small-scale mechanical technology, besides reducing the drudgery of farm work, can increase output per unit of land, in several cases by improving the timeliness of operations, moisture conservation, soil structure and texture, etc. It can reduce areas under fodder and feed crops required to maintain draught cattle that can be partially or wholly replaced by small scale technology. Food and cash crops can then be grown as substitutes in such areas. Gains from biological technology sometimes trigger large farmers demand for mechanical technology.

Two main components of biological technology are breeding new varieties and improvements in cultural practices. The most recent success have been in wheat and rice, grown under relatively good physical production environment. Now there is concern in IRRI that maximum rice yields under best conditions (experimental fields at IRRI) have been constant for 25 years or so. In CIMMYT the same constraint is also being faced for wheat. This should also be a strong reason to develop technologies for diversifying agricultural production to guard against the risks of depending heavily on development of few agricultural commodities. Improved varieties of these two crops for less favourable conditions (e.g., drought prone areas) are scarce. In fact unirrigated agricultural land is about 70 per cent of total arable land and land under permanent crops in the region.

Small farmers are concentrated in drought prone areas and grow primarily coarse or inferior grains, millet, pulses, root and tubers, and other secondary and subsistence crops. Advances in improved technology for these areas, as well as these crops, have been negligible or at best slow. The same is true, to a large extent, of oilseeds. The dearth of appropriate technology is the major barrier to development of these crops (Bhalla, Randhawa and Tyagi, 1989). The small scale livestock enterprises pursued by small farmers suffer from the same weakness. These serious deficiencies in modern technologies have contributed substantially to inter-regional (within a country) income inequality and also, to an appreciable extent, to interfarm income inequality.

In India, the lack of improved varieties have been a major cause for the slow growth in production of coarse grains, pulses and oilseed, compared with the rapid growth of wheat and rice. The former commodities are important for consumption by the poor. Also the import bill of vegetable oils skyrocketed, from a net export of $ 0.4 million in 1961 to $ 690 million in 1987. The breakthrough in oilseed production occurred only in 1988-98. In Pakistan, net imports of vegetable oils also substantially increased, from $ 0.6 million in 1961 to $ 233 million in 1987 (Bhalla, Randhawa and Tyagi, 1989).

The successful impact of the Green Revolution is evidence that appropriate technologies can be developed and adopted by farmers, both large and small. To make it possible, handsight stresses that drastic changes in priorities, location, content and research coverage on biological technology have to be made; greater funds have to be allotted; and more incentives have to be given to scientists in terms of rewarding them (materially and/or by recognition) for creating appropriate biological technologies, which require more effort and time for "difficult" areas and activities than for better physical production environment. In fact, these changes are long overdue, particularly because "an essential condition for success in achieving sustained growth in agricultural productivity is the capacity to generate an ecologically adapted and economically viable agricultural technology in each country or development region" (Ruttan and Hayami, 1984).

Agricultural research has a long time horizon and correspondingly needs a long time commitment. But in the last decade, international as well as national agricultural institutes have increasingly focused on developing new technologies suited to small farmer conditions (location-specific), and for commodities produced by them. The major objective now should be on bringing such proven technologies to the target group as "Economic growth ultimately depends on the flexibility and efficiency of a society in responding to technical and economic opportunities" (Ruttan and Hayami, 1984).

To take full advantage of new technologies for accelerating growth, the integrated role of the six "Ins." is crucial (i) incentives (prices); (ii) inputs (water, fertilizer, equipment, etc.); (iii) innovation (HYV, irrigation, etc.); (iv) information (diffusion of technological knowledge through extension services); (v) infrastructure (roads, markets etc); and (vi) institutions (credit, market, health, education etc.) (Streeten, 1987). I should add the 7th "In" investment in people (human capital).

The combination of these seven "Ins" for small farmers is of particular relevance. Though modern biological technology in neutral, It is non-neutrality of economic, social and political structures or institutions that effect technology's impacts on inter-regional and interfarm income inequalities. Therefore, the undesirable distributional effects of modern technology can be reduced or avoided by rational allocation of funds on development of technology and research itself and through preventive and corrective measures in logistics and substance.

Agricultural Prices

Agricultural prices are the first "In" as they provide incentives to farmers to rationally utilize available technologies and their resources. Most of the countries in the region have adopted agricultural price policy as an important instrument for .... policies have helped small farmers to receive better prices and reduce seasonal price fluctuation; remedial actions against market imperfections. Conclusion of a recent FAO study are briefly below (Bhalla, Randhawa and Tyagi, 1989).

— While several commodities have been covered by agricultural price policy, it has only been adequately implemented for a relatively small number.

— Agricultural price policy has been quite successful for principal food grains; wheat and rice. Innovative technological breakthrough (HYV), supported by incentive prices and the other five "Ins" made the Green Revolution viable by achieving higher growth rates of these commodities, which consequently has led to food self-sufficiency in some countries. Prices policy has also been successful for cotton.

— In contrast, implementation either did not succeed at all or was only partially successful for coarse cereals, pulses, oilseed and secondary crops, because of inadequate support of the seven "Ins", particularly modern technologies. These crops are mainly grown by small farmers.
— Price support policy *per se* can be considered a largely neutral instrument as small, medium and large farmers again according to their marketed surplus. It is the non-neutrality of economic, social, political and technological structures that distorts the impacts of price policy on income distribution. Price policy with uneven distribution of these structures, has contributed to inter-regional and interfarm income inequalities.
— These negative effects on income distribution as stated earlier can be corrected by evolving widespread adoption of HYV for several other crops and improved technologies for livestock activities, particularly those pursued by small farmers, and by developing suitable technologies for regions with less favourable production environments.
— Price support nevertheless has been found to have improved the marketing efficiency of small farmers by reducing or eliminating distress sales immediately after the harvest, when prices are generally low, and by checking market malpractices by private traders.
— In short, agricultural price policy will need to be "restructured" according to the new weight assigned to small farmers in development strategy. The support price policy thus will give due consideration to the produced by small farmers.

Access to Land

Reforms and policies that transfer resources and facilities from the non-poor to the poor face resistance from the non-poor. Land reform is a distinct example. The enthusiasm for land reform of the 1950s and 1960s appears to have withered away. Even improvement in land tenure (tenancy), which was easier to make than land redistribution did not make much headway.

The implementation of the restructuring, in principle, is not tied up to the land distribution structure. Nevertheless, its importance will vary according to the share of small farmers in total number of farmers, total area operated and total agricultural production. In countries where land distribution is largely "asymmetrical", the impact of restructuring will be less. In such countries, there is a strong case to effectively implement land reforms as rapidly as possible. Two successful cases of radical land reform done under different political economies in the early stages of development are cited in Section III. In both countries, a strong determination to fulfil government commitments, particularly for more even redistribution of resources in the short term, was the common thread, despite their differing ideologies. The improved land distribution structures were a major contributor to their achievements in agricultural development.

The small farmer constraint of lack of access to land has been further exacerbated by the fact that there is not much land left suitable for settlement. Land frontiers have been reached in many countries, while in others the investment required to exploit new land is quite high.

A host of changes will have to be made in *logistics* pertaining to the remaining five Ins, (inputs, information, infrastructure, institutions and investment in people) corresponding to changes in *substance* discussed earlier. Instruments to implement these changes will have to be modified. Significant advances have already been made in some countries to bring credit facilities within the reach of small farmers, and consequently, the use of modern inputs by them has appreciably increased.

The main considerations in designing these changes are the speedy transfer of knowledge of new technologies to small farmers, efficient distribution of adequate agricultural inputs, availability of credit and other agricultural services in sufficient degree to bring modern inputs within their reach, existence of market outlets, appropriate incentives and improvements in health, nutrition and education. For example, extension requirements of agricultural development formulated by the restructuring will be different in respect of location enterprises and content. Detailed changes and a *modus operandi* for their implementation should be worked out by each country, depending on conditions relevant to it.

The restructuring fulfils essential conditions and follow the guide for adjustment in agricultural development strategy stated in Section V and the changes suggested in agricultural planning and policies, technology, agricultural price policy and logistics do not envisage any direct transfer of resources from the non-poor to the poor.

VII. MACRO-ECONOMIC VARIABLES AND THE POOR

Linkages of the agricultural sector with the rest of the economy have progressively increased with development. Therefore, the policies in other sectors affect the performance of agricultural sector and vice-versa. For example, some macro-economic variables and policies may adversely affect the poor more unless compensatory mechanisms are adopted.

The debt crisis in the early 1980s in some countries compelled them to undertake adjustment policies to reduce basis imbalances both on external account and in domestic resource use. But the impact of adjustment polices/programmes on the poor was hardly paid any attention. It was UNICEF through its "Adjustment with a Human Face" which brought to focus their implications on the poor. These adjustments have produced mixed results depending on their swiftness and other complementary measures to cushion the adverse effects on the poor. According to the World Development Report 1990, the adjustment measures through effective action in Indonesia and Malaysia contributed to a reduction in poverty. On the contrary Brazil and Costa Rica experienced a sharp increase in poverty when wages fell in 1983 on account of sharply reduced demand. The severe distortions are in sub-Saharan Africa and problems of adjustment and poverty consequently are most grave there. Stabilization in the short run is likely to adversely affect social services, transfers and safety nets that help the poor.

Over valued exchange rate is a deterrent to agricultural development by making exports expensive and by depressing agricultural prices. Exchange rate depreciation and other measures to raise agricultural prices will increase the incomes of some of the rural poor, especially the small farmers, provided demanding reducing impact of restructuring are contained.

Trade

"About one third of the agricultural exports of developing countries appear to be affected by non-tariff barriers. Tariff escalation has a particularly damaging effect on the low-income countries. Industrial countries tariff duties generally increase or 'escalate'—according to the degree of processing" (World Bank, 1990). The generalized system of preference and other such special arrangements have done little to offset the damage done by the much broader range of overtly projectionist devices.

In several developing countries agricultural ministries and institutions dealing with agricultural planning and policies hardly participate or contribute to formulation of macro-economic and trade policies. The assessment of the impact of macro-economic and trade policies on agricultural development and vice versa should be an important task of development planning to avoid contradictions and for creating complementarities between macro policies and agricultural policies and planning. Therefore, a close coordination is needed between those responsible for macro policies and agricultural policies and planning.

VIII. CONCLUDING REMARKS

In brief, agriculture directly or indirectly is the major source of income for the rural poor which forms the bulk of the total poor. Small and marginal farmers constitute a large proportion of total farmers and as well as the total rural poor. Their development is not only ethical but also economically efficient for creating a broad based and long-term sustainable agricultural development and for accelerating overall development also. Since the relative neglect of small farmers was inherent in the agricultural development strategy and the planning system pursued by the developing countries, the remedy should be accordingly addressed there to correct the bias against small farmers. The restructuring of agricultural planning and policies is therefore suggested as a remedy by according appropriate weight to the development of small farmers. It will enable the major proportion of the rural poor to participate more robustly in economic and human development, leading to a broad based sustainable development pattern though it cannot be expected to deal fully with poverty alleviation of agricultural and non-agricultural rural labourers.

For the role of national governments, international agencies and donors in assessing and adopting the restructuring refer to reference (Randhawa and Sundaram, 1990).

Note : Reproduced from *Indian Journal of Quantitative Economics,* pp. 65-99 with kind permission of publisher.

NOTES

1. Definition used by the countries for small farmers have been adopted. They include size, category of land and income levels. For details see Randhawa and Sundaram (1990)
2. This percentage will increase depending upon the strength of the inverse relation between farm size and productivity per unit of land. But some factors like relatively more area under small farms compared to that of large farms in regions with low

productivity per unit of land e.g. dryland are likely to diminish this percentage. The net effect is determined by the impacts of these counter-vailing factors.

3. For details of methodology and results, see Randhawa, Antiporta and Syed (1986).
4. For a complete list of references in this section, see Randhawa and Sundaram (1990).
5. For more details on results, in section 6, see Randhawa and Sundaram (1990).

REFERENCES

Bhalla, G.S., N.S. Randhawa and D.S. Tyagi (1989), "Implementation and Administration of Agricultural Price Policy in Asia", Economic and Social Development Paper, FAO Report No. 83.

Cornia, G.A., R. Jolly and F. Stewart (eds.) (1987), Adjustment with a Human Face, Vol. I (Protecting the Vulnerable and Promoting Growth). A study by UNICEF, Oxford; Clarendon Press.

Donovan, G. (1987), "Broadening production Increase programs to Reach Low Income Farmers", in *Food Policy; Integrating Supply, Distribution and Consumption,* J.P. Gittinger, J. Leslie, C. Hoisington (eds) EDI Series in Economic Development, Baltimore and London; The Johan Hopkins University Press.

Edgren G. and M. Muqtada (1989), "Strategies for growth and Employment in Asia; Learning from Within ", ILO/ARTEP Working papers pp. 32-37.

Enke, S. (1962s), "Industrialization through Greater Productivity in Agriculture", *Review of Economics and Statistics,* 44 (Feb.); 88-91.

Enke, S. (1962b), "Economic Development with Limited and Unlimited Supplies of Labour", Oxford Economic papers 14 (June) 158-172.

FAO (1985), "Toward Improved Multilevel Planning for Agricultural and Rural Development in Asia and the Pacific", Economic and Social Development Paper 52, Rome.

Hirshman, A.O. (1958), *The Strategy of Economic Development,* New Haven; Yale University Press.

Islam Nurul (1990), Horticultural Exports of Developing Countries; past performance. Future Prospects and Policy Issues. Research Report 80 (IFPRI) Washington.

Johnston B.F. and J.W. Mellor (1961), "The Role of Agriculture in Economic Development", *American Economic Review,* 51(4) 566-593.

Krishna, Raj (1962), "A Note on the Elasticity of the marketable Surplus of a Subsistence Crop", *Indian Journal of Agricultural Economics* (July/Sept), 17 : 79-84. (1963, "Farm Supply Response in India-Pakistan: A Case Study of the Punjab Region", *Economic Journal* 73: 477-87, September.

Lewis, W.A., (1954), "Economic Development with Unlimited Supplies of Labour", Manchester, England; *Manchester School of Economics and Social Studies*, 22(2); pp. 139-191.

Longhurst, R. (1987), "Policy Approaches Towards Small Farmers: in Adjustment with a Human Face. Vol. 1, *op. cit.*

Mellor, J.W. and B.F. Johnston (1987), "The World Food Equation: Inter-relations among

Development, Employment, and Food Consumption", in *Food Policy: Integrating Supply Distribution and Consumption"*, J.P. Gittinger, J. Leslie and C. Hoisington (eds.), EDI Series in Economic Development (World Bank), Baltimore and London: The John Hopkins University Press.

Okawa, Kazushi and Rosovsky, Henry (1960), "The Role of Agriculture in Modern Japanese Economic Development", *Economic Development and Cultural Change*, 9/2:43-68, October.

Prebisch, R. (1959), "Commercial Policy in the Underdeveloped Countries", *American Economic Review* 64 (May): 251-273.

Randhawa, N.S. (1988), "Economic Development, Food Security and Poverty: The CORE Adjustment". Paper prepared for the International Fund for Agricultural Development (IFAD), Rome Italy.

Randhawa, N.S., D.B. Antiporta and S. Syed (1986), "Growth, Equity and Poverty in the Far East-Performance of Countries", Internal paper of FAO, Rome (not published).

Randhawa, N.S. and K.V. Sundaram (1990), "Small Farmer Development in Asia and the pacific: Some Lessons for strategy Formulation and Planning," FAO Economic and Social Development Paper in the Philippines and Taiwan: in *Macro Policies for Technology in Developing Countries,* F. Steward (ed.), Boulder, Colorado: Westview.

Riswanul, Islam (ed.) (1987), "Rural Industrialization and Employment in Asia, ILO, ARTEP, New Delhi.

Ruttan, V.W. (1977), "The Green Revolution: Seven Generalization", *International Development Review,* 19(4): 16- 23, August.

Ruttan V.W. and Y. Hayami (1984), "Induced Innovation Model of Agricultural Development", in *Agricultural Development in the Third World,* J.M. Staatz and C.K. Eicher (eds.), Baltimore and London: The John Hopkins University Press.

Schultz, T.W. (1964), *Transforming Traditional Agriculture,* New Heven: Yale University Press.

Seers, D. (1970), "The Meaning of Development", ADC Reprint, New York Agricutlural Development Council.

Staatz, J.M. and C.K. Eicher (1984), "Agricultural Development Ideas in Historical Perspective", in *Agricultural Development in the Third World,* J.M. Staatz and C.K. Eicher (eds.), Baltimore and London: The John Hopkins University Press.

Streeten, P. (1987), "Structural Adjustment: A Survey of the Issues and Options", *World Development,* Vol. 15, 12: 1469- 1482.

UNDP (1990), Human Development Report 1990, New York Oxford Press.

World Bank (1990), World Development Report, 1990: Poverty, World Development Indications, Oxford University Press.

8

A Simplified Approach to District Agriculture Plan Formulation

AUTAR S. DHESI & N. S. BHULLAR

The desirability of district planning has been widely acknowledged as an instrument of decentralising development planning in a large country like India with considerable socio-economic diversity. District planning envisages the stepping up of the rate of economic growth through effective mobilisation and utilisation of local resources, apart from ensuring people's participation in the planning process, which have hitherto been major missing links in Indian planning. By facilitating the diffusion of development, through the realisation of local development potential and people's participation district planning may help in mitigating regional imbalances and social tensions.

For the realisation of the development potential of the district economy, a comprehensive district plan has to be prepared with macro, sectoral and spatial dimensions. For setting macro and sectoral normative targets of growth and attaining the desired structure of the district economy at the end of the plan, the first necessary step is to estimate the development potential of the dominant sector of the economy—agriculture. This paper suggests a simple operational methodology for formulating district agriculture plan with special reference to Jalandhar district.

The agriculture sector plan formulation methodology has been exemplified for a number of reasons. In the first place, agriculture is the predominant sector from which majority of the population draws its sustenance. As such it should be attended to on a priority basis. Secondly, agriculture constitutes the major component of a district plan.

Thirdly, agriculture is the leading sector in the early stages of development with which the development of secondary and tertiary sectors are closely linked. Therefore, once the agriculture development potential is estimated, it is easy to plan the future development of agro-based industries in the district economy. Further, the target growth rates of secondary and tertiary sectors can be determined on a sound basis.

Methodology

The following stepwise procedure, involving simple requisite statistical and analytical tools, has been adopted to exemplify the process of district agriculture plan formulation:

1. An analysis of the existing level and past trends of agricultural development in terms of yield, area, production etc., has been undertaken to identify its strong and weak spots with a view to estimating the potential for further development. In addition to the decomposition analysis, growth rates and index numbers have been computed for the purpose.
2. In consonance with the findings of past performance in step 1, the objectives and strategy of agricultural development plan have been delineated, keeping in view the proposed State and national framework. The emphasis here is on a quantitative analysis.
3. The third step is devoted to estimate the development potential of agriculture in terms of possibility of extension in area (net and gross), shift in the cropping pattern, improvement in yield and increase in production under alternative assumptions. The approach is based on the methodology suggested by FAO and as used by the Planning Commission of India to fix physical targets for different crops.
4. Alternative scenarios of production in the target year regarding various crops have been converted into value terms at constant prices in order to correlate the physical output to gross domestic product of the district and determine the potential growth rate of agriculture.
5. Material balance of agricultural production has been prepared in the light of projected population of the district as well as the estimated seed requirement with a view to studying self-sufficiency in food, possibility of exports or imports and potential for industrial processing.

6. Requirements of investment in agriculture and selected inputs, such as seeds, fertilizers, labour and credit, have been estimated to achieve the targeted production and growth of the district agriculture sector.
7. Lastly, projects schemes have been suggested to realise the plan targets

The secondary data regarding various agricultural variables have been compiled from the Punjab Statistical Abstract. The base year on which the plan has been prepared is 1979-80. The plan has been worked out at constant prices of 1970-71. The terminal year of the plan is 1990-91.

Trends in Area, Productivity and Production of Major Agricultural Crops

We examine the performance of 13 major crops, covering approximately 85 per cent of the gross cropped area, in terms of changes in area, yield and production by computing compound growth rates from 1969-70 to 1979-80, given in Table 8.1.

The area under rice, wheat, dry chilly, potato and masar increased whereas it decreased under mash, linseed, maize, gram, groundnut, rape and mustard seeds, cotton, fruits and vegetables. Obviously, there has been a major shift in the cropping pattern during this period. Rice, wheat, masar, linseed, sugarcane, potato and maize gained in yield whereas gram, rape and mustard seeds, dry chilly, desi cotton and groundnut crops lost in terms of yield. In general, crops which gained in yield benefited from the introduction of HYVs, intensive use of fertilizes, plant protection measures, price support (except potato) and irrigation expansion. The decline in yield of pulses and oilseeds can be explained mainly by the absence of greater improvement in their seeds and price support and their cultivation having been shifted to marginal and dry land.

In production, crops such as rice, wheat, lineseed, potato, sugarcane and dry chilly have registered positive growth rates whereas gram, mash, groundnut, rape and mustard seeds, desi cotton, and maize crops have shown negative growth rates. The highest growth in production has been shown by rice, followed by potato, sugarcane and wheat and the highest negative growth rate is registered by gram followed by rape and mustard seeds and groundnut crops.

TABLE 8.1 : Compound Growth Rate Per Cent Per Annum of Area, Yield and Production of Principal Crops of Jalandhar District (1969-70 through 1979-80)

Sl. No.	Crops	*Area*	*Yield*	*Production*
1.	Rice	19.18* (11.078)	6.42* (7.406)	26.54* (12.965)
2.	Maize	–1.62*** (–1.543)	1.05 (0.646)	–0.24 (–0.172)
3.	Wheat	2.17* (8.367)	1.14** (2.246)	3.33* (7.594)
4.	Principal cereals	3.12* (11.218)	—	5.43* (9.645)
5.	Grams	–4.94** (–2.951)	6.19* (–4.50)	–10.83* (–20.166)
6.	Mash	–4.77 (–0.139)	0.21 (–0.140)	–5.00*** (–1.321)
7.	Masar	0.21 (0.147)	2.67** (2.148)	2.89*** (1.297)
8.	Total pulses	–4.55* (–3.133)	—	–2.55* (–4.613)
9.	Total foodgrains	2.99* (10.899)	—	5.32* (9.521)
10.	Groundnut	–8.29* (–20.166)	–0.33 (–0.228)	–8.59* (–5.517)
11.	Rape and mustard seed	–9.46** (–2.560)	–0.48*** (–1.403)	–9.92*** (–1.954)
12.	Linseed	–1.58 (–1.108)	3.47*** (1.724)	1.86*** (1.205)
13.	Total oilseeds	–9.13* (–12.416)	—	–9.29* (–5.522)
14.	Sugarcane	–1.09 (–0.706)	4.69* (3.304)	3.44*** (1.356)
15.	Dry chilly	4.37*** (1.514)	–1.47*** (–1.167)	2.15 (0.84)
16.	Potatoes	16.19* (12.382)	3.23* (3.442)	19.94* (12.011)
17.	American cotton	–30.00* (–6.57)	—	—
18.	Desi cotton	–4.56* (–3.978)	–3.36*** (–2.155)	–7.63* (–3.117)
19.	Total cotton	–6.12 (–4.890)	—	—
20.	Fruit & Vegetables	–0.45 (–4.168)	—	—

Notes:

1. Figures in parentheses are *t*-values.
2. *Significant at .01 level; ** significant at .05 level; and *** Significant at .10 level.

In short, rice has recorded the highest positive growth rates in area, yield and production. This seems to be a promising crop. In the case of maize, the negative effects of decline in area has not been offset by improvement in yield and, consequently, its production has shown negative growth rate. The growth of area and increase in yield have together contributed to growth of production in wheat, masar and potato. The negative rates of area and yield of gram, mash and groundnut crops have resulted in significant negative growth in their production. In the case of linseed and sugarcane, the negative effects of decline in area has been more than compensated by increase in their yields and, consequently, they have registered positive growth rates in production.

Decomposition Analysis of Increase in Agricultural Production

In the case of individual crops, only the effect of area and yield on production can be ascertained. However, the application of decomposition analysis allows us to bring out the contribution of other factors as well (Minhas and Vaidyanathan, 1965; Misra, 1970). It has been observed that increase in overall agricultural production is the combined effect of increase in area, improvement in yield, shift in the cropping pattern and interaction of cropping pattern and yield. The additive model has been used here.[1]

The results of decomposition analysis in respect of 13 major crops concerning approximately 85 per cent of the gross sown area shows that area effect, yield effect, cropping pattern effect, interaction of cropping pattern and yield effect respectively contributed 28.4 per cent, 28.30 per cent, 25.80 per cent and 17.40 per cent to the increase in production in the pre-plan period. In the plan period further growth potential lies in improving the yield of crop and substituting a superior cropping pattern for the inferior one. The scope for increasing the net sown area in the district is almost nil. However, the gross sown area can be further increased by improving the cropping intensity with better land and water management practices and by adopting the high yielding short- duration crops[2].

Objectives and Strategy of District Agricultural Development Plan

The objectives as well as strategy have to be location specific due to agro-climatic differentiation within a broader framework of societal goals of achieving a higher growth rate, minimising fluctuation in production, diversification, strengthening the national food security system and augmenting the employment and income generating capacity of the rural economy. In this context, the objectives of the district agriculture plan can be the consolidation of the actual gains, a growth rate of 5 per cent, reduction in inter-block disparities, diversification of the cropping pattern in favour of pulses, edible oils and cash crops, proper management of underground water resources, popularisation of mixed farming and modern agricultural practices amongst small/marginal farmers.

Since there is hardly any scope to increase the net sown area, the focus has to be on augmenting the gross sown area by increasing the cropping intensity. The gross cropped area and cropping intensity increased in the past due to increase in the gross irrigated area. Tubewells are the major source of irrigation (90 per cent) in Jalandhar (Table 8.2). Hence, concerted efforts have to be made to improve the efficiency of tubewells by a timely and regulated supply of electricity and diesel. The number of tubewells may be increased in the backward blocks of Shahkot and Nakodar to exploit the ample underground water potential due to their location along the river. Diesel pumping sets need to be replaced by electric sets, wherever possible, to reduce the cost of irrigation. Land levelling, land scrapping and an underground water piping system will facilitate better management of underground water utilisation. There is scope for encouraging the use of HYV seeds, especially in the case of pulses and edible oils, and improving the delivery system of inputs. To augment income generation in the rural economy, the cropping pattern may be shifted in favour of sugarcane, potato, etc. This, however, will require institutional price support and setting up of agro-process units. Further, small and marginal farmers need to be encouraged to adopt mixed farming, i.e., combining crop husbandry and animal husbandry to augment their incomes.

Assessment of Development Potential of Agriculture

The development potential of agriculture is implied in the possibility of extension in area under cultivation, improvement in yield and favourable shift in cropping pattern. These are the major determinants of income and employment in agriculture and are dependent upon a number of factors, such as geographical area, land utilisation, soil type,

TABLE 8.2 : Agricultural Indices of Jalandhar District
Base : Average of Triennium 1969-70 to 1971-92 centered at 1970-71—100

Year	*Index of net sown area (All crops)*	*Index of new sown area (13 crops)*	*Index of gross sown area (All crops)*	*Index of gross sown area (13 crops)*	*Index of cropping intensity (All crops)*	*Index of cropping intensity (13 crops)*	*Index of agricultural production (13 crops)*	*Index of yield (13 corps)*	*Index of land productivity (13 crops)*	*Index of cropping pattern (13 crops)*
1969-70	99.06	98.12	96.31	95.38	97.22	97.21	91.28	95.87	93.05	99.36
1970-71	100.47	100.48	99.08	99.06	98.62	98.59	100.28	101.15	99.80	99.85
1971-72	100.47	101.32	104.61	105.56	104.12	104.18	106.95	100.93	105.47	100.32
1972-73	102.24	103.32	106.62	107.72	104.28	104.26	113.97	105.54	110.33	100.16
1973-74	102.59	103.60	109.14	110.19	106.38	106.25	115.95	103.14	111.94	101.85
1974-75	103.30	103.64	110.64	110.98	107.11	107.08	135.17	113.51	130.43	107.84
1975-76	103.65	105.09	112.66	114.93	108.69	109.36	131.90	111.25	125.52	105.35
1976-77	103.65	105.56	114.17	116.23	110.15	110.11	142.79	111.95	135.25	109.42
1977-78	103.65	105.96	113.16	115.65	109.18	109.14	158.65	120.82	149.72	113.43
1978-79	103.65	108.13	118.44	123.53	114.27	114.24	179.61	121.86	166.12	119.20
1979-80	104.0[illegible]	108.59	117.94	123.11	113.39	113.37	179.87	127.46	165.64	114.53
Compound growth rate	0.46	0.91	1.96	2.42	1.49	1.50	7.04	2.66	6.08	1.83

cropping intensity, irrigation facilities, quality fertilizers and plant protection measures, agricultural implements, human, animal and mechanical power, timely supply and prices of inputs, institutional price policy and farm management skills.

Here we examine the development potential of agriculture under alternative sets of assumptions.

Possibility of Increase in Area

The relevant concept here is the gross sown area. It consists of the net sown area and the area sown more than once. Thus, any increase in area would result from increase in either one or two of its components. In Jalandhar, with 86 per cent of total land (340 thousand hectares) under cultivation, the scope for increase in the net sown area is nil. However, the demand for land foı non-agricultural purposes like residential buildings, commercial establishments and industries may lead to some reduction in the net sown area even after bringing the pockets of marginal land under cultivation. At best, it may remain constant during the next decade—as it has in the Fifth Five Year Plan. Therefore, the only way to increase the gross sown area is by increasing the cropping intensity. Irrigation intensity and cropping intensity have been found to be positively correlated. Irrigation intensity increased due to the gross irrigated area in the district. The net irrigated area increased from 250.5 thousand hectares in 1969-70 to 255.3 thousand hectares in 1969-70 whereas the gross irrigated area increased from 310.3 thousand to 439.2 thousand hectares in the same period. Tubewell irrigation is the major assured source of irrigation (90 per cent). Improvement in supply of electricity as well as diesel can help in increasing the gross irrigated area and, consequently, cropping intensity and the gross sown area.

Cropping intensity has been projected by fitting various growth curves like Modified Exponential, Gompertz and Logistic curves to the time series data on cropping intensity from 1969-70 to 1979-80.[3]

All the curves gave similar results. However, the Gompertz estimate of 177 by 1990-91 with the lowest value of Theil's coefficient (u^2) was accepted.

On multiplying the net sown area by projected cropping intensity, the projected gross cropped area worked out to be approximately 520 thousand hectares, The actual land utilization pattern (for the year

1979-80) and the projected pattern (for the year 1990-91) are given in Table 8.3.

TABLE 8.3 : Projected Land Utilisation Pattern (1990-91)

Sl. No.	*Classification*	*1979-80*	*1990-91*	*Compound growth 1990-91/1979-80*
1	2	3	4	5
1.	Reporting area for land utilistaion	342 (100)	342 (100)	–
2.	Forests	4 (1.17)	5 (1.46)	2.05
3.	Land not available for cultivation	41 (11.99)	43 (12.57)	0.43
4.	Other uncultivated land excluding fallow land	@	@	–
5.	Fallow lands other than fallows	Nil	Nil	Nil
6.	Current fallow	3 (0.88)	0	–
7.	Net area sown	294 (85.96)	294 (84.96)	Nil
8.	Area sown more than once	175	226	2.35
9.	Gross sown area	469	520	0.94
10.	Cropping intensity	160	177	0.92

@ Below 500 hectares.

Note : Figures in parentheses are percentages

Source : Figures relating to the year 1979-80 have been compiled from Statistical Abstract of Punjab, 1980.

Possibility of Improvement in Yield

It depends upon the agro-climatic conditions as well as farm practices. The yield potential of different crops can be estimated by comparing the average yield in the district either with the higher average yield of some other district with similar agro-climatic conditions or with yields observed at progressive/demonstration/experimental farms. The average yields of selected crops in the district have been compared with corresponding average yields in Punjab, the agriculturally advanced adjoining district of Ludhiana, the highest in any district in Punjab and experimental farms, etc. (Gill, 1978, p. 66). Thus four alternative yield potentials have been estimated and are given in Table 8.4.

TABLE 8.4 : Alternative Yield Potentials of Selected Crops of Jalandhar in kilograms

Sl. No.	*Crops*	*In relation to*			
		Punjab	*Ludhiana*	*Best district*	*Experi-mental and the farmers field in Punjab*
1	2	3	4	5.	6
1.	Rice	–	508	508	1465
2.	Maize	–	–	–	1929
3.	Wheat	–	550	550	1614
4.	Gram	–	–	142	814
5.	Mash	–	359	359	–
6.	Masar	–	–	–	–
7.	Groundnut	–	91	275	2110
8.	Rape and mustard seed	–	–	21	698
9.	Linseed	–	–	333	–
10.	Sugarcane	–	613	781	3507
11.	Potato	479	7451	7649	–
12.	Dry chilly	–	–	22	–
13.	Cotton desi	84	–	126	1354

For computing the marginal yield per hectare of selected crops, following equation has been estimated from time series data (1969-70 to 1979-80)

$$O_{it} = + A_{it} + e$$

where

O_{it} = output of the *i*-th crop in *t*-th year.

A_{it} = area under *i*-th crop in the *t*-th year

e = error term,

= is the intercept and

= is the slop co-efficient representing marginal yield (see Singh, 1981, pp. 54-55).

Though the yield performance of Jalandhar is better than that of Punjab, except for potato and desi cotton, there is considerable scope for improvement when set against the other three norms.

Another way to identify crops with increasing yield potential is by comparing the average with the marginal yield. The crops registering a higher marginal yield than the average are given in Table 8.5. All crops, except maize and linseed, have high yield potential. By shifting the cropping pattern in favour of crops with relatively high marginal yield, the total production can be increased. However, the yield of crops depends upon the state of production technology, which is very difficult to project with certainty. So, two yield projection scenarios, as given in Table 8.6, have been made under two assumptions. First, it is assumed that improvement in production technology follows the same trend (FAO, 1962, pp A 29-39; 1971, pp LV-LVI). Second, it is assumed that the best-practised technology at experimental/progressive farm, etc., would soon become the average technology.

TABLE 8.5 : Average and Marginal Yields of Selected Crops (1969-70 to 1979-80)

Sr. No.	*Crop*	*Average Yield (in kgs.)*	*Marginal Yield (in kgs.)*
1	2	3	4
1.	Rice	2764	3415
2.	Maize	1594	700
3.	Wheat	2476	3551
4.	Gram	1137	1554
5.	Mash	534	584
6.	Masar	485	549
7.	Groundnut	939	975
8.	Rape and mustard seeds	613	770
9.	Linseed	615	–33
10.	Sugarcane	5382	6554
11.	Dry Chilly	976	899
12.	Potato	17732	20280
13.	Desi cotton	225	348

Possibility of Shift in Cropping Pattern

A shift in area in favour of high value crops constitutes the third potential of development in the crop sector. During the 1970's the cropping pattern shifted in favour of food grain at the cost of edible oils, sugarcane and cotton in general and within the group of foodgrain against pulses. A shift against edible oils and pulses should be arrested and reversed as a matter of national necessity. In the context of Jalandhar, sugarcane, potato and rice are the promising high value crops. As such, more area can be brought under them. Potato being a short-duration crop has the additional advantage of boosting cropping intensity. In addition, these crops have ample scope for industrial processing and exports.

The cropping pattern depends upon a complexity of factors (soil type, climate agricultural practices and technology, market and

TABLE 8.6 : Projected Yield per Hectare of Principal Crops in Kilograms

Sl. No.	*Crop*	*Scenario-I yields rate*	*Basis of Estimation of rate in Column 3*	*Scenario-II Yield rate*
1	2	3	4	5
1.	Rice	4762	A	5940
2.	Maize	2465	C	4275
3.	Wheat	3000	D	4438
4.	Gram	1093	B	1400
5.	Mash	608	C	875
6.	Masar	692	A	1600
7.	Moong	650	D	1263
8.	Groundnut	1183	C	1350
9.	Rape and mustard seeds	833	C	1105
10.	Linseed	959	A	1250
11.	Sugarcane	8738	A	8333
12.	Dry chilly	978	B	1313
13.	Potato	25214	A	27500
14.	Desi cotton	222	B	1688

Notes :

1. Projected yield of crops in Scenario-I are worked out on the basis of the following methods:

A = Estimated based on Y = A + BT + E

B = Estimated based on eleven years average where there is no trend in yield

C = Estimated based on every highest yield achieved in the district

D = Estimated based on Agricultural Experts' Opinion in District Agriculture Office, Jalandhar.

2. Yields of crops in Scenario-II are adopted from package of Practices for Kharif Crops of Punjab 1981 and for Rabi crops 1980-81 and Farmers' Enterprise Budgets for the same years, Punjab Agricultural University, Ludhiana.

socio-cultural milieu), which makes it difficult, if not impossible to project it. Generally, a simple behavioural approach to a sophisticated optimising approach is used for projecting the cropping pattern.[4] However, the latter technique is more appropriate for a short-run than a long-run exercise (Krishna, 1969, p. 27; Heady, 1961, p. 17). Regarding Jalandhar, the behavioural approach employing regression analysis has been used. This has also been used by the Planning Commission of India and FAO for determining the cropping pattern (Narayana and Parikh, 1979, pp. 1295-96; Planning Commission of India, 1981a, p. 21; FAO, 1962, pp. A 29-39, 1971, pp. LV-LVI). The following three regression equations were estimated by using the time series data (1969-70 to 1979-80) regarding the area under crops.

$$P_{IT} = A + BT + E$$

(1) (PCI, 1981a, p. 21)

$$A_{IGT} = A + (B_1 + B_2T)\, A_{GT} + E$$
$$A + B_1A_{GT} + B_2\, T.A_{GT} + E$$

(2) (Singh, 1981, pp. 54-55)

$$A_{IGT} = A + B_1A_{GT} + B_2\, T + E$$

(3) (FAO, 1971)

P_{IT} stands for percentage share in the gross cropped area of the Ith crop, T for time/year, A_{IGT} for the gross cropped area under the Ith crop in the given year and A_{GT} for the gross cropped area. The results obtained in terms of area projection from the third equation turned out to be similar to the second equation. So, the first and second equations (B and A), along with other suggested practices of FAO (C and D) and the opinion of local experts (F) and the Planning Commission of India (E and G) have been used to allocate the projected gross cropped area amongst selected crops as shown in Table 8.7. Some adjustments in the area under crops have been made so as to conform to the national objective of bringing more area under pulses, edible oil seeds and cash crops (potato and sugarcane regarding Jalandhar). However, the area under cotton is reduced due to its non-suitability to the district.

In the second step in area location, the gross cropped area has been allocated to Rabi/Kharif/Annual (sugarcane and horticulture) crops. The projected cropping pattern is then compared with the base year one in Table 8.8. The projected pattern shows a shift in favour of rice, pulses, sugarcane and potato. Within the cereal group rice gains at the cost of maize. The area under oil seeds, fruits and vegetables and fodder increases in absolute terms, though it remains the same in proportionate terms.

The third and fourth steps in projecting the cropping pattern are related to allocation of area under specific crops into irrigated and unirrigated and HYV and local varieties area, respectively. These steps can be sorted out on the basis of trend growth rates (Narayana and Parikh, 1979, p. 1296). However, in the case of Jalandhar, no such need has been felt. Firstly, because the additional irrigation potential could not be known and hence, the irrigation pattern for the years 1979-80 and 1990-91 has been assumed to be the same[5]. Secondly, the farmers of Jalandhar are quite progressive. Except for maize, for every other crop for which HYV seeds are available the use is already 100 per cent. So, it can reasonably be assumed that farmers will be using HYV seeds in future also.

TABLE 8.7 : Projected Area of Principal Crops

(Thousand hectares)

Sl. No.	*Crops*	*Projected area (1990-91)*	*Basis of projection*	*Adjusted estimated area (1990-91)*
1	2	3	4	5
1.	Rice	143.01	B	140.00
2.	Maize	18.70	A	18.00
3.	Wheat	242.36	A	240.00
	Total (Cereals (1–3)	404.07		398.00
4.	Gram	1.87	A	2.00
5.	Mash	1.66	C	1.70
6.	Masar	0.74	B	0.80
7.	Other pulses	0.73	F	1.50
	Total Pulses (4–7)	5.00		6.00
8.	Groundnut	5.00	F	8.00
9.	Rape & mustard seeds	1.80	D	3.00
10.	Linseeds	0.35	C	0.50
11.	Other oil seeds	0.85	F	1.00
	Total Oil seeds (8–11)	8.00		12.50
12.	Sugarcane	17.47	A	17.50
13.	Potato	22.60	G	22.00
14.	Dry chilly	1.33	A	1.30
15.	American cotton	0.10	E	0.10
16.	Desi cotton	1.40	F	1.40
	Total Cotton	1.50		1.50
17.	Fruit and vegetable	4.00	F	4.00
18.	Foddar	56.33	C	57.00
19.	Miscellaneous	0.03	E	0.20
	Gross Cropped Area	520.00		520.00

A = Projection based on Y = A+(B+BT) AGT+E.

B = Projection based on Y = A+ST+E.

C = Projection based on eleven years average where there is no trend in area.

D = Projection based on average of the last three years where there is sudden rise or fall in trend in area.

E = Projection based on the actual area in 1979-80.

F = Projection based on Agricultural Experts Opinion in District Agriculture Office, Jalandhar.

G = Projection based on ever highest area under the crop.

TABLE 8.8 : Projection Gross Cropped Area under Principal Crops and Cropping Pattern (1990-91)/(1979-80)

Sl. No.	*Crop*	*1979-80 Actual*		*1990-91*		*%age change in area 1990-91/1979-80*	*Compound growth rate 1990-91/1979-80*
		Area (000 hectares)	*Cropping Pattern (%age)*	*Area (000 hectares)*	*Cropping Pattern (%age)*		
1	2	3	4	5	6	7	8
1.	Rice	80.00	17.06	140.00	26.92	+75.00	+5.22
2.	Maize	63.00	13.43	18.00	3.46	–71.43	–12.06
3.	Wheat	218.00	46.48	240.00	46.16	+10.09	+0.88
	Total Cereals	361.00	76.97	398.00	76.54	+10.25	+0.89
4.	Gram	2.00	0.43	2.00	0.38	Nil	Nil
5.	Mash	0.99	0.21	1.70	0.33	+71.72	+5.04
6.	Masar	0.72	0.15	0.80	0.15	+11.11	+0.96
7.	Other Pulses	—	—	1.50	0.29	—	—
	Total Pulses	3.71	0.79	6.00	1.15	+61.72	+4.47
	Total Foodgrains (1 – 7)	364.71	77.76	404.00	77.69	+10.77	+0.93
8.	Groundnut	10.00	2.13	8.00	1.54	–20.00	–2.05
9.	Rape and Mustard Seeds	1.10	0.23	3.00	0.54	+172.72	+9.55
10.	Lin Seed	0.30	0.07	0.50	0.09	+66.67	+4.75
11.	Other Oil Seeds	—	—	1.00	0.19	—	—
	Total Oil Seeds	11.40	2.43	12.50	2.40	+9.65	+0.84
12.	Sugarcane	12.00	2.56	17.50	3.37	+45.83	+3.49
13.	Potatoes	17.90	3.82	22.00	4.23	+22.35	–1.89
14.	Dry chilly	1.34	0.20	1.30	0.25	–2.99	–0.28
15.	American cotton	0.10	0.02	0.10	0.02	Nil	Nil
16.	Desi cotton	4.30	0.92	1.40	0.27	–67.44	–10.74
	Total Cotton	4.40	0.94	1.50	0.29	–65.90	–10.28
17.	Fruit and vegetables	3.32	0.71	4.00	0.77	+20.48	+1.71
18.	Fodder	51.47	10.97	57.00	10.96	+10.74	+0.93
19.	Miscellaneous	2.46	0.52	0.20	0.04	–91.87	–25.63
	Gross Crapped Area (1 – 9)	469.00	100.00	520.30	100.00	+19.87	+8.94

Projected Agricultural Production

We have projected one cropping pattern and the corresponding distribution of area. However, there are two sets of projected yield under two different assumptions regarding production technology. Therefore, two alternative production scenarios are obtained by multiplying this area by two sets of projected yields. These are given in Table 8.9. In Scenario-I, except maize and desi cotton, all the other 11 crops are expected to record significant increase in their production levels. In Scenario-II only maize records decrease in production. But the production of all crops, except sugarcane, is likely to be much higher in Scenario-II then Scenario-I. Increases in area and yield will contribute to increase in the production of rice, wheat, mash, maser, rape and mustard seeds, linseed and potato. In the case of gram and dry chilly increase in production is only due to improvement in yield. Higher production of groundnut becomes feasible with improvement in yield even with decreased area. However, in the case of maize and desi cotton, production declines in Scenario-I. In Scenario-II improvement in yield more than compensates the loss due to decrease in area.

Estimated Growth Rate of Agriculture

The compound growth rates of different crops/groups of crops, along with production are given in Table 8.9. Further, we have prepared an index of agricultural production and allied indices and the growth rate of gross value added to have general view of agricultural development, as shown in Tables 8.10 and 8.11 respectively[6].

The index of agricultural production increases from 179.87 in 1979-80 to 295.24 in Scenario-I and 377.91 in Scenario-II in 1990-91. The corresponding compound growth rates work out to be 4.61 and 6.98 per cent, respectively. The average of the two is about 5.80 per cent per annum. Interestingly, the sum total of compound growth rates of the indices of cropping intensity, cropping pattern and yield come close to growth rate of agriculture production both in Scenario-I (4.55%) and Scenario-II (6.89%). The growth rate of gross value added in the two Scenarios for the period are 4.62 and 6.98 per cent, respectively.

Material Balance of Agricultural Production (1990-91)

The method of constructing material balances has been used to examine the possibility of self-sufficiency, exports and imports and industrial processing (Table 8.12). Food requirements of the projected district population have been worked out on the basis of norms of a balanced diet as prescribed by the Indian Council of Medical Research. Seed requirements of crops have been estimated on the basis of prescribed

TABLE 8.9 : Projected Production Scenarios (1990-91) and Growth of Production of Major Crops

(Production in thousand tonnes)

Sl. No.	*Crops*	*Scenario-I*				*Scenario-II*		
		Production in 1979-80	*Production in 1990-91*	*Percentage change in production 1990-91 / 1979-90*	*Compound growth rate 1990-91 / 1979-90*	*Production in 1990-91*	*Percentage change in production 1990-91 / 1979-80*	*Compound growth rate 1990-91 / 1979-80*
1	2	3	4	5	6	7	8	9
1.	Rice	235	667	+183.83	+8.66	832	254.04	+12.18
2.	Maize	130	44	–66.15	–8.99	77	–40.77	– 4.88
3.	Wheat	629	720	+14.47	–1.24	1065	+69.32	+ 4.90
	Total Cereals (1–3)	994	1431	44.97	3.37	1974	98.59	6.44
4.	Gram	1	2.19	+119.90	–7.39	2.80	+180.00	+ 9.81
5.	Mash	0.5	1.03	+106.00	+6.79	1.40	+180.00	+10.44
6.	Masar	0.4	0.55	+37.50	+2.94	1.28	+220.00	+11.15
7.	Other pulses	–	0.98	–	–	1.89	–	–
	Total Pulses (4–7)	1.9	4.75	+150.00	+8.69	7.46	+292.63	+13.24
	Total Foodgrains (1–7)	995.9	1445.75	–45.17	+3.45	1981.46	+96.96	+6.45
8.	Groundnut	9	9.46	+5.11	+0.45	10.80	–20.00	+ 1.67
9.	Rape and mustard seeds	1	2.50	+150.00	+8.69	3.32	+232.00	+11.53
10.	Linseed	0.2	0.48	+20.00	+8.28	0.63	+215.00	+10.99
11.	Other seeds	–	1.06	–	–	1.06	–	–
	Total Oil Seeds (8–11)	10.2	13.50	+32.52	+2.58	15.81	–55.00	+ 4.06
12.	Sugarcane	63	153	+142.46	+8.40	146	+131.75	+ 7.94
13.	Potato	315	555	+76.19	+5.28	605	+92.06	+ 6.11
14.	Dry chilly	1.2	1.27	+5.85	+0.52	1.71	+42.50	+ 3.27
15.	Desi cotton	0.67	0.31	–46.27	–7.26	2.36	+252.39	+12.13

TABLE 8.10 : Projected Agricultural Indices of Jalandhar District

(*Base : Average of triennium 1969-70 to 1971-72 centered at 1970- 71=100*)

Sl. No.	*Index*	*1979-80*	*1990-91*		*Change 1997-80/1969-70*	*Change 1990-91/1969-80*		*Compound growth rate % per annum 1990-91/1979-80*	
			Scenario-I	*Scenario-II*		*Scenario-I*	*Scenario-II*	*Scenario-I*	*Scenario-II*
1	2	3	4	5	6	7	8	9	10
1.	Index of Net Area Sown All Crops	104.01	104.01	104.01	Nil	Nil	Nil	Nil	Nil
2.	Index of Net Area Sown 13 Crops	108.59	108.54	108.54	+10.47	–0.05	–0.05	–0.004	–0.004
3.	Index of Gross Sown Area All Crops	117.94	130.76	130.76	+21.63	+12.82	+12.82	+0.94	+0.94
4.	Index of Gross Sown Area 13 Crops	123.11	136.43	136.43	+27.3	+13.32	+13.32	+0.94	+0.94
5.	Index of Cropping Intensity All Crops	113.39	125.72	125.72	+ 16.17	+12.33	+12.33	+0.94	+0.94
6.	Index of Cropping Intensity All Crops	113.37	125.70	125.70	+16.16	+12.33	+12.33	+0.94	+0.94
7.	Index of Agricultural Production 13 Crops	179.87	295.24	377.91	+88.59	+115.37	+190.04	+3.61	+6.98
8.	Index of Yield 13 Crops	127.46	174.62	223.51	+31.59	+47.46	+96.06	+2.90	+5.24
9.	Index of Land Productivity 13 Crops	165.64	271.99	348.15	+72.59	+106.35	+182.51	+4.61	+6.99
10.	Index of Cropping Pattern 13 Crops	114.53	123.78	123.78	+15.17	+9.25	+9.25	+0.71	+0.71

TABLE 8.11 : Projected Gross Value Added of Major Crops

Sl. No.	*Crops*	*Ratio of gross value bi-product to gross value of main product*	*Ratio of gross value added to gross value of output (main and bi-product)*	*Gross value of output of main product (1979-80) (Rs. lakhs)*	*Gross value added main and bi-product (1979-80) (Rs. lakhs)*	*Scenario-I 1990-91*		*Scenario-II 1990-91*	
						Gross value of output of main product (1990-91) (Rs. lakhs)	*Gross value of added main and bi-product 1990-91 (Rs. lakhs)*	*Gross value of output of main product (1990-91) (Rs. lakhs)*	*Gross value of added main and bi-product 1990-91 (Rs. lakhs)*
1	2	3	4	5	6	7	8	9	10
1.	Rice	.0126	.65	2397.00	1677.70	6803.04	4477.90	8486.40	85.60
2.	Maize	.0793	.61	780.00	513.20	264.00	173.90	462.00	804.20
3.	Wheat	.1017	.62	4906.20	3351.20	5616.00	3836.00	8307.00	5674.10
4.	Gram	.0096	.80	9.90	8.00	21.70	17.50	27.70	22.40
5.	Mash	.0635	.68	6.00	4.70	13.40	9.70	19.40	14.00
6.	Masar	.0007	.82	3.88	3.20	5.30	4.40	12.40	10.30
7.	Groundnut	.0233	.64	111.60	70.10	117.30	76.80	133.90	87.70
8.	Rape & Mustard Seed	.0063	.71	16.40	12.40	41.00	31.00	54.40	41.10
9.	Lin Seed	.0062	.71	3.24	2.40	7.80	5.90	10.20	7.70
10.	Sugarcane	.0060	.76	692.20	402.20	1438.20	1198.00	1372.40	1143.20
11.	Potatoes	Nil	.60	1791.40	1028.20	3052.50	1831.60	3327.50	1996.50
12.	Dry Chilly	Nil	.77	66.64	60.60	68.50	53.50	93.50	72.00
13.	Desi Cotton	.04446	.71	9.98	7.40	4.60	3.40	35.20	26.10
	Total			10633.94	7136.30	17454.70	11729.50	22342.00	14986.90

Note : Ratio of gross value of bi–product to gross value of main product and ratio of gross value added to gross value of output (main and bi–product) of major crops have been worked out from Farmer's Enterprise Budget for Rabi 1980–81 and Kharif 1981, Punjab Agricultural University, Ludhiana.

TABLE 8.12 : Marginal Balances of Major Crops (1990–91)

Sl. No.	Crops	Food requirement (000 tonnes)	Seed requirement (000 tonnes)	Total requirement (3 + 4) (000 tonnes)	Total production (000 tonnes)		Surplus (+) deficit (–) (000 tonnes)		Potential of industrial processing (000 tonnes)	
					S-I	S-II	S-I	S-II	S-I	S-II
1	2	3	4	5	6	7	8	9	10	11
1.	Rice	50.00	1.85	51.85	667.00	832.00	+615.15	+780.15	665.15	778.30
2.	Wheat	259.00	21.00	290.00	720.00	1065.00	+430.00	+175.00	699.00	1044.00
3.	Maize	27.00	0.36	27.36	44.00	77.00	+16.64	+49.64	43.64	76.64
	Cereals (1–3)	346.00	23.21	369.21	1431.00	1974.00	+1061.79	+1604.79	1407.79	1581.58
4.	Grams	—	a	—	2.19	2.80	—	—	2.19	2.80
5.	Mash	—	a	—	1.03	1.49	—	—	1.03	1.49
6.	Masar	—	a	—	0.55	1.28	—	—	0.55	1.28
7.	Other Pulses	—	a	—	0.98	1.89	—	—	0.98	1.89
	Total Pulses (4–7)	38.00	.14	38.14	4.75	7.46	–33.39	–30.68	4.61	7.32
	Food grains (1–7)	384.00	23.35	407.35	1415.35	1981.46	+1038.40	+1574.41	1422.40	1958.11
8.	Goundnut	—	1.00	—	9.46	10.80	—	—	8.46	9.80
9.	Rape and mustard seed	—	a	—	2.50	3.32	—	—	2.50	3.32
10.	Linseed	—	a	—	0.48	0.63	—	—	0.48	0.63
11.	Other oil seeds	—	a	—	1.06	1.06	—	—	1.06	1.06
	Oil seeds	—	1.3	—	13.50	15.81	—	—	12.20	14.51
	(Oils)	(15.00)	(0.61)	(15.61)	(6.49)	(7.38)	(–9.12)	(– 8.23)		
12.	Potatoes	20.00	55.00	75.00	555.00	605.00	+480	+530	500.00	550.00
13.	Sugarcane (Sugar/Jaggery)	38.00	13.13	51.13	153.00	146.00	+101.87	+94.87	1018.70	948.70

Notes :

1. In Paddy Form (Rice unhusked) Potential of Industrial Processing In S-I and S-II would be 997.73 and 1167.45 thousands tonnes.
2. Estimated 70 thousands tonnes requirement for cereals in terms of rice (in addition to wheat) has been divided into rice (50 thousands tonnes) and maize (27 thousand tonnes, the latter being used as a popular cereals with leafy vegetables of Sarson Sag during winter season in Jalandhar district.
3. Figures below oil seeds in parentheses refer to oil quantity in thousand tonnes.
4. S-I = Scenario I; S-II = Scenario II

seed norms of Punjab Agricultural University (PAU), Ludhiana. The wastage norms have been ignored on the assumption that post-harvest technology of storage would improve. The use of cereals as cattle feed, also, has been overlooked on the assumption that green fodder and oilseed cakes would take care of the cattle feed requirements. Human consumption and seed requirements, constituting the total demand, have, therefore, been juxtaposed with the production of individual as well as group of crops under Scenarios-I and II. Table 8.12 shows that the district would not only be self-sufficient in cereals and food grains but would also export the same to the rest of the country, thereby strengthening the national food security system. The district has to import pulses and oil seeds or products. However, the problem of edible oil for human consumption may not present a major problems due to the possibility of substituting desi ghee for vegetable oils. But demand for vegetable oils for industrial use has to be met by imports.

Cash crops like potato and sugarcane present the possibility of export or industrial processing. New sugar mills can be set up in the Phillaur-Nakodar-Shahkot belt. The additional income and employment opportunities would thus be provided to this backward area of the district. A look at columns 10 and 11 of Table 8.12 gives an indication of the raw materials available for setting up agro-based industries, such as rice husking, rice bran production, flour mills, backeries, oilseed crushing, sugar mills and potato products.[7]

Estimated Input and Investment Requirements

Input requirements depend upon production technology and output level. We have estimated the input requirements on the assumption that the best practice technology in use on the experimental/progressive farms in the State in 1980 would be fully diffused by 1990-91. In this context the physical input requirements of seeds, chemical fertilizer (NPK and total) and manpower on Bullock Operated Farms (BOF) and Tractor Operated Farm (TOF) have been worked out on the Punjab Agriculture University's prescribed norms and are given in Table 8.13. For estimating the Crop Credit Requirements (CCR) the normative crop-loan scale, as adopted in the Jalandhar District Credit Plan (1980-82), has been used. Following this, Rs. 8035 lakhs would be needed to meet the working capital requirement of agriculture. However, for estimating the overall investment requirement of agriculture the norm of incremental

TABLE 8.13: Estimated Input Requirement of Jalandhar District's Agricultural Plan

Sr. No.	*Crops*	*Area*	*Seed*	*Fertilizers*				*Manpower*		
				N	*P*	*K*	*Total*	*BOF*	*TOP*	*CCR*
1	2	3	4	5	6	7	8	9	10	11
1.	Rice	140.00	280	1750	420	420	2590	73.08	66.03	1260
2.	Maize	18.00	36	166	72	41	279	8.41	7.23	162
3.	Wheat	240.00	2112	300	1512	720	5232	88.45	56.59	2160
	Total Cereals	398.00	—	4916	2004	1181	8101	169.95	129.85	3582
4.	Gram	2.00	9	3	4	—	7	0.20	0.19	18
5.	Mash	1.70	3.4	2	4	—	6	0.24	0.18	15
6.	Masar	0.80	2.4	1	3	—	4	0.13	0.10	7
7.	Other Pulses	1.50	2.7	2	6	—	8	0.36	0.30	14
	Total Pulses (4 – 7)	6.00	—	8	17	—	25	1.02	.77	54
	Total Food Grains (1 – 7)	404.00	—	4921	2021	1181	8126	170.97	130.62	3636
8.	Groundnut	8.00	100	12	32	20	64	2.29	2.01	72
9.	Rape and musterd seed	3.00	60	21	9	—	30	0.63	0.53	27
10.	Linseed	0.50	.10	3	—	—	3	0.12	0.11	5
11.	Other oil seeds	1.00	.20	7	3	—	10	0.23	0.18	9
	Total Oil Seeds (8-11)	12.50	—	43	44	20	107	3.27	2.83	113
12.	Sugarcane	17.50	13125	329	—	—	329	35.90	33.57	158
13.	Potato	22.00	5500	275	139	275	689	27.22	20.42	440
14.	Dry chilly	1.30	.16	8	4	4	16	1.80	1.72	12
15.	American cotton	0.10	.18	—	—	—	—	0.04	0.03	1
16.	Desi cotton	1.40	.25	17	—	—	7	0.53	0.45	13
	Total			612	143	279	1041	65.49	56.19	624

TABLE 8.13 (*Contd.*)

1	2	3	4	5	6	7	8	9	10	11
17.	Fruit and vegetables	4.00	—	—	—	—	—	3.06	2.00	80
18.	Fodder	57.00	—	65	167	—	232	25.13	20.42	—
19.	Miscellaneous	0.20	—	—	—	—	—	—	—	—
	Gross Cropped Area (1 – 19)	520.00	—	5644	2375	1480	9506	267.92	212.66	8035

Notes :

1. Area in thousand hectare.
2. Seed and Fertilizers in hundred quintals.
3. Manpower requirement in thousand Standard Year Persons (SYP).
4. BOF and TOF denote respectively Bullock Operated Farms and Tractor Operated Farms
5. CCR denotes Crop Credit Requirements in Rs. Lakhs.

capital-output ratio of 2.5 as adopted in Perspective Model Plan of Punjab State (1977-78), has been used. To achieve a 5.50 per cent growth rate, i.e., for realising an additional gross domestic product worth Rs. 5505 lakhs an additional investment of Rs. 13763 laksh has to be made in the crop sector. Similarly, the investment requirement for different scenarios with growth rates of 4.61 and 6.98 per cent can also be worked out. Alternatively, the rate of growth can be envisaged between the lower and upper limits of growth potential on the basis of mobilisation of resources.

Suggestions for Action Programme

A plan becomes operational through formulation and implementation of schemes/projects. Here we would make some suggestions rather than working out the details of various development schemes. The same can be done by the concerned district officials who are entrusted with such functions. However, with a view to streamlining the process of planning and making it more effective to realise the development potential, the present position of suiting a plan to the schemes should be discarded in favour of suiting schemes to the development plan.

The planning process, starting with scheme planning and ending in development plan in the form of a conglomeration of schemes, can show 100 per cent performance in terms of financial expenditure but much below expectation in physical terms in the absence of an estimated development potential. So, chronologically, the plan should be prepared first and charting the development potential and specific schemes should be formulated afterwards to realise the identified potential.

To recapitulate, it can be stated that the development potential of agriculture in Jalandhar lies in improving cropping intensity, yield and cropping pattern. The suggested schemes should favourably affect these parameters.

The experience of the past shows that improvement in cropping intensity occurred due to increase in the gross irrigated area. The major sources of irrigation are tubewells, both dieselised and energised pumping sets. To increase the gross irrigated area, the efficiency of pumping sets has to be increased. So the scheme should be prepared to provide timely and streamlined distribution of diesel at central villages. Supply of electricity to tubewells can also be boosted by installing a chain of transformers under another scheme within the district. However, the real problem can be solved by generating more power at the State level. This can also help in converting diesel operated tubewells into electric ones. The technicians to repair pumping sets should be encouraged to establish workshops in central villages.

The village panchayats of central villages should be given more grants to build shopping complexes to house such workshops. Shops in focal points can be part of an efficient delivery system of modern agricultural inputs. The crops type, appropriate dosage of fertilizers, right type of seed, etc., should be decided with a view to raising yield by opening a chain of soil testing laboratories in central villages. This will help in evolving the appropriate cropping pattern, also. Agriculture extension services may also be organised at close by central villages rather than at distant urban centres. Agricultural marketing facilities should also be encouraged by establishing marketing yards in central villages. This would eliminate defects in marketing practices. To shift the cropping pattern in favour of high value crops (sugarcane and potato) schemes should be prepared for their timely marketing at remunerative prices. Though the price policy lies under the command of the Centre, yet local initiative in market innovations can fetch higher prices to farmers as experienced under the Apni Mandi Scheme in Punjab regarding vegetables and fruits.

The institutional development for supply of inputs, price support and setting up of agro-based industries needed to facilitate the required transition or shift in agriculture may be worked out by the district planners with the co-operation of expert agencies. In any case, in an agriculturally developed State like Punjab, with a poor natural base the main thrust of industrial strategy should be the strengthening of links between industry and agriculture. The development of industrial sector should evolve around the changing structure of the agriculture sector.

The concerned district official/planners should apply their minds to formulate schemes regarding initiating and promoting three general innovations of agricultural development. These are yield raisers (use of modern inputs), bottleneck breakers (regulated supply of modern inputs) and new enterprises (adoption of superior cropping pattern and cash-market-oriented production). (Kulp, 1977, pp. 13-15). The institutions and instruments falling under their direct command should be favourably exploited. For those beyond their control they may make creative suggestions to the higher-level planning bodies.

Note : Reproduced from *Margin*, April-June, 1989 with kind permission of the publisher.

NOTES

1. The following formula has been used for a decomposition analysis (Minhas and Vidyanathan, 1965, pp. 230-252; Mishra, 1979, pp. 12-13).

   $P_t - P_o = (A_t - A_o)\ W_i Y_{io} C_{io}) + A_t \Sigma W_i C_i (Y_{it} - Y_{io})$
   $+ A_t \Sigma W_i Y_{io} (C_{it} - C_{io}) + A_t \Sigma W_i (Y_{it} - Y_{io})(C_{it} - C_{io})$

   Where P, A, Y, C and W represent the production, area, yield, cropping pattern and the weight used for the crop respectively in base year (0) and current year (t). In the above model the first item on the right hand size captures the area effect, the second the yield effect the third the cropping pattern effect and the last the interaction of yield and cropping pattern effect. Further P_t and P_o have been equated as under (Kahlon, Johal, Singh, 1968, pp. 2-3)

   $P_t = A_t \Sigma W_i C_{it} Y_{it}$
   $P_o = A_o \Sigma W_i C_{io} Y_{io}$
2. A similar picture emerges when we examine time series indices of agricultural development for the period under study with three-year average of area, yield, production from the year 1969-70 to 1971-72 as base. The indices of net sown area, area under major crops, cropping intensity, agricultural production, productivity, yield and cropping patterns are given in Table 9.2. The standard method as used by Economic and Statistical Adviser to the Government of Haryana has been adopted to construct the indices.
3. The method of partial totals has been used for fitting the three growth curves (Croxton, Cowden and Bloch, 1974, pp. 332-338).

   The form estimated equation, value of Theil's Coefficient (u^2) and the estimated cropping intensity for 1990-91 from respective growth curves are as given below:

   (i) Modified Exponential Curve

   $Y = K + AB^x$

   $\hat{Y} = 352.98003 - 207.18379\ (.991432)x$

   $(u_1^2) = .0000797$

   Estimated Cropping Intensity = 177.03791

   (ii) Gompertz Curve $Y = kAB^x$ or

   $\text{Log } Y = \text{Log } k + (\text{Log } A)B^x$

   $\text{Log } Y = 2.4159155 - .252161$
   $(.9791485)^x$

   $(u_1^2) = .0000762$

   Estimated Cropping Intensity = 176.58304

(iii) Logistic Curve $Y = \dfrac{K}{1 + 10^{a+bx}}$

$1/\hat{Y} = 0.0043519 + .002507\,(.96706)^{x}$

$(u^2) = .00008169$

Estimated Cropping Intensity = 176.10006

4. The behavioural approach is based on regression analysis whereas the optimising approach is based on linear programming or its variants like Nerlovian or recursive programming models.
5. Net irrigation intensity, gross irrigation intensity, gross net irrigation intensity, net area irrigated, gross area irrigated, etc., were taken as independent variables separately to explain the behaviour of cropping intensity. It could have been projected provided one could project the above mentioned independent variables. All of them depended on the number of tubewells as 90 per cent irrigated area is irrigated by tubewells. Local geologists officially stated that technically no more tubewells could be installed as the underground water potential has been fully realised, taking the district as a unit. As per instructions of Reserve Bank of India, loan scheme for tubewells installation has also been discontinued. However, in the absence of legislative restrictions farmers are free to install tubewells. The projected figures of tubewells by 1990-91 with linear time trend and quadratic trend turned out to be unreasonable, and thus were not used for projection purposes. However, the gross irrigated area can be increased by improving the efficiency of the existing tubewells, as pointed out by the study conducted by Economic and Statistical Adviser to the Government of Punjab regarding the working efficiency of tubewells.
6. To estimate the potential growth rate of agriculture the following linear regression equation has been used in which the projected agricultural indices for 1990-91 when fitted give the growth rate of 5.65 per cent per annum.

$$\hat{Y} = -103.1473 + 0.9027x + 1.1286X_2 + 0.0163X_3$$
$$(1.6696) \quad (3.5487) \quad (0.368)$$

$R^2 = 0.9483$

$F = 56.0769^{**}$

*Significant at 1 per cent and **Significant at 10 per cent level of probability. In the equation Y, X_1, X_2, X_3, denote index of gross domestic product, cropping intensity yield and cropping pattern respectively.

7. The potential strength of industrial units and employment therein for various agro-based industries can be worked out on the norm of raw material used by an average sized manufacturing enterprise.

REFERENCES

Croxton, F.E. Cowden, D.J. and Bolch, B.W. (1974), *Practical Business Statistics* (Fourth Edition). Prentice Hall of India Private Limited, New Delhi.

Food and Agricultural Organisation of the United Nations, (1962) *Agricultural Commodities Projections for 1970,* Vol. II, FAO, Rome.

Food and Agricultural Organisation of the United Nations, (1971), *Agricultural Commodities Projection for 1970-1980,* Vol. II, FAO Rome.

Gill, K. S., (1978). "Present Status and Future Strategy for Crop Improvements in Punjab' in *Planning for the Growth of the Rural Economy of Punjab,* Punjab Agricultural University, Ludhiana.

Government of Haryana, (1978). A Report on Agricultural Indices, The Economic and Statistical Organisation, Chandigarh.

Government of India, (1976). *Report of the National Commission on Agriculture (Policy and Strategy),* Part II. Ministry of Agriculture and Irrigation, New Delhi.

Government of India, (1977). *Studies on the Structure of Indian Economy and Planning for Development,* Planning Commission, New Delhi.

Government of India, (1981a). *A Technical Note on the Sixth Plan of India (1980-85).* Planning Commission, New Delhi.

Government of India, (1981b). *Sixth Five Year Plan 1980-85 Framework.* Planning Commission, New Delhi.

Government of Punjab (1980), *Sixth Five Year Plan 1980-85 Framework,* Department of Planning, Chandhgarh.

Heady, Earl O., (1961): "Uses and Concepts in Supply Analysis", in *Agricultural Supply Projections—Estimating Techniques and Interpretations,* edited by Heady and Piters, Iowa State University Press.

Krishna, Raj (1969), Agricultural Supply Projections' in the *Proceedings of the 13th International Conference of Agricultural Economists,* Oxford University Press, London.

Kulp, E. M., (1977), *Designing and Managing Basic Agricultural Programmes,* International Development Institute, Indian University, Bloomington.

Minhas, B.S. and Vaidyanathan, A., (1965). "Growth of Output in India: 1951 to 1958-61: An Analysis by Component Element" *Journal of the Indian Society of Agricultural Statistics,* Vol. XII, No. 2.

Narayana, N.S.S. and Parikh, Kirit S. (1979). "Agricultural Planning and Policy of Draft: Will Farmers Fulfil Planners Expectations?" in *Economic and Political Weekly,* Vol. XIV, Numbers 30, 31 and 32

National Council of Applied Economic Research (1972), *Projections of Demand for and Supply of Selected Agricultural Commodities (1975-1991);* 32, New Delhi.

Prahladachar, M. (1979), "Role of Cropping Pattern in Rural Development", in G, Thimmaiah (ed.) *Studies in Rural Development,* Chugh Publications, Bangalore.

Punjab Agricultural University, (1980), *Package of Practices for Rabi Crops of Punjab 1980-81.* Punjab Agricultural University, Ludhiana.

Punjab Agricultural University, (1981), *Package of Practices for Kharif Crops of Punjab,* Punjab Agricultural University, Ludhiana.

Singh, B., (1977), *Agriculture and Animal Husbandry : A Study in Perspective in Training of Personnel or Development Administration* (Course on State Planning with a background of Macro-Planning). Unpublished Sardar Patel Institute of Economic and Social Research, Ahmedabad.

Singh, B., (1981), *Regional Planning, Explorations in Agriculture and Industry,* Oxford and IBH Publishing Co., New Delhi.

United Commercial Bank, (1980), *An Outline of Credit Plan (1980-82) for Jalandhar District,* United Commercial Bank, Lead Bank Office, Jalandhar, p. 48.

SECTION II

LIBERALISATION

9

Agriculture Under GATT

What It Holds for India

ASHOK GULATI & ANIL SHARMA

The long drawn talks on General Agreement on Tariffs and Trade (hereafter GATT), which dragged on for more than seven years have come to an end. The world trade treaty aimed at opening international markets was finally signed in April 1994. In fact eight rounds of negotiations to free international trade have been organised since the inception of GATT in 1947. The first round of GATT to set the rules of international trade was held in Haryana in 1947, which produced a charter for the International Trade Organisation (ITO)[1]. Since then seven more rounds have taken place[2] the eighth in this series which in commonly known as 'Uruguay Round' was initiated in September 1986 at Punta del este, Uruguay. This round has been the most difficult and complex one as it covers new areas like agriculture[3], services, banking, insurance, telecommunications, and so on.

The main objective of GATT treaty is to reform world trade which is highly distorted because of direct and indirect subsidies that flow to various sectors of the economy in various countries of the world. In case of agriculture, specific attempts have been made to measure the degree and extent of distortion in the world trade by some studies like that of Johnson (1971 and 1991), Anderson and Hayami (1986), Islam and Valdes (1990) and Tyers and Anderson (1992). These studies reveal that direct and indirect subsidies, which flow to the agricultural sector, manifest themselves into distorted world price of agricultural commodities. These distorted world prices, in turn, result into a situation of deceptive comparative advantage that leads to inefficient use of world resources, which ultimately leads to efficiency and welfare losses around the world. Therefore, the underlying philosophy of GATT treaty is to correct these distortions in world trade of agricultural commodities with a view of promote efficient allocation and use of world resources.

India is the one of the approvers of this trade treaty. In this context, it becomes important to study the likely impact of GATT on various sectors of Indian economy. The prime focus of this paper, however, is to gauge the likely impact of this agreement on Indian agriculture by analysing the various commitments under GATT in the area of agriculture.

The GATT commitments in the area of agriculture fall under three main categories, namely, market access, domestic support and export competition[4], which are discussed separately in the following Sections I to III. An attempt is also made to present a scenario relating to trade in agricultural commodities in the post-Uruguay round of GATT in Section IV. This section also tries to identify the commodities in which India can gain from international trade. Section V contains concluding remarks of the paper.

I. MARKET ACCESS

Under market access commitment, all member countries of GATT are required to (i) replace all types of non-tariff barriers with tariff barriers,[5] and (ii) reduce the levels of tariffs under a time bound programme. These levels are to be reduced by 24 per cent in case of developing countries and by 36 per cent in case of developed. The period during which these reductions are to be taken up varies from six years in case of developed countries to 10 years in case of developing countries. The least developed countries are not required to undertake these reductions. In addition to these commitments, this measure also calls for maintaining current access opportunities and the establishment of minimum access tariff quota. This minimum access tariff quota is to be established at reduced tariff rates for those basic products where the current market access is less than 3 per cent of domestic consumption. During the implementation period this minimum access tariff quota has to rise gradually to 5 per cent of domestic consumption. However, if there is a surge of imports due to unfair trade practices when these steps are being implemented, the special provisions of the agreement allow any country to impose additional duties. The extent of these additional duties will depend upon the difference between the import price and the trigger price.[6]

The implementation of these measures will not have any adverse effect on Indian agriculture during the implementation period.[7] This is because imports of various agricultural commodities in India are subjected to mainly quantitative restrictions for balance of payment reasons. The special provisions of agreement allow any country to maintain import restrictions up to the end of the implementation period.[8] In any case, even if India does not have BOP cover, that does not mean that India will have to compulsorily import 3 per cent of its domestic consumption.[9] What all it means is that India will allow the imports of these commodities at low levels of tariff. Providing access to the domestic market does not mean that there will be a flood of imports because, (i) if the prices of agricultural products in the domestic market are below their international levels, imports will not be able to compete with the domestically produced products. This is precisely what is true with most of the agricultural commodities in India (with a notable exception of edible oils), and (ii) if the country from where the product is being imported is indulging in dumping, special provisions of the agreement allow us to impose additional tariff barriers as indicated earlier.

This can be explained with the help of import data for the triennium ending 1989-89 (base period), which reveals that except for edible oils, where imports constituted 28.66 per cent of domestic consumption,[10] imports of other agricultural commodities were less than 3 per cent of their total domestic consumption. For instance, imports of rice, wheat and cotton—three major commodities were 0.38 per cent, 1.47 per cent and 0.92 per cent of their total domestic consumption during the base period (triennium ending 1988-89). This clearly shows that there is no adverse impact of the commitment of market access on Indian agriculture. Even in case of edible oils, special provisions allow India to maintain present restrictions till the end of the implementation period, which is 10 years for the reasons explained above and also for balance of payment reasons.[11]

II. AGGREGATE MEASURE OF SUPPORT

Aggregate Measure of Support (AMS) is the annual aggregate value of market price support, non-exempt direct payments and any other subsidy not exempted from the reduction commitment expressed in monetary terms. Thus, to estimate AMS for Indian agriculture, first one works out the level of support for each basic product (like rice, wheat, cotton and so on) that it obtains through (a) market price support; (b) non-exempt direct payments, and (c) other subsidies not exempted from reduction commitments. Support which falls under the purview of non-product specific (like subsidies on various inputs) is totalled into one non-product specific AMS expressed in monetary terms. Policies which have been excluded from reduction commitments are government spending on research, disease control, infrastructure and food security. These also include structural adjustment assistance, direct payments under environmental programmes and regional assistance programmes.[12]

In product specific AMS, the market price support is to be calculated using the gap between a fixed external reference price (fob unit value in a net exporting country and cif unit value in a net importing country) and administered price (domestic support price) multiplied by quantity eligible to receive the administered price. There is not enough clarity in the Final Act whether the quantity eligible to receive administered price is the total quantity of production, or only the marketed surplus.[13] The reference price shall be based on the year 1986-87 to 1988-89. External reference price needs to be adjusted for quality differences as necessary in order to make it comparable with the domestically produced product.

Non-exempt direct payments are to be computed by using the gap between external reference price and administered price or by using budgetary outlays. Input subsidies and other policies like marketing cost reduction measures are to be determined by using budgetary outlays. However, if budgetary outlays do not reflect the full extent of subsidy, the subsidy has to be worked out by using the gap between the subsidised price of input in question and a representative market price multiplied by the quantity used of that input.

If the product specific and non-product specific AMS does not exceed 10 per cent of the total value of agricultural product in case of a developing country (5 per cent in case of a developed country), the country is not required to reduce the product specific and non-product specific support. However, if the AMS exceeds 10 per cent of the total value of agricultural production, it is required to be reduced by 13.3 per cent in case of developing country (20 per cent in case of a developed country) of the value that does not qualify for exemption during the implementation period. The least developed countries have been exempted from these reductions.

India has basically two types of support operations for farmers. First, market price support, which is in the form of minimum support prices announced by the government for different commodities, based on the recommendations of the Commission for Agricultural Costs and Prices (CACP). Second, the support in the form of input subsidies which are on inputs like fertilisers, irrigation, electricity, credit and seeds.

The third type of support which falls under the head of non-exempt direct payments is not given to farmers in India.

The estimates based on support prices reveal that product specific AMS for India (for 17 products out of 22 total products for which India maintains market support programmes) works out to be negative to the tune of (-) Rs. 242 billion. This forms (-)27.74 per cent of total value of agricultural production (excluding forestry and fishery).[14, 15] Non-products specific AMS works out to be Rs. 46 billion, 5.24 per cent of total value of agricultural production (Table 9.1).[16] By adding these two one gets total AMS which stands at (-) Rs. 196 billion, which forms (-) 22.50 per cent of the value of agricultural output during the base period 1986-87 to 1988-89. This indicates the massive amount of 'taxation' that is inflicted on Indian agriculture, which is contrary to the general impression of huge subsidies which flow to this sector.

This negative support (or net taxation) is due to the fact that prices of different crops are fixed by the government below their international levels (barring rapeseed mustard and sugarcane). This is evident from the data presented in Table 9.2, which shows that in terms of per unit output, the highest average negative support is for barley, which was found to be (-)200.30 per cent during the base period.[17] The next highest per unit support (in negative) is found in case of jute, which amounted to (-) 134.92 per cent during triennium ending 1988-89. Other product groups with more than 100 per cent of negative support per unit of output during the same period are maize (-127.26 per cent), sorghum (-119.78 per cent), tobacco (-115.57 per cent) and wheat (-111.98 per cent) respectively. The only product group receiving positive support was sugarcane in which case the subsidy was found to be 14.82 per cent of output.[18,19]

Because of these negative support figures per unit of output, the product specific aggregate measure of support turns out to be negative, i.e. price difference multiplied by total quantity of production. The highest negative product specific AMS was found for wheat which totalled Rs. (-) 85.74 billion. Wheat was closely followed by rice in

TABLE 9.1 : Product Specific and Non-Product Specific Aggregate Measure of Support (TE 1988-89 and TE 1992-93)

(*Rs. Billion*)

Support Type	*TE 1988–89*	*TE 1992–93*
Product Specific AMS		
(a) Based on support prices	–242.25 (–27.74)	–427.89 (–26.33)
(b) Based on farm harvest prices	–151.03 (–17.29)	–295.21 (–18.17)
Non–product Specific AMS	45.77 (5.24)	86.45 (5.32)
(i) Fertiliser subsidy	10.21	24.98
(ii) Irrigation subsidy	17.93	17.15
(iii) Electricity subsidy	15.81	41.54
(iv) Credit subsidy	1.00	1.96
(v) Subsidy on seeds	0.82	0.82
Total product specific and non–product specific AMS		
(a) Based on support prices	–196.48 (–22.50)	–341.44 (–21.01)
(b) Based on farm harvest prices	–105.26 (–12.05)	–208.76 (–22.85)

Notes : 1. Figures in parentheses are percentages to the total value of agricultural production (excluding forestry and fishery).

2. The 17 crops included in product specific AMS are rice, wheat, jowar, bajra, barley, gram, tur, urad, moong, groundnut, rapesed and mustard, soyabean (yellow-and black), cotton, jute, tobacco and sugarcane.

3. In non-product specific AMS, the programmes included are fertiliser subsidy, electricity subsidy, irrigation subsidy, credit subsidy and seed subsidy.

4. According to the Ministry of Commerce product specific AMS during TE 1988-89 is Rs. (-) 244.42 billion. The problems with their estimate have been discussed in endnotes 10, 11, 13, 14 and 15 at relevant places.

which case the product specific AMS turned out to be Rs. (-) 77.23 billion. These two crops were followed by groundnut (-20.49 billion), maize (-16.53 billion) and jowar (-12.52 billion) respectively in that order. The lowest negative product specific aggregate measure of support amounting to Rs. (-) 16.5 billion was obtained for soyabean. However, in case of sugarcane as explained earlier, the product specific AMS turned out to be Rs. 2.41 billion during the base period.

These numbers clearly indicate that India is not required to reduce its support levels as the product specific and non-products specific

AMS works out to be (-) 22.5 per cent of the total value of agricultural output.[20] In fact over the year, the AMS has increased in negative and stood at Rs (-) 341.44 billion during TE 1992-93. This massive increase of 73.77 per cent in absolute value of AMS seems to be the effect of depreciation of rupee against major currencies of the world during the last three years as also in relative prices of different commodities. During this triennium, the highest product specific AMS in negative was found in case of rice which amounted to Rs. (-) 245.98 billion. After rice followed wheat and cotton with product specific AMS totalling Rs. (-) 107.59 billion and Rs. (-27.66) billion respectively. During this period another commodity in addition to sugarcane which has positive product specific AMS is rapeseed and mustard, where the product specific AMS is found to be Rs. 6.09 billion.[21] The relative shares of major crops in total product specific negative support during

TABLE 9.2 : Extent of Support per Unit of Output Based on Support Prices

Crop	*TE 1988-89*			*TE 1992-93*		
	Administered Price (Rs/MT)	*Reference Price (Rs/MT)*	*Product Specific Support (Per cent)*	*Administered Price (Rs/MT)*	*Reference Price (Rs/MT)*	*Product Specific Support (Per cent)*
Rice	2280.00	3520.00	–54.39	3507.46	6834.70	–94.86
Wheat	1670.00	3540.00	–111.98	2383.33	4380.30	–83.79
Maize	1373.33	3121.00	–127.26	2116.67	2525.00	–19.29
Sorghum	1373.33	2945.00	–114.44	2083.33	3427.20	–64.51
Barley	1340.00	4024.00	–200.30	1966.67	2008.05	–2.10
Tur	3350.00	5243.00	–56.51	5550.00	8184.97	–47.48
Gram	2766.67	4128.00	–49.20	4570.00	5563.87	–21.75
Urad	3350.00	4385.00	–30.90	5550.00	6769.83	–21.98
Moong	3350.00	4782.00	–42.75	5550.00	7821.12	–40.92
Groundnut	3966.67	6900.00	–73.95	6583.33	7786.30	–18.27
Soyabean	3033.33	3081.00	–1.57	4566.70	6322.50	–38.45
Rapeseed and Mustard	4350.00	6884.00	–58.25	6766.70	5509.90	18.57
Cotton	16562.00	17333.00	–4.66	24892.00	40734.73	–63.65
Jute	2383.33	5599.00	–134.92	3650.00	6665.60	–82.62
Tobacco	1233.00	2658.00	–115.57	1592.00	3679.00	–131.09
Sugarcane	183.33	156.16	14.82	253.33	156.16	38.36

Notes: Administered prices are minimum support prices.

Reference prices are cif values (except for tobacco in which case reference prices are fob values). Have been derived from Monthly Statistics of Foreign Trade. DGCIS, Calcutta.

Source : GOI, Ministry of Agriculture and DGCIS, Calcutta.

triennium ending 1988-89 and triennium ending 1992-93 are depicted in the Figure.

FIGURE

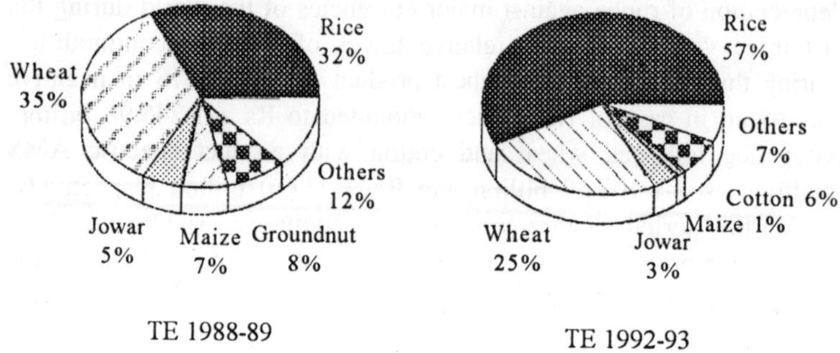

Relative Shares of Different Commodities in Total Product Specific Negative Support

The calculations of products specific AMS for different commodities at farm harvest prices[22] are also somewhat in line with those obtained on the basis of support prices (Table 9.3), but the degree surely differs. The farm harvest prices have been used as an alternative set of prices because support prices have no meaning in the case of those commodities where no support operations are carried on by the government. Besides, farm harvest prices are supposed to be close to the prices which farmers get for their produce.

Though the product specific AMS for 17 commodities was found to be Rs. (-) 151.03 billion, the extent of taxation (both product specific and non-product specific was still quite high, i.e. (-) 12.05 per cent of the total value of agricultural output during 1986-87 to 1988-89 base period.[23] There were only three commodities, namely, gram, tur and sugarcane which showed positive product specific AMS. Similar trends were observed in triennium ending 1992-93, when the total product specific and non-product specific AMS turned out to be Rs. (-) 208.77 billion, which was found to be (-) 12.85 per cent of the total value of agricultural output during this period. The product range with negative AMS narrowed to 8 commodities and with positive support expanded to nine commodities.[24] But three major crops, i.e., rice, wheat and cotton which accounted for 55.52 per cent of total value of agricultural output in 1980s (Bhalla and Tyagi 1989) had negative AMS

TABLE 9.3 : Extent of Support Per Unit of Output Based on Farm Harvest Prices

Crop	*TE 1988-89*			*TE 1992-93*		
	Farm Harvest Price (Rs/MT)	*Reference Price (Rs/MT)*	*Product Specific Support (Per cent)*	*Farm Harvest Price (Rs/MT)*	*Reference Price (Rs/MT)*	*Product Specific Support (Per cent)*
Rice	2828.14	3520.00	–24.46	4302.70	6834.70	–58.85
Wheat	2024.80	3540.00	–74.83	2947.50	4380.30	–48.61
Maize	1870.40	3121.00	–66.86	2818.10	2525.00	10.40
Sorghum	1736.81	2945.00	–69.56	3103.30	3427.20	–10.44
Barley	1812.80	4024.00	–121.98	2987.50	2008.05	32.78
Tur	5550.47	5243.00	5.54	8625.50	8184.97	5.11
Gram	4510.55	4128.00	8.48	6251.70	5563.87	11.00
Urad	3350.00	4385.00	–30.90	5246.70	6769.83	–29.03
Moong	3350.00	4782.00	–42.75	5550.00	7821.17	–40.92
Groundnut	5615.10	6900.00	–22.88	9809.83	7786.30	20.63
Soyabean	3033.33	3081.00	–1.57	7031.67	6322.50	10.09
Rapeseed and Mustard	6194.18	6884.00	–11.14	8274.38	5509.90	–33.41
Cotton	18711.18	17333.00	7.37	25979.90	40734.73	–56.79
Jute	2773.80	5599.00	–101.85	4371.70	6665.60	–52.47
Tobacco	1670.00	2658.00	–59.16	3314.00	3679.00	–11.01
Sugarcane	183.33	156.16	14.82	183.33	156.16	14.82

Notes : 1. Farm harvest prices of urad, moong and sugarcane are not available, therefore, their levels of support are same as under support prices.

2. Farm harvest prices are weighted average prices, weights being shares in total production of those states for which harvest prices are available.

3. Reference prices are cif values (except for tobacco in which case reference prices, are fob values) and have been derived from *Monthly Statistics of Foreign Trade*, DGCIS, Calcutta.

Source : GOI, Ministry of Agriculture and DGCIS, Calcutta.

amounting to Rs. (-) 187.33 billion, Rs. (-) 117.53 billion and Rs. (-) 26.25 billion respectively. Crops with highest positive product specific AMS were groundnut and rapeseed and mustard wherein the AMS was Rs. 15.44 billion and Rs. 13.09 billion respectively.

The above mentioned trends of domestic support in India are contrary to those found in developed countries of the world. Almost all developed countries provide positive support to their cultivators ranging from as high as 72 per cent in case of Japan, 37 per cent in case of European Community and 26 per cent in case of USA (Gulati and Sharma 1992).[25] This is in sharp contrast to India which imposes net tax on its cultivators.

III. EXPORT COMPETITION

In the area of export competition, GATT agreement calls for reducing direct export subsidies to a level of 36 per cent below 1986-88 level in case of developed countries over a period of six years. The quantity of subsidised exports by the developed countries is to be reduced by 21 per cent during the same period. In case of developing countries direct export subsidies are to be reduced by 24 per cent and the quantity of subsidised exports is to be reduced by 14 per cent. These reductions in case of developing countries are to be carried out over a period of 10 years as against six years in case of developed countries. The least developed countries have been exempted from these reduction commitments. Export subsidies included in reduction commitments are direct subsidies paid by the government or any other agency, including payments in kind; payments that are made from the proceeds of levy imposed on agricultural products; subsidies given to reduce the costs of marketing including internal handling, processing, international transport and freight subsidy on export shipments.

In order to see the impact of these commitments on agricultural exports from India, we have to look at the nature and extent of the subsidies. Indian exporters of agricultural commodities do not get direct export subsidy. The only subsidies available to exporters of agricultural commodities are in the form of : (i) exemption of profits from export sales from the income tax, and (ii) subsidies on costs of freight on export shipments of certain products like fruits, vegetables and floricultural products.

Since these payments, by developing countries are exempt from the reduction commitments during the implementation period, the commitments mentioned above will not have adverse impact on agricultural exports from India. In order to encourage the exports of agricultural products these subsidies can be continued for the time being. Once the export supplies become self sustaining during the adjustment period, these can be withdrawn.

It is well documented in literature that most of the developed countries operate export subsidy programmes to enhance their exports. According to one estimate (GATT secretariat 1993) developed country export subsidies amounted to $ 16.4 billion during the base period. These countries would be required to gradually reduce their export assistance programmes. This will open up new market for countries like India, which are efficient producers of agricultural products.[26] Therefore, reductions in export subsidies and subsidised exports of developed countries will have greater implications for developing country exports. Countries like India would benefit not only from improved market access opportunities in the developed and developing countries, but also from the reduction of subsidised exports and trade distorting production incentives in developed countries.

IV. POST-GATT SCENARIO

Having looked at the commitments under GATT in three major areas of market access, domestic support and export competition, the questions that emerge are: (i) what is likely to be the impact of these commitments on world prices and trade of agricultural commodities, and (ii) whether India will be better off once these commitments are fulfilled by the developed countries.

Although, projecting future is a tricky exercise as no one is yet sure of the way how developed countries are going to bring down their AMS, as it is not more commodity specific. Thus, which commodities will bear the burden of adjustment remains speculative. It would be strongly influenced by commodity specific lobbies in developed countries. Nevertheless, one can venture speculating the expected outcome of post-GATT scenario through the judgment of experts in the field. On theoretical grounds one would expect that reduced support levels and rationalisation of trade barriers will increase production and export of various commodities in those countries which have comparative advantage in the production of agricultural commodities. Similarly, the reduction in import barriers will reduce consumer prices in those countries where the tariff barriers are very high, which will generate strong demand and will push up imports. Some empirical modelling exercises show that international prices will rise and so will trade volumes.[27] But the estimates of change in prices and trade vary from study to study. Even if the world prices remain constant, it is clear from the above analysis that prices of various agricultural commodities in India are well below their international levels. India has comparative advantage in agriculture because of its diverse agro-climatic conditions (which are conducive for a variety of crops almost round the year). There is no reason to believe that exports of this sector will not grow with more open world trade policy environment: there are large differences between potential and actual yields, the import intensity of this sector is low compared to other sectors of the economy and labour costs are reasonably low compared to other countries. Our understanding of the situation is that India is likely to gain more from its own policy of liberalising agricultural trade than what may be expected from GATT provisions. The recent trends in exports reinforce these arguments. In fact agriculture has turned out to be a star performer in terms of exports. Agricultural exports increased by 38 per cent in dollar terms during the first half of the year. With the tremendous increase in exports, the export basket of agricultural products has also diversified in the recent years. From tea and mate, which formed nearly 40 per cent of agricultural exports from India in 1960s, exports have diversified to cashew kernels and spices to fish and fish preparations, oil cakes, rice, fruits and vegetables and processed foods.

In the past, agricultural exports from India did not outshine because of : (i) over valued exchange rate of rupee against major currencies of the world; (ii) the external trade in most of the agricultural commodities was and in many cases still is subjected to various controls mainly non-tariff barries like export quotas, canalisation and minimum export prices (MEPs) and (iii) *ad hoc*ism in the export policy, which means export markets are 'residual' markets, i.e., they are approached only in case of country has surplus.[28]

These pervasive controls along with treating export markets as residuals lowered the returns available to the producers. It resulted in diverting the resources away from export competitive crops. All these factors taken together have acted as deterrents and resulted in the lost trade advantage that India has in the area of agriculture. These irritants have outlived their purpose and must be done away with fairly quickly, if India has to take advantage of new post-GATT scenario.[29] The new exim policy of the Government of India does reflect some of the concerns; however, much more remains to be done. First, ministry of commerce should set up an expert group to identify the commodities in which India is likely to gain in post-GATT scenario. This is important because the total cultivable area is almost limited and changes in price ratios will change cropping patterns in favour of those crops which have relative comparative advantage. Second, identify the likely change that the export basket of agricultural commodities is likely to witness, and accordingly prepare the country to improve its competitive edge by working on the fundamentals of various links from production in the farms to exporting in the ships.

In order to find out the precise nature of such shifts in cropping patterns and changes in exports, one may build a large model (linear programming or general equilibrium model) but a very rough and crude idea of this can be had from the relative price ratios prevailing in the domestic market and in the international market. Price ratios of various agricultural commodities in relation to wheat (assuming wheat price equal to 100) in India and the world are given in Table 9.4. A perusal of data presented in table reveals that area under oilseeds is likely to contract, as their relative prices are much lower in the international market than those in the domestic market. This is particularly true for rapeseed-mustard and groundnut.[30] The extent of these variations differs and varies from 31.06 per cent in case of groundnut to 88.80 per cent in case of rapeseed and mustard. Similar changes will occur in case of coarse cereals also, but the impact is likely to be marginal compared to oilseeds. On the other hand, area under fine cereals, rice, wheat and fibre crop, cotton, is likely to expand and they may emerge as important export items along with fish and fish preparations. Thus, the share of oilcakes (particularly non-soya) in agro exports may gradually come down with import liberalisation of edible oils. Our research on the future agro-export basket of the country also reveals that it is likely to be dominated by fish and fish preparations, rice, wheat, tea, tobacco, fruits and vegetables and their processed items reveals that India has high to moderate competitive advantage in most of these commodities (NCAER, 1994). This calls for making additional efforts in terms of devising appropriate policies for these future export items in which the country has competitive advantage.

Therefore, if the exports of agricultural commodities have to increase, the time is ripe to abolish all export controls and regulations like canalisation, export quotas and minimum export prices. Next on agenda should be infrastructural developments in terms of bulk storage and handling facilities at rail heads and sea ports for cereals, fast track facilities, preferably dedicated cargo terminals for perishable commodities (fruits, vegetables and their processed items) at major air-sea ports. Last but not the least, long-term investments in research and extension in yield and quality, eco-friendly production of fruits and vegetables and their packaging material are crucially important for taking advantage from the post-GATT scenario, and sustain higher growth in agro-exports.[31]

TABLE 9.4 : Prices of Various Agricultural Commodities in Relation to Wheat

(TE 1992-93)

Crop	*India (Wheat = 1 00)*	*World (Wheat = 1 00)*
Wheat	100.00	100.00
Rice	147.17	199.43
Maize	88.81	74.68
Sorghum	87.41	74.41
Cotton $	355.25	372.07
Groundnut	248.55	189.64
Soyabean	191.61	177.04
Rapeseed and Mustard	283.92	150.38

Notes : $ = Kapas

1. In case of groundnut, the ratios are for the year 1989, at the international prices for later years are not available.
2. Prices in India are support prices.
3. World prices have been taken from *FAO Quarterly.* The details of these prices are given below. The world prices are for the following varieties :
   Wheat = US HRW No 2 variety, fob US Gulf; Rice = Thai rice 5 per cent of broken variety, fob Bangkok; Maize = US No 2 (yellow) variety, fob US Gulf; Sorghum = US No 2 (yellow) variety, fob US Gulf; Cotton = SMI-1/16 US Memphis variety, cif, Livepool; Sayabean = Soyabean of US origin, cif, Rotterdam; Rapeseed = Canadian 40 per cent cif, NW Europe; Groundnut = Any origin, cif, Europe.

Summary and Conclusions

The impact of GATT commitments in agriculture which fall under three main categories, i.e. market access, domestic support and export competition is explored in this paper. The analysis reveals that India stands to gain rather than lose from trade liberalisation by the GATT members. The domestic support levels in India are negative in most of the agricultural commodities studied here, which is in sharp contrast to the support levels prevailing in developed countries of the world. Almost all developed countries provide positive support to their cultivators and their support levels are quite high (generally more than 10 per cent of total value of agricultural output). In India the product specific and non-product specific AMS (for 17 products out of 22 total products for which India maintains market support programmes) works out to be negative, i.e. (-) Rs. 196 billion, which forms (-) 22.50 per cent of the value of agricultural output during the base period 1986-87 to 1988-89. This indicates the massive amount of 'taxation' that the

Indian agriculture, in reality, is subjected to, contrary to general impression of huge input subsidies which flow to this sector. In fact over the year, this negative support has increased in absolute amount to Rs. (-) 341.45 billion during TE 1992-93, but in percentage terms, has slightly come down to (-)21 per cent.

A few studies that have been carried out in the recent past indicate, that international prices of agricultural commodities will rise and so will their production and trade volumes. Also the variability in world prices will reduce. Though the increase may be a modest one, but it can be said with reasonable degree of certainty that reduced support levels and rationalisation of trade barriers will increase production and export of various commodities in countries like India, which have comparative advantage in the production of agricultural commodities. Diverse, agro-climatic conditions in India and the existing differentials in actual and potential yields, all augur well for exports of agricultural commodities. To cash on the trade opportunities which will emerge from the post-GATT scenario, India will have to change its export strategy of treating export markets as residuals. In addition to this various stringent measures which act as irritants like export quotas, canalisation and minimum export prices must be done away with completely. The analysis also reveals that future export items will be fish and fish preparations, cereals like rice and wheat, tea and tobacco, fruits, vegetables and their processed items. This calls for making additional efforts in terms of devising appropriate policies for these future export access in which the country has comparative advantage.

Note : Reproduced from *Economics and Political Weekly,* July, 16, 1994, 1857-1863.

NOTES

1. In the first round, the 23 countries that founded GATT exchanged tariff cuts for 45,000 products worth $ 10 of trade on annual basis.
2. The second round was held at Annecy (France) in 1949 in which another 10 countries joined and customs duties were reduced for another 5,000 items. Third round was held in Torquay (Britain) in 1950-51 and involved 38 countries. Fourth and fifth rounds were held in Geneva (Switzerland) in 1955-56 and 1960-62 respectively. The sixth round which is known as the "Kennedy round" was also initiated in Geneva in 1964 and was completed in 1967. In this round more than 50 countries accounting for 75 per cent of

world trade participated. Seventh round of GATT negotiations was started in 1973 in Tokyo (Japan) but ended in Geneva in 1979. In this round 99 countries participated in trade negotiations.

3. Agriculture did appear on GATT agenda thrice before this round, but the focus was very narrow.
4. Another area of concern in agriculture which has created fears in the minds of people is the impact of trade related intellectual property rights of farmers and plant breeders. This we feel is basically the area of plant breeders and they should be the right persons to comment on this issue. However, our understanding of this issue is that in *sui generis* system of plant protection, the proposed legislation offers farmers to save, use, exchange and share seeds. The only things which *sui generis* system of intellectual property rights prohibits is the large-scale multiplication and selling of the seed purchased from companies.
5. The level of tariff which will result from this process is not clearly defined in the Draft Final Agreement. Logically, one would think that the level of tariff would be equal to the difference in the domestic and world prices during the base period, 1986-88. But actually, it is not so. GATT secretariat has asked its member countries to propose their respective tariff rates on various commodity groups. India, e.g., has proposed 100 per cent tariff on agricultural commodities, 150 per cent on processed agro-commodities and 300 per cent on edible oils, despite the fact that, on the whole Indian agriculture's AMS is negative
6. The additional duties will be set according to the difference between cif import price and the trigger price. The trigger price shall be equal to 125 per cent, 110 per cent, 105 per cent of the average cif import value of the preceding three years when the market access opportunities are less than or equal to 10 per cent, between 10 per cent and 30 per cent and above 30 per cent respectively. (For details on how additional duties would be determined, see World Trade Centre 1994).
7. Countries like Japan, South Korea and European Community, which has very high tariff barriers and use border measures like variable levies will be required to provide greater market access to other countries.
8. The conditions under which the restrictions can be maintained are (i) imports constitute less than 3 per cent of the domestic consumption of the particular product during 1986-88, (ii) no export subsidies have been provided for the particular product since 1986. (iii) effective production restricting measures are applied to this particular product, and (iv) minimum access opportunities are provided.
9. This apprehension has been expressed in Thomas *et al* (1994).
10. Domestic consumption has been equated with net availability. Net availability includes net production (gross production minus quantity kept for seed, feed and wastage), imports and changes in stocks.

11. Government of India has put imports of palmolein oil under OGL (Open General Licence) at 65 per cent import duty (*The Economic Times,* April 19 1994). NDDB and STC, however, have been allowed to import at 20 per cent import duty.
12. Even input subsidies available to low income or resource poor producers (having operational holdings below 10 hectare) in developing countries are exempt from domestic support commitments (WTC, 1993 page 29 and GOI, Ministry of Commerce). If that is true then almost 80 per cent of the input subsidies will qualify for exemption, because in India operational holdings of 10 hectares or below account for 79.5 per cent of agricultural land.
13. It is worth noting that in 1991, the government estimated AMS for Indian agriculture on the basis for marketed surplus of relevant commodities, while in 1994 the reply given by Indian Government to GATT secretariate included production as the relevant quantity.
14. This is a bit different from the estimate worked out by the Ministry of Commerce. According to their calculations the product specific aggregate measure of support is Rs.(-) 244.42 billion. This is because of the methodological and computational problems which remain in the estimates of the Ministry of Commerce. These have been explained in the latter part of the text or in endontes.
15. Ministry's estimate reveals that product specific support forms (-) 17.40 per cent of the total value of agricultural production, whereas our estimates show that it is (-) 27.74 per cent of total value of agricultural production. It seems they have included animal husbandry and forestry in total value of agricultural production, while support on these components of agriculture has not been included. We have corrected for this and taken crop value only as representing agriculture.
16. Input subsidies for triennium ending 1992-93 have been worked out as follow:

    (i) Fertiliser subsidy has been taken as 50 per cent of the amount shown in government documents, the rest is deemed to be going either to the fertiliser industry or to the feedstock agencies (Gulati, 1990).
    (ii) Irrigation subsidy represents the losses made on working expenses of major and medium and minor irrigation schemes. It does not include the subsidy on capital investment made in these schemes (Gulati, 1994 and Gulati and Sharma 1994).
    (iii) Electricity subsidy has been taken from Gulati and Sharma (1994).
    (iv) Subsidy on farm credit represents only the interest rate subsidy and does not include the loans written-off.
    (v) Subsidy on seeds has been taken from the publications of Government of India.

17. According to the ministry, it is (-) 190.96 per cent. This is again because of the wrong information used for the selected years on prices and production. In their estimates they have used price information for 1987-88, 1988-89 and 1989-90. Actually the price should be for 1986-87, 1987-88 and 1988-89 and corresponding production figures should be for the crop years 1985-86 become available for sale in 1986-87 and so on. Same is true for wheat, gram and rapeseed and mustard, which are all rabi crops.
18. According to the calculations done by the Ministry of Commerce, tobacco also received positive support. But, there appears to be a slip in their estimates. It seems they have used import parity price while comparing support price with external reference price. However India being net exporter of tobacco, it is the export parity price which should be compared with the support price and not the import parity price.
19. Another disagreement between our numbers and the numbers worked out by the ministry appears to be in the case of cotton. In this case, ministry's estimates seem to have compared kapas (seed cotton) prices and the prices of lint. Actually, it is the price of lint which should be compared with import parity price of lint. The like should be compared with like only.
20. India is not required to reduce its support levels under the GATT agreement. However, input subsidies may become unsustainable over time due to budget constraints. The rising burden of subsidies is affecting investment in agriculture. Reduced investment in the agricultural sector has been an area of concern during the 80s. It is believed that this is largely due to the higher budgetary allocations for meeting the expenditure on subsidies.
21. Our feeling is that one more commodity which received positive support during this period could be groundnut. Since, the prices of groundnut in the international market for this period are not available, we have been forced to use 1989 prices due to paucity of information. Due to this reason the AMS turns out to be negative.
22. Farm harvest prices are the weighted farm harvest prices. The weights being shares in total production of a crop in those states for which farm harvest prices are available.
23. The year 1987-88 was a drought year, therefore, farm harvest prices of various crops had abnormally gone up. Probably due to this reason there is such a large difference in these estimates at support prices and farm harvest prices.
24. 1991-92 was again a bad year for coarse cereals. The production of coarse cereals was quite low and because of this reason their prices were very high compared to normal years.
25. These estimates are based on Producer Subsidy Equivalents as estimated by USDA for the period 1981-82 to 1987-88. We do not have access to their AMS for the period 1986-87 to 1988-89, but it is expected the levels of AMS are also quite high.

26. Preliminary estimates by USDA reveal that reductions in export subsidies by developed countries will open up world wheat market for developing countries by 15 to 20 million tonnes.
27. Some studies based on partial and general equilibrium modelling exercises indicate that international prices of agricultural commodities will rise and so will their trade volumes. Also the variability in world prices will reduce. A sample of these models include (i) Tyers and Anderson Mode, (ii) SWOPSIM (Static World Policy Simulation Model), US Department of Agriculture, (iii) IIASA (International Institute of Applied Systems Analysis) Model, and (iv) OECD (Organisation of Economic Co-operation and Development) Model. In a more recent work Brandao and Martin (1993) estimated that international prices increase, but by not very much, from baseline levels (1985-87) e.g., wheat +1 per cent to + 6 per cent; rice - 5 per cent to +2 per cent; coarse grain +2 per cent to +4 per cent; sugar +6 per cent to +12 per cent; oilseeds and pulses +1 to +5 per cent and cotton +1 to +4 per cent. These models vary in their commodity coverage and incorporate different sets of policies and because of these reasons the results of these models also differ, depending upon the assumptions and information they use. But the results are in line with theoretical expectations of trade liberalisation measures.
28. Cotton is a good example of this type of export policy (see Gulati and Sharma, 1994).
29. Some apprehensions have been expressed about the likely impact of such measures on food security of the country (Nayyar and Sen 1994). These apprehensions do not appear to be valid in the present day context, because country has attained self sufficiency in foodgrains production. An inter-ministerial working group set up by the Ministry of Agriculture has found that the current level of foodgrains production exceeds the normative requirements of the population by about 10 per cent (GOI 1994). The group is of the opinion that 180 million tonnes of food grains are adequate to feed the estimated population of one billion in 2000 AD. The report further adds that under nutrition is simply a matter of poors' inadequate access to food.
30. Among oilseeds soyabean perhaps can still survive the impact of changed scenario, because it is more a protein or an oilcake crop than an oil crop. Since the demand for oilcakes is expected to increase both on the international as well as the domestic market, it is quite likely that area devoted to this crop will not decrease.
31. For commodity specific strategies that India should undertake, see NCAER, 1994.

REFERENCES

Anderson, K. and Y. Hayami (1986), *The Political Economy of Agricultural Protection: East Asia in International Perspective.* Allen and Unwin in association with Australia-Japan Research Centre, Australian National University, Sydney.

Bhalla, G. S. and D. S. Tyagi (1989), 'Patterns in Indian Agricultural Development—A District Level Study' Institute for Studies in Industrial Development, New Delhi.

Brandao, Antonia Salzar P. and Will Martin (1993), *Implications of Agricultural Trade Liberalisation for the Developing Countries*, World Bank Working Paper, WPS 1116.

GATT Secretariat (1993), 'An Analysis of the Proposed Uruguay Round Agreement, with Particular Emphasis on Aspects of Interest to Developing Countries', Geneva, November 29.

Government of India, Ministry of Agriculture (1994) : *Report of the Working Group on Foodgrains Requirements by the Year 2000 AD*, Department of Agriculture and Co-operation, April.

Government of India, Ministry of Commerce : *Revised List of Commitments on Agriculture*, New Delhi.

Government of India, Ministry of Finance : *Economic Survey*, various issues.

Goldin, I. and O. Knudsen (eds.) (1990), *Agricultural Trade Liberalisation—Implications for Developing Countries*, OECD and World Bank.

Gulati, Ashok (1990), 'Fertiliser Subsidy : Is the Cultivator Net Subsidised', *Indian Journal of Agricultural Economics*, XLV (1), January-March.

Gulati, Ashok and A. N. Sharma (1992), 'Subsidising Agriculture : A Cross-Country View', *Economic and Political Weekly*, Review of Agriculture, XXXVII (39), September 26.

Gulati, Ashok and Anil Sharma (1994), 'Time to Let the Bales Bail Out the Business', *The Economic Times*, March 7.

——(1994), *Irrigation Cost and Cost Recovery—Case of Minor Irrigation in India*, NCAER-IFPRI Collaborative Study, mimeo.

Islam, N. and A. Valdes (eds.) (1990) : *The GATT, Agriculture and Developing Countries*, International Food Policy Research Institute, Washington, D. C.

Johnson, D. G. (1971) : *World Agriculture in Disarray*, St. Martins Press, New York.

——(1991), *World Agriculture in Disarray*, second edition, St. Martins Press, New York.

Nayyar, Deepak and Abhijit, Sen (1994) : 'International Trade and the Agricultural Sector in India', *Economic and Political Weekly*, pp. 1187-1203, May 14.

Pursell, Gerry and Ashok Gulati (1993) : *Liberalising Indian Agriculture—An Agenda for Reform*, World Bank Working Paper, WPS 1172.

Thomas, P. Sukhpal Singh, Ajit Kanitkar, Sara Ahmed and Micheal E. Johnson (1994), 'Dunkel Text : Implications for Rural Sector', *Economic and Political Weekly*. pp. A42-52, March 26.

Tyers, R. and K. Anderson (1992), *Disarray in World Food Markets : A Quantitative Assessment*, Cambridge University Press, Cambridge, New York.

United States Department of Agriculture (1986) : *World Commodity Markets—Government Intervention and Multilateral Policy Reform*, USDA.

World Trade Centre, Bombay (1994) : *GATT Agreements—Final Text of Uruguay Round*, Academy of Business Studies, New Delhi.

10

Agricultural Pricing, Marketing and International Trade under New Economic Environment

M.S. BHATIA

Agricultural prices in a developing economy are highly influenced by the interaction between producer, consumer and trader groups and their relative effectiveness in influencing Government decision-making. The functions of modern marketing are enormously increased and complicated which, *inter alia,* include to move products from the producer and bring it to the consumer at the right time, in the right form at a reasonable cost and price. The ultimate consumer may be domestic or outside the country and thus marketing is extended to cover international trade. It is because of these linkages, issues relating to agricultural prices, marketing and international trade are closely inter-related and should be discussed together rather than in isolation.

International trade in agriculture is so far being carried on under distorted conditions in which markets were shaped by the "Subsidy War" between the United States and the European Community. The consequence of this trade war was not only influencing these countries but were significant to affect trade in most of the countries including India. The trade practices and market distortions adopted by those countries relating to both price and supply uncertainty in international trade. They also influenced, the markets of most of the countries whether as importers or exporters. The protectionist policy adopted by the major importers were particularly influencing the trade of small countries. The international economic environment is expected to change significantly with the signing of the General Agreement on Tariffs and Trade (GATT) accord and the development of World Trade Organisation (WTO). The protectionist policies followed by various countries have to be changed with rational policies in which subsidy support is to be reduced and quantitative restrictions are to be replaced by moderate tariff restrictions. India has not only signed the GATT Accord, but has also initiated various economic and policy reforms in the recent period to take full benefit of the changed world economic order. Far reaching changes have been made in the trade and exchange rate policies in the past three years. The system of administered discretionary control over imports and exports which was characterised by exceptionally high custom/tariff structure has been abolished. The import duties have been reduced in stages with the maximum duties lowered to 150 per cent in July 1991, 110 per cent in February 1992, 85 per cent in February 1993 and 65 per cent in February 1994. The exchange rate of the rupee is determined by demand and supply conditions in the foreign exchange markets, which is quite stable in the past two years and provides a powerful incentive for exporters. Many restrictions on exports, especially of agricultural exports, have also been removed. All these trade and policy reforms including the GATT Accord are likely to influence prices, marketing and international trade of agricultural commodities in the country. However, it is not yet known how far the new economic environment is likely to influence the parameters of marketing and trade and what are the implications of such reforms. The paper is aimed at examining the issues related to agricultural marketing, prices and international trade under the changed economic environment and the new GATT.

India and World Trade

Before examining the future prospects and implications of the international trade of agriculture in India, it would be important to examine the existing status of its export *vis-a-vis* world trade. Table 10.1 shows the relative share of India in world production and export of principal agricultural commodities.

It would be seen from Table 10.1 that except in a few commodities, like rice, cotton tea, coffee, oilseeds, oilcakes, tobacco and jute, the share of India in total world trade is very small. The share is particularly low in the world export trade of fish, meat and chicken, vegetables and fruits. While India's share in the total world production of many commodities is quite substantial, its share in the export market is relatively very small, particularly of livestock, marine and horticultural

TABLE 10.1 : Share of India in the World Production and Export of Principal Agricultural Commodities, 1992

('000 metric tonnes)

Commodity	*Production*			*Export*		
	World	*India*	*Percentage share of India*	*World*	*India*	*Percentage shae of India*
(1)	(2)	(3)	(4)	(5)	(6)	(7)
Fish production	1,05,125	4,150	3.95	4,87,744	3,807	0.780
Agricultural production	Neg.	Neg.	Neg.	3,57,715	3,198	0.909
Meat fresh + chicken and frozen	1,77,180	2,180	1.23	23,499	112	0.334
Cereals	19.52,224	1,99,585	10.22	38,809	373	0.961
Wheat	5,63,649	55,084	9.77	19,017	3	0.016
Rice	3,52,068	73,372	20.84	5,123	370	7.226
Potato	2,68,492	15,500	5.77	1,468	0.4	0.27
Pulses	57,455	12,850	22.36	2,230	16	0.72
Tomato	70,433	5,000	7.09	1,910	0.2	0.01
Onion	28,233	3,500	12.39	675	55	0.15
Oranges	57,048	1,900	3.33	5,806	12	0.20
Bananas	49,630	7,000	14.10	3,122	0.1	4.09
Apples	40,346	1,200	2.97	2,589	31	0.14
Coffee	5,919	200	3.38	618	11	1.78
Sugar total	1,15,939	14,400	12.43	1,060	11	1.04
Tea	2,479	703	28.36	2,212	370	16.73
Soyabean cake	83,228	2,153	2.50	577	40	6.93
Groundnut cake	9,872	3,444	34.88	92	30	32.15
Tobacco unmanufactured	3,965	538	6.75	5,856	163	2.78
Cotton lint	18,430	2,195	11.91	8,619	330	3.83
Jute fibres	3,135	1,260	40.19	132	30	2.27
Natural rubber	5,304	351	6.60	3,640	8	0.22

Neg. = Negligible

products for which quite substantial surpluses are available. As India is already in the export market for these commodities, there is a need to have proper assessment of the availability of these items for export as well as identification of export markets.

Structure of Export Trade of India

In order to assess the possibility as well as direction of required export trade, it is important to study the existing pattern and structure of export trade of India. The principal agricultural exports of India in selected years since 1960-61 to 1992-93 are given in Table 10.2.

Table 10.2 shows that the major commodities of export from India included fish and fish preparations, oilcakes, tea and mate and rice. These commodities together accounted for about 55.4 per cent of total agricultural exports during 1992-93. Among these, tea is traditionally an export commodity whereas oilcakes, rice and fish preparations are relatively new items of export. The relative share of traditional export commodities of spices, sugar, raw cotton and tea in the total agricultural exports has declined over the years. This is primarily because India is facing tough competition for some of the traditional export items from its neighbouring countries due to its limited world trade. The relatively higher growth on export of some of the new commodities showed that their demand in the international market is high and India is placed at a better comparative advantage in exporting these commodities. An important aspect of agricultural export trend is that it supports the classical theory of economic development and trade which suggests that at initial stages of development, the share of agriculture or primary production in the total exports of the country is high. This is reflected in the export pattern of India over the years which shows that the share of agricultural and allied products in the total exports has decreased from 44.23 per cent in 1960-61 to 30.65 per cent in 1980-81 and then to 17.61 per cent in 1992-93. Despite India having a comparative advantage in the production of some of the commodities like meat and meat preparations, fresh fruits, vegetables, processed foods, etc. their exports are at present very low. Not only that their share in the country's total agricultural exports is very low but India's share in the world trade of these commodities is almost negligible being 0.01 per cent in the case of tomato. 0.003 per cent in bananas and 0.167 per cent in apples. Even in major export items of fish and fish preparations India's share in the world export trade is just 0.80 per cent. The major issue for consideration is to examine the circumstances that has so far been restricting the expansion of export of new items.

Domestic and World Prices

The relative level of domestic and world prices would indicate the export competitiveness[1] and possibility of export and import of

TABLE 10.2 : Principal Agricultural Exports of India

(Quantity in 000 tonnes)
(Value in Rs. crores)

Commodity group	*1960-61*		*1970-71*		*1980-81*		*1990-91*		*1992-93*	
	Quantity	*Value*	*Quantity*	*Value*	*Quantity*	*Value*	*Quantity*	*Value*	*Quantity*	*Value*
(1)	(2)	(3)	(4)	(5)	(6)	(7)	(8)	(9)	(10)	(11)
Coffee	19.7	7.0	32.2	25	87.3	214	86.5	252	141.4	376
Tea and mate	199.2	14.0	119.1	148	229.2	426	199.1	170	168.1	977
Oilcakes	433.8	14.0	878.5	55	886.0	125	2,477.8	609	3,678.9	1,545
Tobacco	47.5	16.0	49.8	33	9.3	141	87.1	263	88.3	474
Cashew Kernels	43.6	19.0	60.0	57	32.3	14	55.5	447	62.7	349
Spices	47.2	1.7	46.9	39	84.2	11	103.3	239	128.7	393
Sugar and molasses	99.6	3.6	473.6	29	970.0	40	191.0	36	485.1	354
Raw cotton	32.6	12.0	32.1	14	131.6	165	374.0	846	65.7	182
Rice	(–)	(–)	32.8	5	726.7	224	505.0	462	1,085.4	976
Fish and fish preparations	19.9	5.0	32.6	31	69.4	21.7	158.9	960	210.8	1,743
Meat and meat preparations	(–)	1.0	(–)	4	(–)	56	(–)	140	(–)	257
Fruits, vegetables and pulses	(–)	6.0	(–)	12	(–)	80	(–)	216	(–)	366
Miscellaneous processed foods (including processed juices and jams)	(–)	(–)	(–)	(–)	(–)	36	(–)	213	(–)	373
Agricultural and allied products	(–)	284.0	(–)	487	(–)	2.057	(–)	6,317	(–)	9,457
All commodities	(–)	642.0	(–)	1,535	(–)	6,711	(–)	32,553	(–)	53,688
Agricultural and allied products as per cent of total exports	(–)	44.23	(–)	31.72	(–)	30.65	(–)	19.4	(–)	17.61

(–) Not available.

agricultural commodities in the country. Relatively low prices in the domestic market indicate that the commodity has comparative advantage in the export of that commodity if the international prices are higher than the domestic prices plus transport and other handling charges (free on board (f.o.b.) price). Against this, if international price plus insurance and freight charges (cost, insurance and freight (c.i.f.) price) is lower than the domestic price, then the country is placed at a disadvantage in its production. However, comparison of domestic prices with the world prices is an exercise fraught with a number of problems. These problems arise from the fact that the quality of the products whose prices are to be compared are very often not sure and known, the transport cost involved is not known and the volatile nature of world prices which vary considerably over time and space does not depict the real situation as they are often highly subsidised. Yet the comparison between the domestic and world prices provides an useful idea about the possibility of export of these commodities. The levels of prices that prevailed in the domestic markets of India and in the international markets for some of the principal agricultural commodities during 1992 are as shown in Table 10.3.

An examination of Table 10.3 shows that the ratio of domestic price to world price during 1992 is significantly lower than that in the case wheat, rice, maize, cotton, jute, tea, coffee, rubber, tobacco, pepper and oilcakes and horticultural products like potato, mango and banana. In all these commodities, India was placed at an advantageous position to export as the world prices were relatively higher than the domestic prices. Against this ratio of prices of edible oils being significantly higher than one showed that India is comparatively at a disadvantageous position in the international trade of edible oils as the world prices are almost 40 per cent of the wholesale prices in India. The ratio of prices of sugar was at variance by 28 per cent from unity which shows that the situation in this case is at margin which has to be watched to make appropriate decision as to whether to export or import these commodities. The export competitiveness examined here is slightly different than that of Nominal Protection Coefficient (NPC)[2] which is used to measure the degree of protection to domestic producers.

Movement of Indian and World Prices

The ratio of domestic to world prices prevailing at a point of time

TABLE 10.3 : Domestic and World Prices of Principal Agricultural Commodities in 1992

(per tonne)

Commodity	*World price*		*Domestic price*		*Domestic/ world price ratio*
	World price (US $)	*CIF price (Rs.)*	*Wholesale price*	*Border price*	
(1)	(2)	(3)	(4)	(5)	(6)
Wheat	151	5,242	2,750*	3,050	0.58
Rice	306	9,730	4,050*	4,350	0.43
Maize	104	3,890	—	—	—
Groundnut oil	609	18,505	36,096	36,300	1.96
Soyabean oil	430	13,322	30,250	30,550	2.29
Sunflower oil	450	13,901	32,000	32,300	2.32
Coconut oil	578	17,897	50,490	53,490	2.99
Copra	381	11,903	33,654	33,900	2.55
Cotton	1,852	54,503	35,495	36,000	0.66
Jute	358	11,236	3,830	4,130	0.37
Sugar	200	6,661	8,208	8,508	1.28
Tea	1,995	58,644	36,763	37,263	0.64
Coffee	1,499	44,280	38,477	38,777	0.87
Rubber	2,188	64,233	24,071	24,371	0.38
Tobacco	3,918	1,14,334	39,668	39,968	0.35
Pepper	2,066	60,700	27,315	27,615	0.45
Banana	564	17,202	2,557	2,957	0.17
Mango	640	19,693	4,500	4,800	0.24
Potato	116	4,228	1,850	2,180	0.52
Soyabean meal	206	6,835	—	—	—

* Minimum support prices.

depicts the situation of comparative advantage as it existed at that point. However, whether India is to continue to enjoy the relative advantage in the prices of these commodities for export would depend on the relative movement of Indian and world prices. If prices in the domestic market in India are rising faster than the world prices, then the export of the commodities becomes relatively less remunerative or unprofitable unless the local currency is devalued through re-adjustment of exchange rate proportionately. The rate of increase in Indian and world prices of major agricultural commodities is shown in Table 10.4.

It would be seen from Table 10.4 that the world prices during the period 1980-92 have shown a decline in most of the commodities except in a few commodities like cotton, jute, tobacco and some horticultural products in whose case there was some marginal increase. Against this, the wholesale prices of various agricultural commodities in India have

TABLE 10.4 : Compound Rates of Growth of Prices of Principal Agricultural Commodities During 1980-92

(per cent per annum)

Sr. No.	Commodity	World prices	Indian prices	
			Wholesale price	Minimum support price
(1)	(2)	(3)	(4)	(5)
1.	Wheat	–1.77	6.91	5.79
2.	Maize	–2.21	7.11	5.23
3.	Rice	–1.99	7.12	6.50
4.	Cotton	5.42	9.46	7.12
5.	Jute	0.57	7.94	7.48
6.	Soyabean	–1.21	10.90	6.10
7.	Groundnut oil	–1.23	9.98	NC
8.	Tea	–1.20	9.60	NC
9.	Coffee	–6.05	10.41	NC
10.	Sugar	–2.91	5.32	6.92*
11.	Pepper	–0.27	10.36	NC
12.	Banana	3.45	8.55	NC
13.	Tobacco	0.74	7.20	5.61

NC = Not covered under minimum support prices.
* Relates to sugarcane

increased in the range of 7 to 10.5 per cent per annum during the same period except sugar in which case that increase is relatively less. The minimum support prices (MSPs) of agricultural commodities have also shown an increase of almost similar magnitude in the range of 8 to 10 per cent per annum during the period 1980-92. Relatively higher increase in the prices of agricultural commodities in the domestic market than in the world market affects the export of these commodities as it reduces the competitiveness and profitability for exporters. However, the competitiveness and profitability of agricultural commodities was projected by correcting the exchange rate through devaluation of the Indian rupee which has shown a growth rate of –8.5 per cent per annum during the period 1981 to 1992. The dollar has shown appreciation with respect to the rupee at the rate of 9.4 per cent per annum. This shows that the adjustment of the exchange rate through devaluation of the Indian rupee has been able to offset the adverse effect of higher increase in their domestic prices for export. However, the decline in the world prices of some of the major commodities and continued higher increase in the domestic prices adversely influence the future prospects of agricultural exports.

Prospects of Future Prices

The most important variables that will decide the future of the country's export performance are the relative increase in the domestic to world prices and competitiveness of the country's exchange rate. If the past is any indicator of the future, it would be appropriate to assume that the world prices of most of the agricultural commodities in the future are not likely to show any significant upward trend. Instead of declining trend of the past, it may show a relatively stable position in the future. Assuming the world prices as stable and Indian domestic market prices continued to show the same increase as observed during the period 1980-92, the emerging prices scenario for Indian and world prices for 1997 and 2000 A.D. appears to be disturbing as the ratio of the Indian to world prices is likely to increase sharply. The ratio which was less than one in most of the commodities in 1992 is likely to become positive and much greater than one which could render most of the agricultural commodity/products uncompetitive and non-profitable to export in the world market. The crucial factor that will decide the country's future export performance, at least in the short run, is the competitiveness in the country's exchange rate. Unless the required adjustments in the exchange rate of the Indian rupee with respect to major currencies are made, the higher rate of inflation as compared to increase in world prices may render the export of agricultural commodities unprofitable and thus affect future export performance adversely.

GATT Accord on Agriculture and Export

India is one of the 120 member countries who have signed the GATT Accord in December 1993. The Dunkel Draft Treaty commonly referred to as the DDT which has been the basis of agreement is a most debated issue but no definite consensus has emerged on this. It would, therefore, be desirable to study what the GATT Accord is and what would be its implications for international trade in agriculture for India.

GATT Accord on Agriculture

The long-term objective of the GATT Accord on agriculture is to establish a fair and market oriented agricultural trading system by initiating a reform process through negotiation, commitment on support and protection and establishment of strengthened and more operational effective GATT rules and discipline. The objective is to provide for substantial progressive reduction in agricultural support and protection sustained over an agreed period of time resulting in corrective and preventing restrictions and distortions in the world agricultural market. Thus it provides for achieving specific commitment in the area of domestic support, make access, export competitions and on sanitary and phyto- sanitary issues. The agreement also provides for greater liberalization by developed countries particularly for trade in tropical agricultural products to provide better access to developing countries. It also provides special provisions for food security and environment and also treatment to developing countries.

There are four major provisions of the GATT Accord relating to agriculture, viz., (1) reduction in domestic support, (2) market access, (3) Trade Related Intellectual Property Rights (TRIPS) and (4) Sanitary and Phyto-Sanitary Provisions. Implications of these provisions to Indian agriculture particularly with respect to trade are discussed hereunder.

Reduction in Domestic Support

The GATT Accord provides that all domestic support in favour of agricultural producers with the exception of measures that have been exempted from reduction, shall be reduced from the years 1995 to the year 2001 by 20 per cent if the support provided in the base period of 1986 to 1988 is over and above the limits fixed for different nations. It has been provided that there will be no requirement to reduce support if the Aggregate Measure of Support (AMS) and/or for Product Specific Support is below 5 per cent of total value of production for developed countries. The *de minimis* level for developing countries like India has been set at 10 per cent. The reduction commitments are to be expressed and implemented in terms of AMS or in terms of equivalent commitments, where the calculation of AMS is not practical and have to be implemented in equal instalments. The AMS has to be calculated separately for each basic product taking into account all product specific support and for the total agricultural production, taking into account non-product specific support.

While calculating and reducing the domestic support, the measures under Government service programmes not involving direct payment to producers or processors such as research programmes, pest and disease control measures, training, extension and advisory services, inspection services for health safety, grading and standardisation, marketing, infrastructural services, public stock holding for food security, domestic food aid, payment for relief for natural diseases, structural adjustment assistance provided through producers retirement programmes and investment aid are not to be considered.

Aggregate Measure of Support

In order to examine the implications of reduction in domestic support for India, the Government has calculated AMS for the non-product specific and product specific subsidies. While calculating non-product specific subsidies it was observed that the support provided under area development programmes like drought prone area programme, dry farming, desert development programme, soil conservation, integrated watershed management and other programmes like crop insurance, assistance to small and marginal farmers, research, extension, etc., are generally under the exempted category and may be excluded. Input support programmes, however, were the major items which require adequate scrutiny. It was felt that under input support programmes also, the supply of agricultural implements at subsidised prices falls within investment subsidies generally available to agriculture which are exempted from reduction commitments in respect of developing countries. Subsidy inherent in supply of seeds is very small or negligible in India. Thus the major input items which fall in support measures are (1) fertilizes, (2) agricultural credit, (3) electricity, and (4) water for irrigation supplied to farmers at subsidised rates. Non-product specific AMS as calculated on these items by the Government for reference base period of 1986-88 for submission to the GATT is shown in Table 10.5.

It is seen from Table 10.5 that total subsidy support provided to Indian agriculture in the reference base period is estimated at 2.9 per cent of the total value of agricultural output. The GATT Agreement also provides that low income and resource poor farmers would be exempted from subsidy reduction commitments. It will not be inappropriate to assume that all farmers having holding under two hectares (small and marginal) in India are covered by the exemption.

As these farmers account for about 30 per cent of the total area and rebate is given to total subsidy for these farmers, thus the AMS for agriculture will be about 2 per cent only. This shows that in the near future and medium-term, India is unlikely to be required to take any commitments to reduce the generally available subsidies in agriculture.

TABLE 10.5 : Non-Product Specific Aggregate Measure of Support (AMS) in Agriculture in India during 1986-88

(Rs. crores)

Item of support	*1986-87*	*1987-88*	*1988-89*	*Average*
(1)	(2)	(3)	(4)	(5)
Fertiliser	65.52	512.22	1,777.69	741.00
Electricity	1,089.84	1,735.23	1,924.51	1,582.00
Credit	73.42	112.84	106.56	97.60
Water for irrigation	26.17	26.72	27.56	26.80
Total	1,119.91	2,387.01	3,836.32	2,447.40
Value of agricultural output	72,990.00	80,932.00	1,02,028.00	85,317.00
Aggregate measure of support in per cent	1.53	2.95	3.76	2.90

Product Specific Support

There are three programmes which provide product specific support, viz., crop oriented programmes for increase in production, consumer subsidies and market price support. The first two programmes are not covered under support reduction commitments. Therefore, the only item to be considered for product specific support is the market price support. The new GATT provides that market price support shall be calculated using th gap between fixed external reference price and the applied administered prices multiplied by the quantity of production eligible to receive the applied administered price. The budgetary payment made to maintain this gap such as buying and storage cost shall not be included in AMS. The fixed external reference price shall be based on the years 1986 to 1988 and shall generally be the average f.o.b. unit value for the product concerned in the net exporting country and the average c.i.f. unit value of the product concerned in a net importing country in a base period. In India administered price support is being provided to 21 agricultural commodities, viz., paddy, wheat, jowar, bajra, maize, ragi, barley, gram, tur, urad, moong, groundnut, rapeseed and mustard, soyabean, sunflower, safflower, cotton, jute, tobacco and copra for which the MPSs are fixed and that of sugarcane for which statutory minimum prices are announced. As per pricing policy of the Government, the entire produce that is offered for sale in market is to be purchased at the MSPs. Hence, the entire marketable surplus quantities of various commodities and not the total production, is eligible to receive the price support. Taking into consideration the average annual production and marketable surplus ratios[3] of different commodities during the reference base period of 1986-88, the value of price support of various commodities and its proportion of total value of eligible production at external reference price were estimated and these are presented in Table 10.6.

TABLE 10.6 : Production Specific Market Price Support, 1986-88

(price per tonnes)

Commodity	*World price (US $)*	*CIF price[4] (Rs.)*	*Minimum support price per tonne (Rs.)*	*Gap or support per tonne (Rs.)*	*Annual average produc-tion (million tonnes)*	*Eligible produc-tion for support (million tonnes)*	*Value of support (Rs. crores)*
(1)	(2)	(3)	(4)	(5)	(6)	(7)	(8)
Paddy	247.2	3,648	2,280	–1,368	63.08	28.28	–3,869
Jowar	110.0	1,842	1,373	–469	10.52	3.45	–162
Bajara	106.0	1,790	1,373	–417	5.20	1.73	–72
Maize	106.0	1,790	1,373	–425	7.18	1.95	–83
Ragi	—	—	—	—	—	—	—
Wheat	125.3	2,042	1,740	–310	48.20	25.28	–784
Barley	68.3	1,293	1,383	90	1.67	0.45	4
Gram	—	4,297	2,983	–1,314	4.67	1.96	–258
Urad	—	4,124	3,350	–724	1.24	0.76	–55
Moong	—	4,498	3,350	–1,148	1.05	0.63	–72
Tur	—	5,332	3,350	–1,982	2.36	1.20	–238
Groundnut in shell	217.5	3,139	3,967	828	7.13	5.87	486
Rapeseed	219.3	3,281	4,350	1,069	2.88	2.67	285
Sunflower	263.3	3,860	3,967	107	0.44	0.40	4
Soyabean	242.6	3,587	2,633	–954	1.11	1.00	–95
Cotton	1,368.0	18,398	16,899	–1,499	1.25	1.22	–183
Jute	277.1	4,041	2,384	–1,657	1.19	1.19	–197
Tobacco	3,506.3	46,537	12,330	–34,207	0.44	0.37	–1,265
Copra	301.5	4,362	14,000	9,638	7.39	3.26	3,161

It would be seen from Table 10.6 that market support for most of the agricultural commodities in India during the reference base period of 1986-88 was negative except for oilseeds crops of groundnut, rapeseed, sunflower and copra and marginal for barley for which the AMS was positive. Sugarcane is a special case as it is not an internationally traded commodity. However, if we compare the average c.i.f. price of sugar with statutory minimum price of sugarcane along with conversion cost of sugarcane to sugar or the domestic price of sugar, it is observed that the AMS was marginally positive for sugarcane in the reference period. The aggregate value of price support provided to various commodities[5] adds up to (–) Rs. 3,393 crores which is 12.1 per cent of the total value of eligible annual production of Rs. 28,007 crores as c.i.f. prices. It is, therefore, evident that, on the whole, Indian agriculture is being taxed rather than being subsidised.

According to Dunkel proposals, the product specific and non-product specific AMS were to be considered separately. However, as per the new GATT 1994, the two can be clubbed together. Though our non-product specific AMS is also within the permissible limits, after clubbing it with product specific support, our consolidated AMS becomes negative. India is thus not required to undertake any reduction commitments with regard to domestic support.

Market Access

One of the most important articles of the new GATT is the provision of the market access. The basic objective of this clause seems to facilitate the movement of agricultural products from developed surplus countries to the relatively less developed and comparatively disadvantaged and food deficit countries. According to the provision of market access, the member countries have to regulate their tariffs and quantitative restrictions in such a way that they have to allow certain access in the domestic market to the exporting countries, which is currently assessed at 3 per cent of the base period domestic consumption in the first year of the implementation period and thereafter, increased by 0.4 per cent of the corresponding domestic consumption in the base period per year for the remainder of the implementation period. In case the member countries have already been providing access (importing), then it has to maintain the minimum access opportunity already in effect at such times and increase the access opportunities by 0.4 per cent of the corresponding domestic consumption in the base period per year for the remainder of the implementation period. The minimum access thus arrived in the final year of the implementation period shall be required to be maintained thereafter. This shows that once the market access commitments have been made through the reduction of tariffs, the member countries would not be allowed to revert or to undertake any measures except only under certain special safeguards/provisions provided in the Treaty. The concessions under which the provisions of the article may be invoked are if—(i) the volume of imports of that product extending the custom territory of the participant countries granting the concession during any year exceeds a trigger level equal to 125 per cent of the corresponding average quantity during the three preceding years for which data are available or 125 per cent of the minimum access opportunity, whichever is greater, or but not concurrently, and (ii) the prices at which import of that product may enter the custom territory during the concession as determined on the basis of c.i.f. import price expressed in terms of its domestic currency, falls below a trigger price equal to the average 1986 to 1988 reference price for the product concerned.

The requirements of the Dunkel Treaty are to remove quantitative restrictions and allow access to domestic market for agricultural imports with certain exemptions to those countries having balance of payments problems like that in India. It is being argued that since India is having balance of payments difficulties, the provisions of the market access clause would not influence the import restrictions being used by the country. There are two important aspects of market access clause: (a) current market access clause, and (b) market access as per Dunkel Treaty for the future. In regard to (a) it is mentioned that the current access opportunities on terms equivalent to those existing shall be maintained as part of the tariffication process. The current access opportunities shall be no less than the average annual import of quantities for the years 1986 to 1988. It is not very clear whether a country facing balance of payments difficulties will be exempted from the requirement of giving minimum and current market access even if it is allowed to maintain quantitative restrictions due to balance of payments difficulties. Since India imported large quantities of edible oils as also sugar during the base year of 1986-88, the clause relating to maintaining current market access can drain millions of foreign exchange whenever international prices are significantly lower than the domestic market despite our having substantially improved production. The import of such a magnitude may result in a fall in domestic prices and adversely affect the producers and production in the country thereafter.

In regard to market access for the future, apart from the fact that it is the International Monetary Fund (IMF) which decides according to the GATT whether member countries are having balance of payments problem or not, the concerned provisions in the Treaty are so made that it will be difficult for the country to use these provisions to continue without the policy of quantitative restrictions on imports. If the policy is changed from quantitative restrictions to tariffication for regulation of imports it would be essential to examine the implications of these shifts in the policies. This would require the likely reduction of subsidies by various countries as also its impact on international prices. The implications of such a shift in policies, therefore, at this stage are difficult to explain. However, as Dunkel Treaty provides for only 20 per cent reduction of subsidies during the implementation period 1995 to 2001 and the existing level of subsidies in some of the developed countries including the U.S.A., Canada, European Economic Community is so high that even after the proposed reduction in subsidies, it would not be inappropriate to assume that it would not have any significant impact on international prices. The going trend in the foreign and domestic market prices would, therefore, indicate the likely impact of the market access clause of the Dunkel Treaty.

Trade Related Intellectual Property Rights

The new GATT provides for effective and adequate protection of property rights but at the same time ensuring that measures and procedure to enforce intellectual property rights do not become barriers to legitimate trade. The term 'Intellectual Property' refers to all categories of intellectual property including copyright and related rights, trade mark, patents, etc. According to the provisions of the GATT, patent shall be available for any invention weather products or process in any of the fields of technology, provided that they are new, involve inventive step and are capable for industrial application. It provides that members may exclude patentability inventions, which is necessary to protect human, animal or plant life or health or to avoid serious prejudice to the environment, provided that such exclusion is not made merely because the exploitation is prohibited by domestic law. It also provides that members may exclude from patentability of (a) diagnostic, therapeutic and surgical methods for the treatment of humans or animals, and (b) plants and animals other than the micro organisms and essentially biological processes for the protection of plant or animals other than non-biological and micro-biological processes. It further states that, however, the members shall provide for the protection of plant varieties either by patents or by an effective *sui generis* system or by any combination thereof.

The major provisions of the TRIPS which have implications for Indian agriculture are those influencing agricultural chemicals like pesticides or those affecting plant varieties. As per the Agreement, the country has the option to protect its own plant varieties. The Government is preparing a *sui generis* legislation through which it is proposed to protect the traditional and 'inalienable' rights of the farmers to save, use, exchange, share and sell seeds on a non-commercial basis. The proposed legislation would allow researchers the right to free access to protected material or breed, to develop new variety without the authorisation of original breeder.

This shows that the *sui generis* system that the country is proposed to enact would fully safeguard the farmers' and researchers' privileges and the provisions of the TRIPS of the new GATT in no way will have an adverse impact on our agriculture.

Sanitary and Phyto-Sanitary Measures

'The New GATT provides that no membership will be permitted for adopting and permit the measures necessary to protect human, animal or plant life or health, subject to the requirement that they are not applied in a manner which would constitute a measure of arbitrary or unjustifiable discrimination between members where the same conditions prevail or disguised restrictions on international trade. The measures shall be based on international standards, guidelines and recommendations where they exist except or otherwise provided in this Agreement. In order to have transparency members shall notify changes in their sanitary and phyto-sanitary measures and provide information on these measures as per the provisions of the Agreement.

As the implementation of the sanitary and phyto-sanitary measures require adequate technical expertise, technical infrastructure or resources, the absence of these facilities may have adverse impact on exports. Although, the Agreement provides that the least developed country members may delay application of the provisions of the Agreement for a period of five years following the date of entry into force of Multilateral Trade Organisation (MTO) with respect to their sanitary and phyto-sanitary measures. The other developing countries have been given a grace period of two years. However, these provisions are to be implemented in such a way that it should not affect the export of agricultural products from the developing countries where facilities and infrastructure for such standardisation are not adequate.

New GATT and Growth in Export

The new GATT 1994 through reduction in tariff and increase in market access is likely to provide upward push to the rate of growth of world export trade. The past rate of growth of world export has been around 6.5 per cent per annum during the eighties. The rate of growth of export trade has been different in developing and developed countries as would be seen from Table 10.7.

TABLE 10.7 : Growth of World Export Trade

(million US dollars)

Region	*1970*	*1981*	*1990*	*1992*	*Growth rate per cent per annum during 1981-92*
(1)	(2)	(3)	(4)	(5)	(6)
World	3,15,100	19,73,800	34,87,851	38,46,257	6.25
Development market economy	2,23,500	12,28,100	24,39,299	28,18,070	7.84
Developing countries	57,900	5,66,000	8,00,156	10,28,186	5.58
Asia	25,590	3,59,580	8,85,390	11,67,953	11.30
India	2,026	8,373	18,215	19,563	8.02
India's share in per cent to					
(a) World	0.64	0.42	0.52	0.51	—
(b) Asia	7.92	2.32	2.05	1.67	—

An examination of Table 10.7 shows that the rate of growth of export trade in India was relatively low as compared to that of the world during the seventies as its share in the total export was reduced in 1981. During the eighties, the rate of growth of export trade of India was higher than that of the world as its share in the world export during 1990 has shown improvement over that in 1981. The rate of growth of export trade in India has been very much lower than that of Asia as a whole as India's share as per cent to total export from Asia has decreased over the years. The average rate of growth of export trade during the period 1981 to 1992 in the world was observed as 6.25 per cent per annum whereas it was 11.3 and 8.02 per cent per annum in Asia and India respectively. With the new GATT, the rate of growth of world export trade is likely to go up to about 7 per cent per annum and the total world export which was about US $ 3,846 billion in 1992 is expected to reach about US $ 7,070 billion in the year 2001.

Gains from New GATT

It would be difficult to quantify the potential gains to the world export trade and income resulting from liberalisation and greater predictability of trading environment in the wake of the new GATT. New trading environment would lead to opening of markets across the globe and thus would provide greater opportunities for diversification of exports and markets that are essential to sustained growth of export trade.

While gains from the new GATT would be undoubtedly considerable, it would not be spread evenly across the member countries. The countries which were more flexible to capitalise new market opportunities would benefit most. If India can plan its export strategy properly and diversify the export trade in the future, it can easily raise its rate of growth of export trade from the existing level of 8 per cent to about 10 per cent per annum from 1995 onwards. Thus increased rate of export trade would result in the annual export of about US $ 42 billion which otherwise at the existing trend is likely to be about only US $ 38 billion in the year 2001. Thus India is likely to gain in export trade by about US $ 4 billion annually by the turn of the century from the new GATT.

Future Strategy

In order to take the maximum benefits from the new world trade environment, it would be essential to properly assess the available export surpluses of various commodities in the country and to give greater emphasis to production strategy for the commodities for which the country has greater comparative advantage. In view of the increasing population pressure, the country may not have large surpluses of average quality of foodgrains in the long run. Therefore, focus will have to be on the high value products like Bâsmati and other superior varieties of rice and value added processed products. The rate of growth of production of fruits and vegetables in the country has been around 4.5 per cent per annum which is higher than the rate of growth of population of about 2.11 per cent per annum during the period 1981-91, which may provide good export surpluses of these products. If property handled, stored and processed, export of fruits and vegetables may be expanded significantly as the country has significant cost advantage in its export.

The new technology in the case of cereals has helped in reducing its cost of production in real terms. There is a need to improve the technology for other agricultural commodities, particularly in the case of fruits and vegetables which would help in reducing the cost of production, as also for processing and production of value added products. It may also be essential to create the required infrastructure through appropriate investment both in public and private sectors which would facilitate to provide storage, packing and marketing facilities. In view of the higher rate of increase in prices in the domestic market, it would be essential to make appropriate adjustments in the exchange rates with major currencies to sustain the growth in export.

Note : Reproduced from *Indian Journal of Economics*, Vol. 49, No. 3, July-September, 1994 with kind permission of the publisher.

NOTES

1. The export competitiveness was measured as ratio of domestic border price and c.i.f. price expressed in local country. This measure is slightly different than the Nominal Protection Coefficient in the sense that it brings both international and domestic prices at border point in the same currency.
2. Nominal Protection Coefficient (NPC) is the ratio of domestic prices (preferably farm harvest price of MSP) to border price, i.e. c.i.f. price convened to domestic currency at official exchange rate. It is expressed as:

$$\text{NPC} = \frac{\text{Actual farm gate price}}{\text{C.I.F. price in domestic currency}}$$

3. The marketable surplus ratios for estimation of eligible production for market price support were used as estimated in the Report of the Sub Group on *Estimation of Marketable Ratios of Agricultural Commodities*, Government of India.
4. C.I.F. prices were estimated by adjusting international prices in the reference year

with insurance and freight charges as obtained from the *FAO Outlook Report* and local shipping companies. For gram, *urad, moong* and tur, C.I.F. prices were estimated on the basis of unit value of imported quantities in the reference period.

5. *Ragi* and safflower could not be considered for estimation of product specific price support as international reference price for these commodities were not available. These commodities are, however, very minor in terms of production and support.

REFERENCES

Bhagwati, Jagdish and T.N. Srinivasan (1993), *India's Economic Reforms,* Ministry of Finance Press, New Delhi, July.

Food and Agriculture Organisation of the United Nations (FAO) (1987), Agricultural Price Policy : Issues and Proposals, FAO Economic and Social Development Series No. 42, Rome.

FAO (1993), *FAO Quarterly Bullein of Statistics,* Vol. 6, No. 4.

FAO (1993), *FAO Production Year Book,* Vol. 46, Statistical Series No. 112, Rome.

FAO (1993), *FAO Trade Year Book,* Vol. 46, Statistical Series No. 115, Rome.

FAO (1993), *Selected Indicators of Food and Agricultural Development in the Asia Pacific Region*, 1982-92, Regional Office for Asia and Pacific (RAPA), RAPA Publication, No. 1993/26, Bangkok.

FAO (1994), ''Global Information and Early Warming System on Food and Agriculture'', *Food Outlook,* No. 5/6, June.

Ganesan, A.V. (1994), ''Uruguay Accord: An Opportunity', Seminar on GATT Accord: India's Strategic Response, Indian Institute of Foreign Trade, New Delhi, March.

GATT Secretariat (1993), Final Act Embodying the Results of the Uruguay Round of Multilateral Trade Negotiation, Trade Negotiation Committee, Geneva, December.

Government of India (1984), Report of the Sub Group on Education of Marketable Surplus Ratios of Agricultural Commodities, Ministry of Agriculture, New Delhi, November.

Government of India (1993), ''Economic Reforms—Two Years After and the Task Ahead'', Discussion Paper, Department of Economic Affairs, Ministry of Finance, New Delhi.

Government of India (1994), Economic Survey, 1993-94, Ministry of Finance, Economic Division, New Delhi.

Nayyar, Deepak and Abhijit Sen (1994), ''International Trade and Agricultural Sector in India'', In G.S. Bhalla (ed.), *Economic Liberalisation and Indian Agriculture,* Institute for Studies in Industrial Development, New Delhi.

Petit, Michel (1988), ''Agricultural Trade Confrontation between the United States and the European Community: A challenge to Our Profession'', *Agricultural Economics, (Journal of the International Association of Agricultural Economics),* Vol. 2 No. 3, November.

Salvtore, Dominick (1984), *Theory and Problems of International Economics,* Second Edition, McGraw Hill Book Company, New York.

United Nations (1994), *UNCTAD Year Book*, United Nations Conference on Trade and Development, Geneva.

United States, Department of Agriculture (1987), *The International Trade Research Consortium, Agricultural Trade Modelling, The State of Practice and Research Issues,* Economic Research Services, USDA Report No. AGES-861215, Washington, D.C. USA.

World Bank (1994), *World Developing Report 1994: Infrastructure for Development*, Oxford University Press, Delhi, May.

11

Farmers in Market Economy

Would Farmers Gain through Liberalisation?

V.M. RAO

I. INTRODUCTION

Liberalisation is expected to provide a powerful thrust to growth and modernisation of the economy including agriculture. Even some of the farmer-spokesmen themselves appear to be optimistic about liberalisation though the reforms in this direction would involve eventual phasing out of subsidies for the farmers and exposing agriculture much more thoroughly then now to the market forces—domestic and world. It is indeed likely that there are parts of agriculture such as the green revolution areas or the newly emerging corporate activities in the fields of horticulture, floriculture, etc., for which liberalisation could be a boon (Rao, 1994). This is also true of the impact of liberalisation on the overall growth and diversification of agriculture. However, as Dantwala reminded us many years back, given the skewed distribution of land and other agricultural resources, a situation which favours better and efficient use of resources concentrated in 'large farms' would not necessarily benefit the farming community consisting predominantly of 'small farmers' (Dantwala, 1959). The purpose of this paper is to take a look at the prospects of such farmers in the new policy regime for agriculture focused on increased marketisation value addition and exports.

The context in which the farmer operates now is different in many major respects from that which prevailed during the 1950s when Dantwala drew the distinction between the 'small farmer' and the 'small farm'. Technological change has transformed some parts of agriculture and is likely to spread further though its pace remains both sow and unsteady. The farmer's access to irrigation, modern inputs, credit and extension has improved. He has achieved some progress in reducing his dependence on village trader and money lender by shifting over to organised markets and institutions. The point is that the farmer (typically, small, poor and unorganised) has now opportunities open to him to increase his output as also to widen his contacts with the markets and the world outside his village. However, it would not be easy for him to make the transition from the survival-oriented traditional attitude and modes of behaviour which still prevail among a large number of farmers to the ruthlessly competitive environment of the modern markets dominated by powerful organised groups. Obviously, the farmer would remain vulnerable in the new environment until he acquires the capacity to withstand its pressures.

The gains of a farmer from liberalisation would depend on: (a) the increase in production achieved by him which would be determined by his access to resources and his efficiency and skills as a producer, and (b) the returns obtained by him in the market determined by his ability to comprehend and respond to market signals and to get a 'fair' price.

There is extensive evidence in the areas where agricultural growth has taken place in India that not only does the small farmer contribute to growth but he is also seen to achieve higher productivity of land than his larger counterpart. In fact, the relatively better performance of the small farmer in this resepct is often cited as an argument in favour of radial programme for redistribution of land. On the other hand, it is also generally observed in the same areas that the economic status and the lifestyle of a small farmer do not differ much from that of the agricultural labourer. An inference which could reconcile these two apparently contradictory findings is that, even when he achieves increases in production, the small farmer is not able to bring about steady improvement in his economic conditions and become upwardly mobile. The purpose of this paper is to present some illustrative evidence to suggest that the prevailing market environment could be one of the factors hindering the progress of the small farmer. It is easy to see that the tilt towards liberalisation which is the objective of the current policy reforms could lead to a serious situation in the agricultural sector if it promotes growth but allows only a small part of it to reach the numerically large group of small farmers.

The plan of the paper is as follows, Section II takes a look at the monthly Wholesale Price Index (WPI) data for the 1980s for selected commodities. We believe that the trends and the month by month changes in the prices of commoditie give useful insights into the working of the commodity markets. Our attempt is to bring together clues reflecting the difficulties that the typical farmers is likely to face in operating in the commodity markets. The section also contains a brief assessment of the government interventions to help farmers and the scope for such interventions in the emerging phase of liberalisation. The choice of the eighties for this exercise helps us in assessing the situation as it existed just prior to the launching of reforms for liberalisation in the early 1990s. Liberalisation is expected to give a stimulus to agriculture through increases in value addition activities and exports. Value, addition and exports increase the 'distance' between the farmer and the final consumer of his produce by lengthening the chain of markets intervening between the two. More important, the gains of the farmer from value addition and exports would depend on the interaction between the relatively competitive commodity markets, on the one hand, and the markets for manufactures and exports, which tend to be oligopolistic, collusive and heavily protected, on the other. Section III considers the implications of this situation for the farmer. The final section enumerates for discussion the main propositions emerging from the findings of the paper.

II. COMMODITY MARKETS

The monthly WPI data used in this section cover the period of little over a decade from April 1982 to September 1992. This gives 126 monthly observations and it is hoped that the results based on them would reflect well the features of the commodity markets to which they relate. Nine major crops have been selected for analysis. Crops are grown in agro-climatic environments ranging from relatively stable and favourable to adverse and subject to large fluctuations. The crops also differ in demand conditions and the attention they receive from the government. Our attempt has been to select crops which could bring out the wide range of market situations prevailing in agriculture.

Table 11.1 presents the growth rates in WPI (per cent per month). Column (2) shows the growth rates in nominal WPI and column (3) for WPI deflated by WPI of 'All commodities'. If we take the table, particularly column (3), as indicating the signals provided by the market, the favoured crops were pulses, oilseeds and cotton as against cereals. Do the prices prevailing in the market and changes in them over time convey the same signals? Given our data, this can be checked by comparing the frequency and magnitude of positive month-to-month changes in prices with the frequency and magnitude of negative changes. Table 11.2 shows, for both series, the percentage of number of positive changes to total month-to-month changes and the percentage of sum of magnitude of positive changes to the sum of magnitude of all changes. The sum of magnitude of all changes, positive and negative, divided by the total number of month-to-month changes gives the Average Absolute Change (AAC) which serves as an index of amplitude of fluctuations.

TABLE 11.1 : Growth Rates of Prices during April 1982 to September 1992

(per cent per annum)

Crop (1)	*Nominal WPI (2)*	*Deflated WPI (3)*
Rice	0.57	–0.04
Wheat	0.58	–0.04
Cotton	0.77	0.15
Groundnut	0.78	0.16
Rapeseed-mustard	0.72	0.11
Jowar	0.70	0.08
Bajra	0.62	0.01
Gram	0.57	0.26
Tur	0.92	0.31

Source: Based on data compiled from the Office of the Economic Adviser, Ministry of Industry, Government of India, New Delhi.

A clear indication given by Table 11.2, particularly by the changes occurring in the relative prices, i.e. the deflated WPI, is that an observer of price changes is likely to notice only the fluctuations in prices rather than the trend concealed by fluctuations. It can be seen that, in terms of both frequency and magnitude, the positive and the negative changes have about the same weight in the total changes occurring in the deflated WPI of the selected crops. As regards nominal WPI, the predominance of positive changes stands out clearly in the case of WPI of 'All commodities' with positive changes according for 83 per cent of frequency and over 90 per cent of magnitude. However, similar clarity is not observed in the case of selected crops, excepting rice in whose case the positive changes in nominal WPI have a weight of three-fourths. It is also

TABLE 11.2 : Month to Month changes in Prices During April 1982 to September 1992

	Normal WPI			*Deflated WPI*		
	Percentage of positive changes			*Percentage of positive changes*		
Crop	*Average absolute change (AAC)*	*Frequency*	*Magnitude*	*Average absolute change (AAC)*	*Frequency*	*Magnitude*
(1)	(2)	(3)	(4)	(5)	(6)	(7)
Rice	2.43	76.8	74.9	1.28	51.2	52.6
Wheat	4.66	65.6	60.4	2.88	54.4	49.2
Cotton	5.32	56.0	60.0	3.26	48.8	51.5
Groundnut	5.41	59.2	62.0	3.35	52.8	52.7
Rapeseed-mustard	6.39	57.6	59.0	3.93	53.6	51.9
Jowar	3.53	64.8	68.7	2.19	56.0	55.1
Bajra	4.52	63.2	60.9	2.91	54.4	50.4
Gram	5.49	57.6	59.4	3.60	48.0	51.5
Tur	6.22	64.8	63.4	3.81	54.4	53.7
All commodities	1.17	83.2	93.3	—	—	—

Source: Based on data compiled from the Office of the Economic Adviser, Ministry of Industry, Government of India, New Delhi

observed in Table 11.4 that the crops favoured by the growth rates in Table 11.1 cotton, oilseeds and pulses—have higher AAC values, indicating larger amplitude of fluctuations than fluctuations in the cereal prices. It is necessary to remember that the price fluctuations in individual markets are likely to be even more severe than those in the WPI at the nation level. It is indeed true that the farmers respond to price changes as shown, for example, by quick shifts of area between crops which are close substitutes in production when the ratio of their prices changes. The point is whether their responses over time to price changes remain consistent with the signals provided by the national market.

Interestingly, the changes in crop areas, yields and production which actually occurred during the eighties i.e. the period covered by Tables 11.1 and 11.2 appear to have been influenced as much by the government interventions as by market signals. Table 11.3 shows the growth-rates in area, production and yield for two periods, (i) 1967-68 to 1980-81 which may be called the first green revolution decade and (ii) 1980-81 to 1991-92, i.e., the eighties. In the case of oilseeds, there was a marketed increase in the growth rates of area, production and yield during the eighties as compared to the preceding period. It was mentioned above that this crop group was favoured by the market but an even more powerful influence on the performance of oilseeds since the mid-eighties has been the Technology Mission and the market intervention opertions by the public agencies. A heartening feature of the growth in oilseeds production has been that it occurred in the agriculturally backward areas of states like Madhya Pradesh and Rajasthan.

TABLE 11.3 : Growth Rates of Area, Production and Yield

(per cent per annum)

Crop	*1967-68 to 1980-81*			*1980-81 to 1991-92*		
	Area	*Production*	*Yield*	*Area*	*Production*	*Yield*
(1)	(2)	(3)	(4)	(5)	(6)	(7)
Rice	0.76	2.22	1.46	0.55	3.69	3.13
Wheat	2.94	5.64	2.61	0.33	3.59	3.25
Cotton	0.08	2.62	2.54	–0.63	3.24	3.89
Oilseeds	0.24	1.04	0.79	1.07	4.71	3.60
Coarse Cereals	–1.03	0.62	1.67	–1.68	0.06	1.77
Pulses	0.45	–0.39	–0.83	0.08	1.42	1.34

Source: Government of India (1993).

In comparison with oilseeds, the performance of pulses, which received little policy attention, has been quite modest though, possibly in response to the market signals, some improvement in pulses did take place in the eighties while their production and yield had actually decreased during the preceding period in the wake of the surge in the production of wheat. In the case of cotton, its area decreased in the eighties but there was a marked rise in the growth rate of its yield and production between the 'green revolution' decade and the eighties. Like oilseeds, cotton also benefited from policy interventions to help its production as well as marketing though, considering the decline in its area during the eighties, the interaction between the policy support and the favourable market has apparently been much more effective in certain selected areas and not uniformly in all cotton growing regions.

As regards cereals, the market-disfavoured coarse cereals witnessed practically stagnant production and share reduction in area during the eighties. It may be mentioned here that the coarse cereals started losing area in the wake of green revolution and the process gained momentum during the eighties. However, it is worth noting that the coarse cereal yield registered improvement in both the periods though the growth rate was modest. The point is that with more research and more effective price support it may be possible to get a breakthough in coarse cereals which are believed to have considerable market potential yet to be exploited. As regards rice and wheat, with limited scope for expansion in area, the growth in production during the eighties was much more yield-based than was the case in the green revolution decade. Rice and wheat form the foundation of the food security system. Procurement of large quantities of these crops by the public agencies at minimum support prices, provision of subsidised irrigation, fertiliser and other inputs and strong research-cum-extension programmes have had the effect of maintaining favourable conditions for their increased production. The decline in their relative prices during the eighties observed in Table 11.1 was a result of deliberate policies to keep the price of foodgrains low. During the eighties, policy interventions can be regarded as having played a much more dominant role in the case of rice and wheat as compared to their role in the case of other crops.

In the light of experiences gained so far, two broad propositions can be made about the role of government interventions during the emerging phase of liberalisation. The interventions to provide infrastructures, to organise research and extension and to regulate supplies of inputs critical for agricultural growth have been, on the whole, very helpful in promoting agricultural development. Contituation and further strengthening of these interventions would need priority attention in the future too. While the open and implicit subsidies occasioned by them would have to be reviewed and, also, an attempt would have to be made to confine their benefits to areas, groups and crops which, left to themselves, would seriously lag in development, the interventions on the production side would have to remain as an essential component in the agricultural strategy. On the other hand, interventions by the public agencies in the commodity markets seem to have had mixed results. The support provided by them was important in the case of rice and wheat but coarse cereals have benefited little from these interventions. With a few exceptions, public agencies have had meagre success in reducing fluctuations in prices, in disciplining traders and in making markets farmer friendly in their operations.

While even in a market-guided economy the Government would have to intervene in the commodity markets in situations of price collapse or of extreme scarcity of critical commodities, it would be desirable in the long run to leave the markets as free as possible as ineffective interventions could do more harm than good to the farmers besides inhibiting their own individual and collective efforts to cope with markets. It would be far better for the government to find ways to support these efforts of the farmers to become self-reliant rather than supplant them with direct interventions which, more often than not, serve only to expose the shortcomings of public agencies as market operators. The government initiative could go a long way in reducing the fluctuations in production through extension of irrigation, evolution of varieties capable of withstanding adverse conditions and improvements in cultivation practices. Similarly, better roads and market infrastructures, arrangements for wider and quicker dissemination of market intelligence and effective enforcement of provisions for open auctions and fair marketing practices would be of help to the farmer in realising better returns in the market. It is also to be expected that with growing production and increasing participation in the commodity markets, the farmer would gradually improve both his understanding of market operations and ability to make effective use of such understanding in his own decisions and actions.

III. VALUE ADDITION AND EXPORTS

Starting with the base of agricultural commodities, value addition through modern industrial processing and exports help in generating further rounds of increases in income and employment from which the farmer would also benefit through better return to his produce owing to larger demand. It is indeed true that value addition and exports may turn out to have the potential to transform agriculture and to bring about substantial improvement in the economic conditions of the farmer. However, it seems unlikely that this process would either be quick or automatic for the reasons briefly described below.

Having remained for long in a protected environment, the Indian industries are believed to be seriously lagging behind their competitors in both technology and economic efficiency. Even if the policies for liberalisation are seriously pursued, the industries may need several years and sizeable investments to catch up with the competitors. However, if one goes by the current fairly vociferous lobbying for 'level playing field', the industry may instead choose the easier alternative of trying to prolong the era of protection under a different label. Significantly, while liberalisation is now three years old, the industrial recession is still reported to be continuing. The interesting point to note is that the cost of providing protection to industries is probably being borne to a large extent by agriculture and the rural sector. Gulati's studies show that during the period 1986-87 to 1988-89 ''massive amount of taxation'' has been imposed on agriculture "which is contrary to the general impression of huge subsidies which flow to this sector'' (Gulati and Sharma, 1994), it is to be expected that the industrial elities will deploy their considerable skills in lobbying for continuation of this cushion. It may be mentioned here that many developing countries have succumbed to the temptation of getting industrialised quickly through taxation of agriculture and in the process succeeded only in damaging their agriculture (Rao, 1993).

As regards exports, the ''massive amount of taxation'' on agriculture implies that many agricultural commodities have comparative advantage for export. However, two major obstacles in the way need to be noted. First, despite the official pronouncements about liberalisation the policy with respect to agricultural exports continues to operate with the pre-liberalisation perspective of permitting such exports only in years of domestic surpluses. It has been pointed out by many observers that such a policy seriously hampers the efforts of exporters to establish themselves as dependable suppliers of quality products. The recent experience with cotton exports have demonstrated well the damage that an ad hoc policy can cause (Government of India, 1994). Second, it would be possible to form an assessment of the eventual scope for Indian exports only after the new arrangements for multilateral trade get stabilised. Some have argued that being a large country ''India's entry in the world trade would significantly affect world prices.... This is more so because of the glaring imperfections and interventions in agricultual markets'' (Bhalla, 1994; Nayyar and Sen, 1994). According to a recent OECD study, ''international trade in food will likely rise by as much as 100 per cent over the next decade, with most of the increase geneated by OECD member countries. Developing nations thus face a reduction in food self-sufficiency and consequent rises in their food import bills'' (Brown and Goldin, 1992). This assessment goes against the current optimism in India about the scope for export of food grains like rice and wheat. While our rice exports appear to be picking up, the recent policy initiatives for selective wheat exports have yet to show results. Also, given the requiremetns of food security, it is unlikely that the trade in rice and wheat would ever be fully liberalised.

Considering the reasons mentioned above, a plausible hypothesis for the medium-term future is that the increases in value addition activities and exports would be modest in relation to the thrust needed by Indian agriculture as a whole but certain favoured areas, groups and crops may come up fast and make dramatic gains. There might be a replication of what happened during the green revolution; an early phase of handsome rewards to enterprising farmers followed by a broad-basing of the process along with levelling off of returns to late entrants. It is obvious that the farmer groups that we have in mind are likely to join the process in the later phase and, as a result, make only modest gains.

While assessing the gains of the farmer from value addition and exports, it is necessary to take into account a further adverse factor affecting his prospects which is best explained through an illustration. Consider Table 11.4 which shows the behaviour over time of the prices of cotton along with the prices of cotton-based manufacturers. The period covered by the table is September 1990 to March 1993 which was marked by a sharp rise in the Wholesale Price Index (WPI) for 'All Commodities' from 180.9 during September 1990 to 230.7 in September 1992; the price level appears to have stablised after 1992. The point to be noted is the striking contrast observed in the behaviour of cotton prices as compared to that of the prices of cotton manufacturers. The WPI for raw cotton rose from 141.3 during September 1990 to 251.5 in September 1991—an increase of nearly 80 per cent over a year—compared to an increase of 16 per cent in the WPI for 'All commodites' . Almost 90 per cent increase in the WPI of raw cotton occurred after the peak marketing period—September to December —which means the beneficiary of the rise in prices was the trader rather than the farmer. After September 1991, while the general price level continued to increase, there was a fall of nearly 20 per cent in the WPI for raw cotton. Thus, in the context of an inflationary rise in the general price level, the relative price of raw cotton increased sharply but without necessarily benefitting the farmer during the first year and, then, decreased equally sharply over the 18 months from September 1991 to March 1993. It is instructive to compare the behaviour of cotton prices with that of cotton manufacturers. The WPI of handloom cloth increased steadily over the entire period but the increase was only 9 per cent as compared to 30 per cent increase in the general price level. In constrast to the farmer and the handloom weaver, the mills and powerlooms—the organised and modernised parts of textile industry—seem to have benefited from the inflationary phase with the relative prices of their products showing some improvement rather than an erosion. The general point that we would like to suggest based on Table 11.4 is that in a value addition process involving an unorganised producer (in this case, the farmer) who is a 'price-taker' and an organised manufacturer (in this case, the mills and powerlooms) who is a 'price-fixer', the advantage in terms of share in the returns to the process is likely to lie with the latter.

TABLE 11.4 : Wholesale Price Indices of Raw Cotton and Cotton Manufactures (September 1990 to March 1993)

(1981-82 = 100)

Month	*Raw cotton*	*Cotton cloth (Handloom)*	*Cotton cloth (Powerloom)*	*Cotton cloth (Mill)*	*Cotton yarn*	*All commodities*
(1)	(2)	(3)	(4)	(5)	(6)	(7)
September 1990	141.3	201.6	169.6	157.6	187.3	180.9
December 1990	153.3	201.6	174.9	162.1	190.6	186.6
March 1991	171.9	200.6	192.7	166.6	195.4	191.7
June 1991	219.6	202.7	193.1	177.9	240.6	198.4
September 1991	251.5	202.7	200.0	184.1	241.2	210.4
December 1991	242.1	202.7	196.9	185.6	241.2	213.2
March 1992	231.1	213.1	196.9	196.9	258.2	217.7
June 1992	233.1	216.6	209.5	198.3	248.1	224.1
September 1992	217.7	216.6	214.1	199.7	254.2	230.7
December 1992	204.8	216.6	218.9	201.5	254.2	231.4
March 1993	202.3	218.5	226.8	205.8	252.2	232.5

Source: Government of India (1993 b).

A similar unequal relationship between the farmer and the agro-based manufacturer seems to exist in other crops too. Taking the case of groundnut, the following observations of Kahlon and Tyagi are worth nothing: "Given the regime of acute shortage of edible oils, the major problem with oilseed prices... is their volatile character... The violent intra-seasonal fluctuations produce a disparity between the prices realised by the grower and those prevailing in the subsequent months which is often far in excess of the carrying cost of the commodity.... This apart, there are wide spatial differences in the prices of oilseeds and edible oils as between different market centres.... It is the market imperfections which tend to heighten the spatial margins beyond that normally justified by freight and other incidentals... With the price parity between seed and oil, in the case of oilseeds in general and groundnut in particular, moving in favour of the end product...., the grower does not seem to have fully benefited from the high market prices of edible oils'' (Kahlon and Tyagi, 1989).

It is only through group action that the farmer would be able to achieve a measure of parity with the manufacturer-purchaser of his produce belonging to the organised industrial sector. The variant of group action which immediately springs to mind is co-operative marketing which, by and large, had a somewhat indifferent record so far. However, when the farmers notice profitable new crops which have to go through value addition and exports, they often show remarkable capacity to bring into being reasonably well-functioning cooperative marketing bodies. Consider, for example, the case of soyabean cultivation in Rajasthan. Over the period 1982-83 to 1991-92, soyabean area in Rajasthan increased from a little less than 12,000 hectares to over 1,70,000 hectares, production from a little over 6,000 tonnes to 1,30,000 tonnes and yield from around 500 kg per hectare to over 1,100 kg. It is reported that soyabean, which was introduced in Kota district in Rajasthan in 1980-81, ''has revolutionsed the agriculture of Kota and adjoining districts of the state''. A recent field study done in Kota district shows that 83 per cent of soyabean growers sold their produce within the village to the purchase centre set up by the Oilseed Growers Co-opertive Society. A suggestive finding of this study was that both the net price receieved by the soyabean grower and his share in the price paid by the processor were higher for the growers who sold their produce to the purchase centre of the co-operative society within the village as compared to the price and share received by the growers who sold their produce in the Kota market (Agarwal and Sharma, 1994). No generalisation is possible from a single field study but its findings seem to be consistent with the hypothesis that the farmers would have to acquire capacity for group action to benefit fully from the activities in which they have to directly or indirectly interact with parties with much stronger bargaining power and organisational structure.

IV. PROSPECTS FOR FARMERS—SOME PROPOSITIONS

The propositions emerging from the findings of this paper are enumerated below for convenience in discussion. The propositions relate to the farmers forming the major part of the framing community, viz., the farmers who are small, poor and unorganised.

There are indications in areas of agricultural growth that the small farmers to achieve increases in production and in productivity of land. However, the crucial requirements for the farmers to gain from liberalisation is that, in addition to achieving increases in production, they should be in a position to operate in the commodity markets with good understanding of market dynamics and adequate bargaining power. Considering, the prevailing market environment, it is doubtful that the farmers would be able to meet this requirement. In fact, the available evidence seems to show that in areas of agricultural growth the small farmers have an economic status not much better than that of the agricultural labourers.

Liberalisation implies guidance of producer decisions by markets rather than by government interventions. However, the changes in crops which occurred during the eighties appear to have been influenced as much by the latter as by the former. The wide fluctuations in commodity prices suggest that reading the market signals is a difficult and confusing task. As regards the future, the interventions to help the farmers with infrastructures, research extension and critical inputs and those designed to reduce fluctuations in commodity markets would have to remain as priority items on the policy agenda. However, it would be desirable to limit the direct interventions by the government in the commodity markets to exceptional situations like price collapse or extreme scarcity of critical commodities.

Some of the major policies pursued so far cast a shadow on the prospects for value addition and exports. The policies for promoting industries through restricting imports and taxing agriculture have had the effect of impairing their competitive strength. Similarly, the policies with respect to agricultural exports continue even today to permit exports only when there are domestic surpluses. Even if it is assumed that there would now be a thorough overhaul in such policies, it would need several years for the industries to gather strength and for the exports to establish themselves in the international markets and to adjust to their structure and dynamics. Hence, the prospects appear to be that, considered against the vast expanse of Indian agriculture, value addition and exports would remain modest in pace and spread in the medium-term future. In such a scenario, the farmer groups that we have in mind would tend to remain on the periphery of the growth process.

Another feature of value addition and exports is that they involve transactions between the farmer and organised industries having capacity to influence market prices to their own advantage. Both cotton and oilseeds, crops holding considerable promise for backward areas, appear to have market structures in which the farmer bears the burden of fall in prices but does not receive his due when the prices are high. Such asymmetry is likely to be a rule rather than an exception when farmers deal with organised groups and the two parties differ markedly in bargaining power.

Considering the fluctuations in commodity prices, factors constraining the pace and spread of value addition and exports and the market influence of organised groups involved in these activities, it is clear that the policies for liberalisation would place the farmer in a market environment which could be hostile both to his efficiency and welfare. He would have to overcome the problems facing him primarily through his own efforts—by producing and selling more, by getting acclimatised to markets and above all, by building up and operating professionally competent bodies which can fight his battles in the market place. Crucial to his success would be the supportive interventions by the government and a strategy which seeks to promote growth by developing agriculture and not by discriminating against it.

Note : Reproduced from *Indian Journal of Agricultural Economics,* Vol. 49, No. 3, July-September, 1994.

REFERENCES

Agarwal, N.L. and J.L. Sharma (1994), ''Promoting Agribusiness: Soyabean Marketing Problems in Rajasthan''. *The Bihar Journal of Agricultural Marketing*, Vol. 2, No. 1, January-March.

Bhalla, G.S. (1994), "Introduction", in G.S. Bhalla (Ed.) (1994), *Economic Liberalisation and Indian Agriculture,* Institute for Studies in Industrial Development, New Delhi.

Brown, M. and Golden (1992), The Future of Agriculture Developing Implications, Organisations for Economic Corproation and Development, Paris.

Dantwala, M.L. (1959), ''Small Farmers, Not Small Farms'', *Indian Journal of Agricultural Economics,* Vol. 14, No. 3, July-September.

Government of India (1993 a), *Agricultural Statistics at a Glance,* Directorate of Economics and Statistics, Department of Agriculture and Co-operation, Ministry of Agriculture, New Delhi.

Government of India (1993b), *Compendium of Textile Statistics,* 1993, Office of the Textile Commissioner, Bombay.

Government of India (1994), *The Report on the Price Policy for Kharif Crops of the 1994-95 Season,* Commission for Agricultural Costs and Prices, Ministry of Agriculture, New Delhi, March.

Gulati, Ashok and Anil Sharma (1994), "Agriculture under GATT: What It holds for India", *Economic and Political Weekly,* Vol. 29, No. 29, July 16.

Kahlon, A.S. and S.S. Tyagi (1989), *Agriculture Price Policy in India*, Allied Publishers, New Delhi.

Nayyar, D. and A. Sen (1994). "International Trade and Agricultural Sector in India", in G.S. Bhalla (Ed.). *Economic Liberalisation and Indian Agriculture*, Institute for Studies in Industrial Development, New Delhi.

Rao, V.M. (1993), Government Inventions in Agriculture: A Review of Studies in Developing Countries", *Indian Journal of Agricultural Economics,* Vol. 48, No. 4, October-December.

Rao, V.M (1994), "Agriculture and Liberalisation: Some Implications for Development Policies", *Economic and Political Weekly*, Vol. 29, Nos. 16 and 17, April 16-21.

12

GATT : Implications for Agriculture with Special Reference to Punjab

KARAN SINGH, D.S. SIDHU, S.S. GREWAL, JOGINDER SINGH, D.R. SINGH, O.P. SAHNI & M.M. VERMA

I. INTRODUCTION

GATT has become one of the major concerns of debate amongst the political leaders, media and the public in the country for the last one year. Most of the controversy seems to be based on political expediencies than on any objective analysis. The document is a voluminous one on a complex and technical subject, and very few have had the opportunity to go through the original document and the revised drafts or even the final agreement. Most of the debate, the comments have, therefore, been constrained and are inconclusive.

GATT, an international forum was established in 1948, initially with 23 countries including India as original signatories to over see the smooth conduct of world trade, i.e. to promote the world trade by reducing barriers to it, such as custom duties and quotas, with some 'special and differential' treatment to developing countries.

Traditionally, GATT was concerned with issues related to trade in merchandise, i.e. the goods sector only. Its jurisdiction stopped at the national borders thus keeping domestic policies and matters other than goods trade beyond its orbit. The first seven rounds of GATT, confined to this paradigm only, ended with the Tokyo Round in 1979.

The eighth round of GATT Negotiations, which began in 1986, when its Direction General was Arthur Dunkel (hence the Dunkel Text) began at Uruguay, Punta Del Este (hence also called Uruguay Round).

There was a major change from the previous rounds by including new areas such as agriculture, textiles, investment, intellectual property rights, and services. The expanded scope of GATT has made it much more sensitive than ever before especially by touching issues related to the domestic policies of member countries. The eighth round of GATT has been finally signed on 15.4.1994 by 125 countries, including India and will be effective w.e.f. July, 1995. This is a commitment on the part of each signatory to give all other signatories, the Most Favoured Nations (MFN) status. The long-term objective is "to establish a fair and market-oriented agricultural trading system" through "substantial progressive reductions in agricultural support and protection over an agreed period of time" with specific binding commitments on market access, domestic support and export competition.

The agreement is to facilitate free international trade in goods and services by removing/minimising different impediments. The member countries have committed themselves to conduct their trade policy in a more transparent manner and in an open trading environment. It envisages for liberalisation at a greater speed and will contribute to more effective survelliance and to strict observance of multilateral rules and disciplines. And in order to supervise the goibalization operations of various economies and settle their mutual disputes, the GATT henceforth will be replaced by WTO (World Trade Organisation) which will have independent machinery for consultation, evaluation, implementation, negotiations and punishment.

There are many general arguments in favour of free international trade based on the economic philosophy of unquestioned faith in the market forces. The free trade promotes specialisation in the areas of comparative advantage. This in turn leads to the most efficient allocation of the productive resources. There will be cost effectiveness encouraged all round with each country moving towards specialisation in producing (and exporting) goods which it can do more economically (cost effectively) and from the increased export earnings, importing the other goods in which the country does not have the comparative advantage. Thus, free trade is shown to contribute to good to all concerned. The total gain in global income has been estimated to the tune of as much as $200 billion.

Another important benefit of promoting free trade is the increase in business confidence which would in turn trigger off greater investment, economic growth and development and new employment generation. The developing countries facing a slow growth rate of capital formation or even a declining growth in real capital investment particularly in agriculture, such as India, are already pursuing a number of structural adjustment programmes to correct the imbalance and increase the economic growth rate. With the process of trade liberalisation, the consumers will also have a greater choice due to increased competition; so will be the farmers for the improved production technology. These will augment to improve the growth rate of the economy.

The state intervention in monopolising the market (for some commodities) has been over-stretched, and in fact, as argued by many, many a times becomes counter productive. Inter-zonal restrictions and other non-tariff measures* often generate more disincentive signals for the farmers which are proposed to be removed and converted to tariffs. Above all there have been enough provisions built into the agreement to safeguard the interests of the developing countries.

*The non-tariff measures include :
— quantitative import restrictions,
— variable import levies,
— discretionary import licensing,
— voluntary export restraints,
— minimum export price,
— domestic inter-zonal restrictions, etc.

II. GATT AND AGRICULTURE

In the early 1980s, an increasing awareness and documentation of the distortions in the world markets for agricultural goods emerged. It has been observed that :

(i) In general rich industralized countries provide more support for agriculture relative to other sectors;
(ii) Importers of agricultural commodities tend to be more protectionist than exporters; and
(iii) Exporting developing often tax their agricultural sector, both directly by interventions within the sector and indirectly by protecting their urban industrial sectors.

The massive domestic subsidies given by industrialized countries to their agricultural sector have been causing distortions in international trade in agriculture. In fact in various meetings since 1986, the strong proponents of including agriculture in GATT, namely U.S.A., Canada and some of the exporting developing countries have been negotiating hotly with the opponents which included EEC, Japan, Korea and net importing countries. There was a wide gap between the early positions of proponents of complete elimination of farm programmes (the U.S. 'zero' option) and a stage by stage negotiation process of opponents. After a number of expert discussions and international meetings at various levels, a new compromise (Dunkel Compromise or Draft Final Act) came up in December, 1991 which formed the core of talks till the final version was discussed in December, 1993.

In India, being predominantly agricultural country, much of the controversy centres around agriculture. The major agriculture related issues in GATT are :

(i) Reduction in domestic subsidies;
(ii) Reduction in export subsidies;
(iii) Provision of market access leading to reduction in tariffs;
(iv) Agreement on sanitary and phyto-sanitary measures;
(v) Agreement on textiles; and
(vi) Trade intellectual property rights (TRIPS).

Reduction in Domestic Subsidies (Domestic Support Commitment)

The "Total Aggregate Measurement of Support" (AMS) which means the sum of all domestic support provided in favour of agricultural producers (generally called as subsidies) is to be reduced.

This has been considered necessary to allow the free play of market forces. The reduction is to be effected if the Aggregate Measure of Support (AMS) consisting of product-specific and non-product specific subsidies in agriculture exceed 5 per cent of the value of the total output in developed countries and this limit is 10 per cent in case of developing countries; India is in the group of developing countries. If the AMS exceeds this limit, the subsidies are to be reduced by 20 per cent and 13 per cent respectively. So far as India is concerned, we are as yet quite below this cut off point. Total AMS has been estimated at a maximum of 6 per cent by the Ministry of Commerce, Government of India. Thus, the apprehension that subsidy shall be withdrawn by being a part of the GATT is totally misplaced rather there is a scope to give some more subsidies, if need be. Already the country has adopted a policy of slashing subsidies to contain the fiscal deficit. Then there are some important exemptions for the developing countries. Resource poor and low income farmers are exempted from the provision. Further, the developing countries are allowed to continue with the system of building up of buffer stocks of foodgrain for food security purposes and also can give food subsidies to the rural and urban poor but the condition here is that the stock building shall have to be made at the prevailing market price and the quantum of stocks shall be made according to pre-determined targets. In the matter of domestic subsidies, it is most of the developed industrial nations that have to slash their subsidies to a larger degree. For instance, the overall Producer Subsidy Equivalent (PSE), which is one variant of Aggregate Measure of Support (as provided in GATT) in some of the developed countries has been as follows :

U.S.A. 26.17 per cent	Canada 33.50 per cent
Europe 37.00 per cent	Japan 72.50 per cent

Reduction in Export Subsidies

The developed countries which are providing huge export subsidies and dumping their products cheap in the importing countries have an obligation to reduce the subsidies both on value basis and quantity basis. On value basis they are required to reduce the subsidies by 36 per cent in 6 years and on quantity basis by 21 per cent in 6 years. This limit is 24 per cent and 14 per cent for developing countries and the period is 10 years.

In the matter of reduction of export subsidies, again, India is not affected because the country is not providing any export subsidy in agriculture. Rather, there is some scope to provide these subsidies in India.

Market Access

The market access provision applies to countries that at presently maintain strict import restrictions like Japan, South Korea, etc. In such cases, non-tariff measures are to be converted into tariffs and after tariffication, these are to be reduced by the same quantum as given earlier in case of export subsidies. The least developed countries are exempted from this. Such countries are further required to allow a minimum access of 3 per cent of their domestic consumption to the foreign exporters. This is to be increased to 5 per cent in 6 years period. However, if there is surge of imports due to unfair trade practices by some exporting countries when these steps are being implemented, the special provisions of the agreement allow any country to impose additional duties. India at the moment, because of adverse balance of payments position, is exempted from the market access provision.

Agreement on Sanitary and Phyto-Sanitary Measures

As regards Sanitary and Phyto-Sanitary measures, this is a non-tariff measure which the countries can invoke to refuse imports from different countries on the plea that imports could be harmful to human, animal or plant health. This would require the exporting countries to maintain the desired quality standards as per international norms. India and Punjab have to be equally more conscious now about maintaining the quality standards acceptable in the world market in order to penetrate the international market and increase its share of exports.

Agreement on Textiles and Clothing

The textiles sector remained beyond the scope of GATT in the first seven rounds. Even in the eighth round, separate treatment has been given to textiles and clothing since it was expected that for eventual integration of this sector in GATT would require longer period. This consideration was based on the two basic factors. Firstly, the autonomous adjustments in the industry cannot be made in the short-run and secondly, the prevailing Multifibre Arrangement (MFA) is to be phased out.

Under the MFA which came into being in 1974, India's quota of textiles is fixed with main buyers such as USA, EEC, Canada, Finland, Norway and Austria. For example India's quota with USA is fixed at 500 million squaremetre of cloth. This quota of tofs and yarn, fabrics, made up textile products and clothing is to be phased out under GATT to the extent of 16 per cent in the first 3 years from January 1995 to December, 1998, by 17 per cent in the next 4 years and by 18 per cent in still next 3 years. Thus, 51 per cent of the existing quota will come in the open market in 10 years. The remaining 49 per cent quota is to be abolished in the 121st month of the Agreement. As a result of this, the growth rate in textile trade with the quota countries is to increase by 16 per cent, 25 per cent and 27 per cent every year during the 3 stages respectively.

Another salient feature of GATT agreement on textiles is that Textile Monitoring Body (TMB) would be set up to monitor the trade commitments, reviewing the progress and providing special treatment to small and new traders and exporters of under-developed countries.

Trade-Related Intellectual Property Rights

Coming to Intellectual Property Rights (TRIPS) Agreement, it covers several categories of Intellectual Property namely : (i) Copyright and Related Rights, (ii) Trademark, (iii) Geographical Indications, (iv) Industrial Design, (v) Layout Designs (Topographies) of Integrated Circuits, (vi) Protection of Undisclosed Information, (vii) Control of Anti-Competitive Practice in Contractual Licenses, and (viii) Patents.

It is in this sector of the GATT that the advanced developed countries have a distinct advantage and a decisive lead over the developing and the least developed countries and the subject of intellectual property rights has been brought in the gambit of GATT on the insistence of the industrial world led by USA.

All the member countries have an obligation to bring the legislative measures in line with the provision of GATT. Our existing laws meet some of the requirements of TRIPS (Annexure I) but in so far as plant patents are concerned, our existing law is significantly different from TRIPS Provisions where Product Patenting is essential. The existing Indian Patent Act, 1970 provides for process patenting and not the product patenting in (i) Food (ii) Pharmaceuticals and (iii) Chemical sectors. Patents are granted for seven years in the case of food and pharmaceutical sectors an for fourteen years in the chemical sector, whereas the TRIPS provide protection for 20 years. In India, grant of patent also provides for automatic license of right without authorisation of the patent holder whereas TRIPS provide for compulsory licence only on a case to case basis on merit for a particular period and the patent holder has a right to appeal. The patented product can be locally produced or can be imported. Transitional period of five years has been provided for the developing countries. An additional period of five years has been provided where no product patenting is available in a given country.

The Plant Breeder's Rights (PBR's) protect the right holder from other agencies that ought to produce or sell and seeds of the protected cultivar. PBR will not protect against other firms that used this cultivar in a breeding programme and backcrossed the desirable gene into their own cultivars. Use for research does not require permission. Internationally accepted 'SUI-GENERIS' system of PBR has been developed (UPOV 1961/1968, 1972, 1978, 1991) and is known as International Convention for the Protection of new Varieties. The 1991 Convention grants protection of 20 years as compared to 15 years for plants and 25 years instead of 18 years for trees. It also limits the use of farm saved seed to the farmers own land (owned, rented or leased). Farmers are prohibited to sell branded seeds cut can save, exchange and share the seeds. At the same time the farmer will be entitled to a suitable compensation from the breeder of a protection variety and/or dealer in the event of failure in stated performance of seed/propagating material of a protected variety. Hence, the breeders shall put in their maximum efforts to get only those things patented which really have an edge over the existing materials. Breeders have the right over the harvested material and even on the product made from the harvested material if the seed is obtained without the authority of the breeder. This would affect the importing of protected seed or the product directly derived from the harvested protected seed from countries that do not have breeders right into countries that do have rights.

According to TRIPS agreement, the patenting of micro-organisms and micro-biological processes is essential. However, due to complexity of their nature, the micro-organisms have not yet been defined but shall include genetically modified organisms. Sub-cellular material like genes, gene sequence, plasmids and replicons will be included. Biotechnology techniques are now increasingly being applied in agriculture, industry, food, medicine and ecology and may yield products and processes which will be patented

The member countries shall provide for the protection of plant varieties by patents or by effective 'SUI-GENERIS' system or by any combination thereof. Thus, the choice of the system of protection of plant varieties is left to each country.

III. GATT AND INDIAN AGRICULTURE

The impact of GATT on Indian agriculture is being debated seriously with a great concern. The debate is important for an economy like that of India, where the agriculture sector still carries the burden of supporting two-third of its population and contributes about 30 per cent of GDP (Gross Domestic Product). The production actively is carried on some 97 million farms averaging to tiny holdings of 1.69 hectares only, though independently but highly dependent upon the public support for input price and distribution incentives as well as little marketed surplus again market supported and where more than 30 per cent of population living below the poverty line also looks towards the publish support for getting the foodgrains at affordable prices. We take all the issues one by one.

Domestic Subsidies

The two types of subsidies in Indian agriculture are :

(a) *Non-Product specific subsidies* : These are given for all crops, which in the case of India are the subsidies given for fertiliser, irrigation, electricity, seeds and credit. The calculations made by the Ministry of Commerce, Government of India, show that these non-product specific AMS works out to be 2.9 to 6 per cent depending upon the assumptions made.

(b) *Product specific subsidies* : These are given for specific crops in the form of minimum support price and is calculated with reference to the international price for the commodity. There are 20 agricultural products for which India provides minimum support price There will be subsidy only if the support is at a level higher than the international market price of that product but in case of India, as many as 17 commodities are supported at prices lower than the c.i.f. prices. It is only in case of sugarcane, groundnut and tobacco that there is a subsidy (but it is still less than 10 per cent). In case of other 17 agricultural commodities, the minimum price support, being less than the comparable international price, imbibes a NEGATIVE subsidy.

It may be again mentioned that for the purpose of calculating the total subsidies given to farmers i.e., the "total AMS", both the product and non-product subsidies are to be totalled together, and reductions to be made, only if the "Total AMS" exceeds 10 per cent. In case of India, in fact, the sum of the two comes to be even "negative", earlier estimates being minus 2.33 per cent; the latest ones (since rupee has been considerably devalued since then) coming to be more than minus 20 per cent. And if the allowance is to be made for "low income or resource poor farmers" (who by any norm in India are going to be enormous, absolutely as well as proportionately), the negative magnitude of subsidies in Indian agriculture will widen further.

It also needs to be mentioned that in the wake of liberalisation and structural adjustment policy of the Government of India, the subsidies on fertilisers, irrigation, electricity, etc. are being contained to reduce the fiscal deficit of the government rather than as any prelude to GATT.

Export Subsidies

None of the subsidies listed in the GATT as "export subsidies" apply to Indian agriculture because these are not provided at the moment. The only export subsidy provided by India is the exemption of export profit under Section 80-1111 C of the Income Tax Act which is not one of the listed subsidies. It is also worth noting that developing countries are free to provide three of the listed subsidies, which are :

— reduction of export marketing costs.
— internal transport and freight charges
— international transport and freight charged

And India, in order to promote exports, have been provided with this opportunity.

Market Access

The provision to allow market access to the extent of three per cent of the domestic consumption, to be gradually increased to 5 per cent, does not apply to India because of adverse balance of payments situation. Secondly, "to allow imports", i.e. to open the market does not mean "compulsory imports". And thirdly, when the international scenario changes with increase in the world market prices, our negative product specific subsidy in case of major agricultural commodities will itself be a sufficient check.

It is also worth mentioning that the Public Distribution System (PDS) will not be affected in India. The provision of foodstuffs at subsidised prices with the objective of meeting food requirements of urban and rural poor in developing countries on a regular basis at reasonable price is allowed. Food purchases by the government for public distribution system shall be made at current market prices and the financing and administration of the aid shall be transparent. Public stock holding for food security purposes shall correspond to pre-determined targets related solely to food security. The process of stock accumulation and disposal shall also be financially transparent. Food purchases by the government for food security shall also be made at current market prices and sales from food security stocks too shall be made at no less than the current domestic market price for the product and quality in question.

The distribution of foodgrains through Public Distribution System (PDS) in India, at present, is little above 20 million tonnes per annum which is less than 10 per cent of the total demand (consumption) of foodgrains in India compared with more than 30 per cent people still living below the poverty line. The proper targetting of the PDS is already an on-going subject with the Indian economy. And the operations for food security purposes (buffer stocks) through the Food Corporation of India and other public agencies are already transparent. Again, although the public agencies buy the food stocks only at the minimum support prices, the farmers are free to sell at higher market prices if offered by the open trade, and thus in fact, the minimum support prices are the open market prices during the post-harvest season.

Agreement on Textiles

India is one of the major exporters of cotton, its exports are constrained due to quotas fixed under the existing arrangements but with the GATT, quotas being gradually phased out, India's share in the world market in case of export of cotton will increase substantially.

The total export of cotton during 1992-93 was Rs. 12498 crores which accounted for 25 per cent of the total exports of the country. This is likely to boost upto over Rs. 20,000 crores during 1994-95. Since long, we have suffered due to quota system and along with Pakistan and Indonesia have been pressing for its abolition. Though the period for phasing out of MFA is long and slow, yet it will fetch us better prices for our textiles which are already quite competitive in the world trade in spite of restrictions.

Textiles make use of cotton as a raw materials apart from wool, silk, synthetics, rags, etc. India with 7.5 million hectares area under cotton is at the top in the world but ranks fourth in terms of production. China and USA top in cotton production. India's contribution to the world trade in textiles is to the tune of 5 to 6 per cent. However, modernisation of industry is very essential to capture the world market still more in coming years. The production of cotton in India has been fluctuating. This is partly as a result of high percentage area under cotton being rainfed in the country and high attack of pests and more so due to fluctuating prices of cotton. On the other hand, the mill intake has been consistently increasing. Thus, the exportable surplus has varied widely. For example, during 1989-90 and 1990-91, exports were 13.70 and 11.90 lakh bales which declined to only 0.77 lakh bales the very next year. This uncertainty in the world trade of raw cotton has serious repercussions on the cotton prices in the international and national markets and in turn on the stability in production.

Trade Related Intellectual Property Rights

In the developing countries like India, only recently private research has started and that is largely confined to the field of hybrids. The major objective of protection is to encourage private and public investment on varietal development research by providing exclusive right to the Breeder/Institution so as to supply seeds on superior new varieties to the farmers which increase production and productivity. The costs of variety improvement programme are increasing and are likely to further increase because of the requirement of Distinctness of New Varieties from the existing cultivars for purposes of identification and registration. The breeder has not only to introduce some agronomically desirable features but also a distinct morphological essential character in the New Variety to qualify for protection. Thus, the seeds of the protected improved genotypes may become expensive but there is a self-regulating check as farmers will buy only those seeds that bring them more returns. For most of the crops, the cost of the seed is a very small proportion of the total cost of raising the crop. Furthermore, there is provision of compulsory licensing of the production of the seed of the protected variety if the Breeder is found to be artificially creating the shortage of the seed and is charging high price. Issue is what will happen if we do not go in for this protection? Obviously, other agencies (within and outside the country) will be free to use and exploit the new varieties and germplasm developed by the public institutions at the public cost but the availability of improved materials will be restricted to the public institutions even for research purposes. However, there is one advantage also that once such technology is legally imported in the country, then there is no restriction in its free use by any agency. Furthermore, the protection rights will definitely provide motivation and incentives for hard work by the scientists of the public institutions and there will be healthy competition between the private and public enterprises. The availability of exclusive right to a plant invention invariably determines whether or not the invention will be commercially successful. In course of time, India can develop its own marketable technologies, protect them and export them and also use locally. Indian seed industry will develop. Collaborative research may increase and India may be able to export the seed of the protected varieties both of the Indian or Exotic origin.

There is a notion that the research in developed nations/multinationals is much ahead of us. They may be having an edge in this but we also do have a huge trained and talented manpower in agriculture and with sincere efforts can hope to come upto the standards comparable with the best in the world. What we presently lack is the resources, the cumbersome and time-consuming official procedures and the absence of incentives to the performing scientists and disincentives to the non-performers. Once these impediments are corrected and clear directions are set in order, our research system has the potential to produce much better results; the need is to exploit it.

Proposed Sui-Generis System for Indian-Draft Plant Varieties Act, 1993

— Ensures to the Breeder of a New Plant Cultivar certain rights regarding its use by others such as that production of the seed, offering for sale-marketing, export and import will require the authorisation of the Breeder.

— Only new varieties developed and released in India/imported but not utilised for more than 4 years/6years (other crops/trees) in other countries will be protected.

— New variety must differ from other known varieties for atleast one distinguishable essential character. It should be : New, Distinct, Uniform and Stable.

- — Applies to : Self-pollinated variety, Hybrid, Parental line, Cross-pollinated variety, Composite, Synthetic, Multiline/components of Multiline not included even though a multline is more homogeneous and morphologically more uniform than the composites/synthetics.
- — Farmers have a right to save, use, exchange, share and sell propagating material from seed obtained/decended from seed obtained except for commercial purposes.
- — Use for research purpose (breeding, studying, characterization) will not require permission of the Breeder.
- — Certification of seed for sale of the protected variety essential through the official seed certification agency.
- — Sample seeds to be supplied to the National Bureau of Plant Genetic Resources (NBPGR).
- — Old variety released/notified under the Seeds Act will be protected as Extant Variety for the period of 15/18 years from the date of release/notification.
- — All sorts of germplasm will be protected and exchanged on certain terms and conditions through NBPGR/Gene Bank.

The Draft Act needs some change particularly for the rights of the germplasm and use of resources generated through protection, constitution of the authority and participation of States.

Prospects of India's Agriculture

The production and exports of rice and wheat, in the world market, Asian market and India are given in Table 12.1. The international trade in wheat is far higher at 22 per cent of its production compared with 3.7 per cent only for wheat. The relative Asian export market is much smaller for wheat. Out of 121 million tonnes of wheat in the international market, Asia accounts for only 6.3 million tonnes which is less than 6 per cent of production in Asia. The major exporters of wheat are U.S.A., Canada, Australia and France and the major importing countries are USSR, China, Japan, Italy, Germany and Belgium (Table 12.2). Similarly, in case of rice, the major exporting countries are Thailand and U.S.A. (Table 12.3). The case of India is a marginal one at the moment; in 1991, India exported only 0.5 million tonnes of rice and 0.7 million tonnes of wheat. In fact, during the latest five years, India is still a net importer of foodgrains. The foodgrains production is around 180 million tonnes whereas the demand is estimated at around 200 million tones for 1995. In case of oilseeds, India will remain an importer (Table 12.4) and in case of cotton, it will continue to be a major exporter (Table 12.5).

TABLE 12.1 : Export Trade, 1991

		(Million tonnes)	
		Rice	*Wheat*
World :	Production	347	551
	Export	12.8	121
	%	3.7	22
Asia :	Production	170	109
	Export	6.5	6.3
	%	3.8	5.8
India :	Production	74	55
	Export	0.5	0.7
	%	0.7	1.3

Source : *UNCTAD Commodity Yearbook*, 1993, United Nations.

TABLE 12.2 : Major Exporting and Importing Countries for Wheat

	Million tonnes					
Country	*1986*	*1987*	*1988*	*1989*	*1990*	*1991*
Exports						
Canada	16.38	22.62	20.33	11.49	18.17	23.51
United States	26.45	32.60	42.27	38.27	28.75	32.48
France	15.41	16.48	16.93	17.08	19.34	19.73
Germany	2.62	2.74	3.83	4.63	2.82	3.47
Italy	1.63	1.41	2.23	2.75	1.78	2.47
United Kingdom	4.09	4.22	2.03	3.39	4.56	4.13
Australia	16.17	14.87	12.28	10.59	11.63	12.02
Argentina	4.08	4.23	3.68	4.42	6.04	5.73
Saudi Arabia	0.59	1.35	2.06	1.61	1.27	2.00
Hungry	1.68	1.28	1.79	1.43	1.16	1.33
Imports						
Belgium Luxamburg	1.29	1.31	1.62	1.23	1.73	2.33
Germany	2.52	1.87	1.77	2.18	1.85	1.37
Italy	5.28	4.64	4.92	5.65	4.71	6.53
Netherlands	1.51	1.61	2.67	1.79	3.25	1.79
United Kingdom	1.73	1.74	1.65	0.87	0.89	0.77
Japan	5.62	5.48	5.72	5.58	5.47	5.69
USSR (Former)	16.02	18.40	21.42	14.43	15.39	19.92
China	6.35	13.84	14.69	15.08	12.64	12.44

Source : Same as Table 12.1.

TABLE 12.3 : Major Exporting and Importing Countries, Rice

	Million tonnes					
Country	*1986*	*1987*	*1988*	*1989*	*1990*	*1991*
Exports						
United States	2.39	2.47	2.26	3.06	2.47	2.24
Pakistan	1.32	1.27	1.21	0.85	0.74	1.20
Thailand	4.52	4.44	5.27	6.31	4.02	4.33
China	0.95	1.02	0.70	0.32	0.33	0.66
Vietnam	0.16	0.12	0.09	1.42	1.62	1.00
Imports						
Developing	8.85	8.23	8.12	9.63	9.11	8.94
Countries and territories						
America	2.11	0.87	0.70	1.18	1.43	1.80
Africa	2.97	3.05	2.44	2.88	2.73	3.24
Asia	3.51	4.10	4.79	5.35	4.71	3.62

Source : Same as Table 12.1

TABLE 12.4 : India's Production of Oil Seeds and Imports of Vegetable Oil

	Million tonnes		
Year	*Production*	*Consumption*	*Imports*
1982-83	11.0	12.5	1.5
1983-84	12.5	14.5	2.0
1984-85	14.0	16.0	2.0
1985-86	13.0	15.0	2.0
1986-87	12.0	13.8	1.8
1987-88	14.0	16.9	2.9
1988-89	17.0	18.5	1.5
1989-90	18.0	19.5	1.5
1990-91	19.0	20.0	1.0
1991-92	20.0	21.0	1.0

Note : Figures have been converted from the bar-diagrams and rounded.

Source : Price Prospects for Major Primary Commodities, 1990-2005, IBRD, Figure 16, p.141

With the reduction in domestic subsidies, export subsidies and tarffis, the exports from developing countries including India shall become more competitive and we would be in a position to a export more. Prices of agricultural commodities are also likely to register an increase. The same thing will happen with textiles where we are even better placed. The agricultural exports accounts for about 16 per cent of the total country's exports. Textiles, on the other hand, account for about 25 per cent of the total exports of the country. With the phasing out of quotas, the quantum and value of exports are likely to register further increase but the major

TABLE 12.5 : Cotton Exports from India

(Lakh bales)

	Opening Balance	*Crop Production*	*Imports*	*Total Supply*	*Mills Intake*	*Extra Factory Intake*	*Exports*
1980-81	25.60	78.00	—	103.60	76.80	3.40	7.00
1981-82	16.40	84.00	0.50	100.90	73.50	4.00	3.70
1982-83	24.80	83.00	—	107.80	75.00	4.70	7.00
1983-84	25.40	72.50	—	100.60	80.10	4.70	3.50
1984-85	16.00	101.50	1.00	118.50	86.50	4.70	1.80
1985-86	28.80	107.00	—	135.80	86.60	5.10	4.50
1986-87	39.50	95.00	—	134.50	95.20	5.50	14.30
1987-88	20.00	90.00	3.00	113.00	92.00	5.50	0.40
1988-89	18.70	106.00	2.30	127.00	98.00	6.00	1.00
1989-90	22.00	135.80	—	157.80	104.50	6.50	13.70
1990-91	33.00	117.00	—	150.00	108.00	7.50	11.90
1991-92	22.64	119.00	3.00	144.64	103.09	8.00	0.77
1992-93	32.78	130.00	0.73	163.51	112.00	8.00	17.88
1993-94	25.63	—	—	—	—	—	—

Source : G.O.I. Directorate of Economics and Statistics.

question that arises at this stage is that the very opening of external markets shall not automatically lead to increase in our exports unless we are able to capitalise upon these opportunities by making suitable structural adjustments in the system. India should look for export markets in the Asian and Gulf countries in order to economise on freight charges. The foremost requirement is to increase our productivity, quality standards, packaging and other things consistent with the requirements of competitive markets. The country shall have to increase its investment in research and development, human resource improvement and the related infrastructure. Fortunately, the country has great untapped potential particularly in the agricultural sector. To harness this potential, we shall have to ensure that the necessary capital flows into the sector. It should not be difficult particularly when we have already opened our economy to external investment. Instead of concentration of primary production alone, we shall have to concentrate more on agro-processing industries for value addition.

IV. GATT AND PUNJAB AGRICULTURE

Punjab is a progressive agricultural state. It agriculture sector is more dominant, still contributing about 47 per cent of the state GDP compared with about 30 per cent in India. Punjab has an export surplus of wheat, rice, cotton, milk and livestock products and there is a scope in case of some fruits, vegetables and flowers (Table 12.6). In fact, the export surplus of rice of Punjab of about 6 to 6.5 million tonnes is equivalent of the exports of Asia and in case of wheat, the Punjab's export surplus of 7 to 8 million tonnes is even more than that of the Asia's exports of little above 6 million tonnes (Table 12.6.)

TABLE 12.6 : Punjab's Export Surplus

	(Million tonnes)		
	Production	*Consumption*	*Surplus*
Wheat	12-13	4-5	+7-8
Rice	7+	0.5	+6-6.5
Cotton	2.4	1.0	+1.4
Milk	5.6	3.6	+2.0
Fruits	0.7	—	+
Vegetables	1.5	—	+
Flowers	1.5	—	+

It is being feared that when India is deficit, Punjab will not be allowed to export and hence it will not benefit the Punjab farmers. It is also being feared that Punjab which uses about 10 per cent of India's fertiliser consumption and about 11 per cent of pesticides, etc. will be at a disadvantage due to slashing of subsidies drastically. The argument is misplaced; GATT is an Agreement between the countries not with the individual states in each country. Secondly, due to this Agreement, the domestic prices of agricultural produce will be in relation to the international prices, and on this account, Punjab stands to gain. The prices of wheat in the world market f.o.b. Gulf and of Punjab wheat f.o.b. Bombay (since the international freight and handling charges to major importers of wheat from India, namely, USSR, China and Japan will be about the same from Gulf or Bombay) are given in Table 12.7. It shows that inspite of the decline in world market prices, which is more than compensated by the devaluation of Indian rupee, more particularly after the liberalisation process, the price of Punjab wheat f.o.b. Bombay has been less than the world price of U.S. wheat f.o.b. Gulf by as much as Rs. 444 per tonne in 1989, Rs. 878 in 1992 and Rs. 362 in 1994. Further in the wake of implementation of GATT, the world market price of wheat may increase by, say, about 10 per cent, i.e. shifting the wheat price to U.S. $ 150 per tonne which at the current exchange rate of about Rs. 32 per US $ gives f.o.b. Gulf price of wheat at Rs. 4800 per tonne compared to Rs. 3840 f.o.b. Bombay of Punjab wheat, i.e. a margin of Rs. 1000 per tonne which shows ample scope to profit provided we increase the productivity and production to become a real export surplus wheat economy.

TABLE 12.7 : Comparative Position of Punjab and World Market Prices, Wheat, 1989 to 1994

Year	*(Per tonne)*			
	World market price U.S. Wheat for f.o.b. Gulf		*Punjab wheat price f.o.b. Bombay*	*Net margin*
	U.S. $	*Rs.*	*Rs.*	*Rs.*
1989	170.1	2464	2020 (1730)	+444
1990	136.8	2278	2130 (1830)	+148
1991	128.9	2313	2560 (2250)	-247
1992	151.1	3698	2820 (2500)	+878
1993	142.0	3686	3580 (3250)	+106
1994	134.0	4202	3840 (3500)	+362

Notes:

1 Figures in the parentheses are procurement prices of wheat.

2. The exchange rate is the average for the year computed from total export in US $ terms and in India Rs terms from Economic Survey, Government of India, 1993

Sources: For World Market Prices same as of Table 12.1

In case of rice, Punjab is even better placed. The price of super fine varieties of rice f.o.b. Bombay is less than the Thai rice f.o.b. Bangkok (Table 12.8). This is not withstanding the fact that Thai rice is 5 per cent broken and of poor quality which is often discounted to the extent of as much as 70 per cent in comparison with the superfine varieties grown in Punjab. And we export basmati which is super superfine.

TABLE 12.8 : Comparative Position of Punjab and World Market prices, Rice, 1989-1993

	(per tonne)			
Year	*Thai Rice f.o.b. Bangkok*		*Punjab Rice f.o.b. Bombay*	*Net margin*
	U.S. $	*Rs.*	*Rs.*	*Rs.*
1989	320.3	4640	3385 (2050)	+1255
1990	287.2	4782	3750 (2250)	+1032
1991	314.2	5638	4138 (2500)	+1500
1992	287.4	7034	4754 (2900)	+2280
1993	232.0	6022	5570 (3500)	+452

Source : For World Market Prices, same as of Table 12.1.

Notes :

1. Figures in parentheses are procurement price of superfine rice.
2. Same as of Table 12.7.

In case of cotton, Punjab which has 60 per cent surplus exportable production, even India is much better placed, so is Punjab. The price of cotton in the world market was as much as Rs. 1339 per quintal in 1992-93 for oriens/texas qualities which is comparable to our J-34 variety cotton (Table 12.9). The national wholesale price of J-34 was only Rs. 1020 per quintal. Similar, has been the case for most of the years. Thus, on an average, the prices of cotton in the international market as compared to that in the national market has been higher by about 25 per cent. It is expected that linkage of these two prices would result in rise in the return to cotton growers by about Rs. 300 crores in the Punjab State alone. Situation of long staple cotton grown mainly in the southern part of the country shows still wider gap (Table 12.10). The value addition to the commodity through manufacturing in textiles and clothing should make it still more remunerative.

Somc shifts in the cropping pattern in Punjab, keeping in view the export prices and prospects, may be summed up as follows. The area under quality wheat (Durum Varieties), quality rice (basmati) and cotton will increase. Marginal shifts towards horticultural crops like vegetables and flowers are also expected. Punjab also has a distinct advantage in exporting livestock products, particularly good quality buffalo beef to the Gulf market. It may also be pointed out that there is no comparative advantage in the cultivation of oilseeds particularly sunflower.

TABLE 12.9 : Price of J-34 (Medium staple) Cotton

	(Rs./qtl. of Kapas)		
Year	*Wholesale Price*	*International Price Orieans/Texas*	*Minimum Support Price*
1980-81	455.50	667.90	304.00
1981-82	493.00	522.43	N.A.
1982-83	397.75	576.12	380.00
1983-84	510.25	748.03	400.00
1984-85	512.25	748.39	410.00
1985-86	441.25	679.52	425.00
1986-87	482.75	663.12	430.00
1987-88	837.00	821.76	440.00
1988-89	671.00	814.04	500.00
1989-90	713.00	1140.08	570.00
1990-91	947.00	1312.74	620.00
1991-92	1323.75	1339.00	695.00
1992-93	1020.00	1339.17	800.00

TABLE 12.10 :Price of MCU-8 (Long staple) Cotton

	(Rs./qtl.)		
Year	*WP MCU-8 Tripura T.Nadu (Kapas)*	*International Price Giza (Kapas)*	*Minimum Support Price (Kapas)*
1980-81	630.00	1041.58	304.00
1981-82	590.00	913.08	380.00*
1982-83	645.00	1006.65	380.00
1983-84	695.75	1352.18	527.00
1984-85	635.00	1513.94	535.00
1985-86	608.00	1246.81	535.00
1986-87	890.00	1297.26	540.00
1987-88	889.60	1700.21	550.00
1988-89	1066.00	2299.94	600.00
1989-90	1024.00	2889.96	890.00
1990-91	1029.00	3045.26	750.00
1991-92	1253.00	2335.20	840.00
1992-93	1891.25	2995.95	850.00

*Recommended by the Agr. Prices Commission but not announced by GOI.

The Punjab Agricultural University is already seized of the problem and is initiating steps to modernise its research programmes to become competitive with the outside world. Some policy issues are given in attached Statement-A. It is being realised that the potential consuming markets particularly in our neighbourhood have to be explored. The details of the Marketing Cell being proposed to be set up in the PAU are being worked out. Based upon the price environment outside, one of the envisaged mandate of the Marketing Cell, PAU shall be to provide the guide to reformulate our production strategies. The University is also in the process of reorienting its extension programme to educate the farming community in line with the requirements of the external market.

Overall, the country including the State of Punjab shall be a gainer from being a member of GATT provided we capitalise upon the opportunities that come our way. But we shall have to come up to the world standards to meet the challenges that liberalised trade would throw open. The marketing system and infrastructure need to be tailored towards an export-oriented one, particularly the transportation will have to be modernised. Punjab should be immediately linked to a sea-port like Kandla through good road and rail system, besides activating a modernised international airport. A plane-load of fresh flowers and vegetables from Punjab taking off to the Gulf market can land back in a few hours as a plane-load of dollars.

STATEMENT-A

GATT AND PAU : SOME POLICY DECISIONS

A. Research

— Registration of Extant Varieties developed by PAU.
— Co-sharing of the rights of the PAU/Breeders, some system of rewards/bonus/royalty has to be developed. Scientists to be designated as breeders/role of other scientists to be defined.
— Policy on licensing/sale of exclusive/non-exclusive rights of PAU to official and private organisations.
— Reorienting of Research Record keeping system, secrecy of records, restricted visits to the guarded experimental area and security of seeds in the seed production chain. Flow of funds to crop improvement research has to increase in order

to compete with the multinationals. Hybrid breeding in all crops including vegetables, flowers, fruit trees has to be strengthened.

— Characterisation (Finger-printing by molecular techniques) of the potential material in the pipeline and genetic stocks.

— Cell at the Directorate of Research level dealing with all the above relevant issues may be designated as Variety Production Rights Monitoring Cell.

B. Marketing Cell : Details are being worked out.

ANNEXURE I

TRIPS and India

— Computer software is protected as a literary work under Indian Copyright Law since 1983.

— A Bill is pending before the Indian Parliament for providing protection to performers, sound recordings and broadcasting organisations.

— For Trademarks, Trade Secrets and Industrial Designs, our laws require only minor incremental changes which are consistent with our policies.

— For Integrated Circuits, process of enacting a law has already been initiated before signing the GATT.

INDEX

Accelerated agricultural growth, 69-70
Access to land, 239-40
Acreage price elasticity, 93-85
Action programme, suggestions for, 268-69
Adult literacy, 213, 216-17
Agarwal, N.L., 330
Aggregate Holding Tax (AHT), 27
assessment, 141-42
classification, 156-58
Aggregate Measure of Support (AMS), 279, 280, 282, 284, 306-12, 337
Aggregate Reteable Value (ARV), 138
Agreement on taxation, 344-45
exporter of, 286-90, 298-300
future prices, 305
international trade of, 287-88, 291, 298-300
wholesale prices, 150
world prices and trade of, 178-79, 287-90
Agricultural development, 105
productivity, 108-109, 121-22
strategy, 28, 111, 217-19
Agricultural export, 288-90, 327
policies for, 331-32
Agricultural goods, world markets for, 336
Agricultural holding tax, 27, 137
constitution questions, 155-60
incidence of, 148-50, 152-54
land revenue rationalisation, 160-64
modified scheme, 140-46
recommend by Raj Committee, 107
yield and incidence, 148-52
Agricultural implements and machinery, 41-42
Agricultural income tax, 134-36
Agricultural labourers, 205-207
Agricultural planning and policies, 231, 234-36
development strategy, 234-35
restructuring of, 204-240
Agricultural price policy, 24, 26, 99-130, 238-40, 297-310
implementation of, 106
package of, 99
politics of, 109-112
principles, 103-108
Agricultural Prices Commission (APC), 72, 73-74, 100, 112, 116, 155
instruments, 103
objectives, 101-102
recommendations, 74, 103
Agricultural production, decomposition analysis of, 129-50
material balance of, 260-61
Agricultural taxation, 134, 137-39
income tax, 147-48
Agricultural technology, 235-37
Agriculture,
cooperative and banking support for, 45-46

direct taxes on, 134
development potential of, 250
unimodel pattern of, 232
estimated growth rate, 260
implecation for, 334-40
incentive policies, 53-54
institutional credit for, 63-64
private investment in, 65-66
under GATT, 277-90
Agro-climatic regional planning, 25, 253
Agro-processing, 42-53
Alagh, Yoginder K., 25, 35
Alleviation rural poverty, policies for, 204-40
Anderson, K., 277
Andhra Pradesh Land Revenue (Addl. Assessment) and Cell Revision Act, 153, 155
Antiporta, D.B., 212
Apni Mandi Scheme, 269
Aquaculture, 44
Assessment, 27, 38, 140, 155-60, 321
revision of, 145-46
Average Absolute Change (AAC), 322

Bagchi, Amaresh, 27, 134
Bangladesh, 206, 257
Bardhan, P., 68
Bhaduri, A., 49
Bhagwati, J., 85
Bhalla, G.S., 237, 238, 284, 327
Bhatia, M.S., 30, 297
Bhullar, M.S., 29, 245
Biological and chemical technique (land saving), 236
Brahmaputra Flood Control Board, 38
Brown, M., 327
Bullock Operated Farms (BOF), 265
Bureau of Industrial Costs and Prices (BICP), 40, 43
Bureaucracy, 67, 68
Butter oil production, 190-93

Canal irrigation system, 21, 26, 63, 235
Cash Crops, 142, 236
Charan Singh (former Prime Minister), 112
Cheap food policy, 125
China, 206, 210, 217, 345
Command Area Development (CAD), 37
Commercial agriculture, 52
Commercial Crops, 85, 116, 122
Commercialization, 94
Commission for Agricultural Costs and Prices (CACP), 26, 101, 112, 281
Commodity markets, 325-26
Composite index, 209, 210, 212-14
Concentration index, 85
Constitutional questions, 153-60
legislative competence, 153-60
Constraints, 101, 236, 240
Consumer price index, 65
Controversy, 334
Cornia, G.A., 220
Correlation coefficient, 80, 211
Cost of production, 106-108, 112-113
Cost reducing new techniques, 108
Cotton export, 350, 354
Crash schemes for rural employment, 23
Criticism, 134
Crops, 321-22
cesses, 152, 162-64
prices, 23
Crop Credit Requirement (CCR), 265
Crop loan manual, 45-46
Crop production, 247-48
programmes, 53
Cropping intensity, 252
Cropping pattern, 255, 257-59
Cummings, J.T., 91

Curb price movements, 118

Dairy development, 43-44, 194-96
Dairy farming, economics of, 197-200
Dantwala, M.L., 319
Debt crisis, 241
Decentralisation, 234, 235
Democratisation, 216
Development villages and small scale industries, 23
Dharam Narain, 90
Dhesi, A.S., 29, 245
Dhindsa, K.S., 21
Dietary Energy Supply (DES), 208
Disparities, 22, 50, 123
District agricultural development plan,
  formulation of, 245-268
  methodology, 246-47
  objectives and strategy of, 249-50
District rural industries project, 23
Domestic consumption, 278-79
Domestic subsidies, 336, 342-43, 349
  reduction in, 337-39
Domestic support, 306-307
Domestic and world prices, 300-302
Donovan, G., 236
Double cropping, 226
Drought prone area programme, 23
Dunkel, Arthur, 334
Dunkel draft treaty, 305, 311, 312
  proposals, 310

Edgran, G., 220
Edible oils,
  crisis, 24, 27
  butter oil cost production, 191-92
  domestic production, 184-85, 186-87
  imports of, 182-83, 186-87, 289
  international trade, 302
  precautions and pre-requisites, 193-96
  problems, 190, 196
  price distribution, 184-86
  public distribution, 186
  viable alternative, 189-90
Eighth Plan, 35, 39
Employment, 35, 46-47, 125, 226, 232, 250
  assurance scheme, 23, 38
  strategy, 222
Enke, S., 208
Excise duty, 42
Existing land revenue, 139-40
Exports, 22, 85, 227, 305-320
  competition, 286-87, 306
  markets, 344-48
  subsides, 29, 68, 286-90, 338, 343-45
  trade, 348
Exporting and importing countries, 347-48

Farm incomes, 104
Farm mechanisation extension, 41
Farm size, 225-26
Farmers agitation, 112
  prospects for, 330-32
Faster industrial growth, 222
Fertilizer prices, 117
  subsidies, 58-60
  use, 52-53, 61
Fifth Five Year Plan, 252
Financial constraints, 101
Finance Minister, Budget Speech, 42-43
Fishery development, 44-45
Floriculture, 30
Fluctuations, 112-14
Food and Agricultural Organisation (FAO), 232
Food Corporation of India (FCI), 23, 344
Food insecurity problems, 231

Food problems, 100
Food security system, 325, 338
Food subsidies, 23, 120
Foodgrains,
  economy, 99
  fluctuations, 113-14
  international markets, 118
  output and procurement, 50-51
  price policy, 111-12, 121-22
  production, 35-36, 50-52, 112-15, 124-25
  supply, 65-66
Food-price management, 65
Foreign exchange management, 67
  market, 298-99
Free international trade, 335
Future prices, prospects of, 305
Future strategy, 315-16

Gallerson, M.W., 62
Ganga Flood Control Commission, 38
General Agreement on Tariff and Trade (GATT), 252, 277, 280-287, 290-92, 297, 305-310
  accord on agriculture, 305-310
  aggregate measure of support, 279-85
  agreement on phyto-sanitary measures, 339-40
  agreement on sanitary, 339
  agreement on textiles and clothing, 333-40, 344-45
  export competition, 286-87
  market access, 278-79
  post-GATT scenario, 287-91
  provisions of, 306-307
  WTO agreement, 24, 29
GATT and Indian agriculture, 244-45
GATT and PAU, policy decisions, 355-56
GATT and Punjab agriculture, 351-55
Gini-coefficient, 208
Gompertz curve, 252
Govt. Stock policy, 119
Green revolution, 30, 52, 65, 68, 77, 90, 111, 115, 125, 220, 237, 238, 323-324
Grewal, S.S., 334
Gross Area Sown, 170, 252
Gross Cropped Area, 25, 58, 247, 250, 257
Gross Domestic Product (GDP), 324, 351
  growth rate, 208, 209, 215, 216, 221
Groundwater developments, 37-38
Gulati, Ashok, 29, 177, 285
G.V.K. Rao Committee, 46

Haryana Land Holdings Tax, 162-64
Hayami, Y., 237
High Yielding Varieties (HYV), 40, 226
Hirshman, A.O., 218
Horticulture, 30
  production, 227
Human development index, 212-14
Human development,
  average per farmers of, 215-16
  pattern of, 212-16

Imports, 119
  restrictions, 279, 311-12
Imports and stocks, 118-20
Incentives prices, 103-104
Income distribution, 123-25
India,
  export trade structure of, 299-300
  agricultural export, principal of, 299, 301-302
Indian Council of Agricultural Research (ICAR), 40, 43, 46
Indian-Draft Plant Varieties Act, 346-47
Indian Patent Act, 340

Indonesia, 209, 228
Input-seeds, 39-40
Input subsidies, 58-66, 279-82
Intellectual property, categories of, 340
Integrated Rural Development Programme (IRDP), 23, 37
Inter-crop price movements, 90
  ratio index, 92-93
International Convention for Protection of new varieties, 341
International Fund for Agricultural Development (IFAD), 232
International Labour Organisation (ILO), 222
International Monetary Fund (IMF), 312
International Trade Organisation (ITO), 277
Inter-regional disparities, 121
Inter-state disparities, 50
Inter-state price dispersion, 84-87
Inter-zonal restrictions, 336
Investment requirement, 265-66
Irrigation, 58, 65-66
  development, 122-23
  intensity, 292
  subsidies, 26, 58-59
Islam Nurul, 227, 277

Jalandhar district, 29
  agricultural indices of, 250-51
  projected agriculture indices, 262
  yield performance of, 254
Jawahar Rojgar Yojna, 23
Johl, S.S., 27, 182
Johnson, D.G., 277
Johnston, B.F., 218
Jolly, R., 220

Kahlon, 329
Kandla sea port, 355
Karan Singh, 334
Karnataka Urban Land Tax Act, 159
Kerala Building Tax Act, 153, 155, 158
Kharif crops, enterprise budgets for, 202
Krishna Raj, 26, 55, 72, 90, 219, 256
Krishnaji, M., 26, 99

Labour saving cultivation techniques, 55
Land classification, 145-46
Land distribution structure, 239
Land reforms, 110, 239
Land rateable value, 151-53
Land revenue retionalisation, 160-64
Land taxation, 159-60
Land utilisation, 100
  pattern, 252-53
Land and water resources, development and management, 36-39
Lewis, W.A., 218
Liberalisation, 24, 319-30, 323, 325, 335
  policies for, 326-27
Liberalising agricultural trade, 288
Livestock improvement, 21
Longhurst, R., 228
Low-income countries, 214-15, 217
Low investment policy, 65

Macro-economic variables, 240-42
Madras Urban Land Tax Act, 153, 155
Major agricultural crops, 247-49
  gross value of, 263-64
  productivity and production of, 247-49
Malaysia, 214, 228
Market access, 278-79, 310-12, 338-40, 343-44
Market economy, 319-30
Market prices, 114-17, 120
  support, 280, 290
Marketing cell, 355, 356

Marketing and international trade, 297-310
Mechanical technology (labour-saving), 236-37
Mellor, J.W., 218
Mexican wheat, 76
Minhas, B.S., 249
Minimum Export Prices (MEPs), 288, 289
Minimum Support Prices (MSPs), 304
Mitra, A., 67
Modern technology, 36, 226
Most Favoured Nations (MFN), 335
Multi-fibre Arrangement (MFA), 339
Multilateral Trade Organisation (MTO). 314
Multiple cropping, 226
Mysore Building Tax Act, 153, 155, 157

NABARD, 45
Narayana, N.S.S., 256
National Bureau of Plant Genetic Resources (NBPGR), 347
National Commission on Agriculture, 148
National Council of Applied Economic Research (NCAER), 58
National Rural Employment Programme (NREP), 23, 35, 37
National Seeds Corporation, 40
Nepal, 217
New GATT, gains from, 315
New HYV technology, 53, 82, 238
New agricultural strategy, 99
New economic environment, 297-300
New economic policy, 22
New seeds policy, 40
Nominal Protection Co-efficient (NPC), 302
Non-product specific AMS, 280-85
Non-tariff measure, 336
NSS data, 150, 157

OECD, 327
Oilseed Growers Co-operative Society, 330
Oilseeds production, 186-88
Okhawa, Kazushi, 219
Open market prices, 120
Operation Flood, 43
Operation holding, 142-43
Optimal land use planning, 36
Out-input price ratio, 122

Paddy, cost of production, 56
Pakistan, 206, 210, 212, 215, 228
Panchayat ghars, 46
Parity, 147
PEPSU, 161
Peasant cause, 111
Periodical assessment, 140-41
Pesticides, 39-41
Philippines, 209, 228
Planning Commission, 25, 41, 44, 100, 246, 256
Plant Breeder's Rights (PBRs), 340
Plasticulture, 44-45
Political factors, 109-10
Political mobilisation, 117
Political will, 107
Post-GATT scenario, 287-90
Poverty, 23, 65, 123-24, 204-208, 241
  alliviation programme, 194, 214-16
  incidence of, 204-205
  performance of countries, 208-210
  reducing strategy, 206
  trickle down effect, 211
Poultry farming, 43
Power subsidies, 62-63
Price dispersion, 73
  fixation, 116
  fluctuation, 113-14, 323
  incentive policy, 102-103, 105

movements and relations, 112-120
rationalisation scheme, 25
strategy, 39
stabilisation, 102-104
subsidisation, 105
support, 54-56
programme, 125-26
Principal crops, production, 170
projected area, 256-58
Private investment, 99-100
Procurement prices, 75-78, 84-88, 93-96, 105-110, 112-17, 120-22, 123-24
Procurement/output ratio (PQR), 81, 85, 88
Producer Subsidy Equivalent (PSE), 338
Product Specific AMS, 280-84
Product specific support, 308-310
Projected Agricultural Production, 259-60
Public Distribution System (PDS), 21, 23, 27, 101, 103, 106, 113, 120, 124, 125, 231, 343, 344
Public investment, 67-68
Punjab,
agro-climatic environment, 55-56
Apni mandi scheme, 269
cropping pattern of, 353
export surplus, 357
labour-saving techniques, 55
older irrigation system, 62
Punjab Agricultural University (PAU), 31, 190, 265, 355

Quinquennial assessment, 141

Rabi crops, enterprise budgets for, 201
Raj Committee, 157, 158, 164
features of, 137-39
recommendations, 139
scheme, 140-43
Randhawa, N.S., 28, 204, 212, 224, 233, 238, 242
Rao, V.K.R.V., 50, 68
Rao, V.M., 30, 319
Rateable Value per hectare of Land (RVL), 138-39
computation of, 170
schedules, 139, 141, 146-48
Raychaudhuri, G.S., 26, 72
Regional benefit, 55-56
Regional disparities, 24, 48-50, 69, 124
Regional inequity, 50-52
Remunerative prices, 111
Republic of Korea, 212, 214
Restructuring agricultural development planning, methodology, 230-40
Rosovsky, Henry, 219
Rural development schemes, 37
Rural disparities, 110
Rural electrification, 62
Rural industrialisation, 235
Rural Landless Employment Guarantee Scheme (RLEGP), 23, 35, 37
Rural poverty, 52, 217-20
Rural-urban migration, 229
Ruttan, V.M., 227, 237
Ryotwari system, 50

Sahni, O.P., 334
Sanitary and Phyo-sanitary measures, implementation of, 313-14
Schultz, T.W., 219
Second Five Year Plan, 220
Self-pollinated crops, 40
Share-croppers, 46
Sharma, Anil, 29, 277, 285
Sharma, Anju, 21
Sharma, J.L., 330
Sharad Joshi, 112
Singh, D.K. 334

Singh, Joginder, 334
Sidhu, D.S. 334
Small formers, characteristics of, 207
Small formers and landless labourers, 224
Small and Marginal formers, 205-207
Soil conservation, 37
Special foodgrain production, programme, 39
Sri Lanka, 209, 215, 228
Staatz, J.M., 220
State incentive policies, 55
State Seeds Corporation, 40
State revenues, 49
States, fertilizer use, 61-62
Stock-building operation, 114
Stock management policy, 119
Storage facilities, 23
Strategies, 35-38
Streetan, P., 238
Subbarao, K., 25, 48, 55
Subsidies, 21, 27, 39, 53-54, 69-70, 122, 277-80, 310-12, 324-28
Subsidized institutional credit, 63-64
  programmes, 55
  public funds, 69
Subsidy war, 297
SUI-GENERIS system, 341, 346-47
Sundaram, K.V., 224, 242
Support prices, 55-56, 103-104
  policy, 57
Syed, S., 212

Tank irrigation, 39
Tariff cscalation, 241
Taxation, 65, 68, 290, 327
Taxation Enquiry Commission (TEC), 161
Technological modernisation, 22
  variables, 95
Terms of Trade (TOT), 117
Thailand, 206, 214
Tokyo Round, 334
Tractor Operated Farm (TOF), 265
Trade and licensing policies, 67
Trade Related Intellectual Property Rights (TRIPS), 29, 306, 337,, 345, 357
  agreement, 340-41
  provisions of, 312-13
Travancore-Cochin Land Tax Act, 153, 155
TRYSEM scheme, 41
Tyagi, D.S., 237, 238, 284, 329
Tyers, R., 277

Unemployment, 23
UNICEF, 232, 241
Uniform purchase prices, 116
Unimodal development, 232
Urban poverty, 205
Uruguay Round, 277, 334
USA, soybean production, 189

Vaidyanathan, A., 249
Value addition and exports, 326-30
  prospects of, 333
Vanaspati industry, 184, 186
Verma, M.M., 334

Warabandi, 38
Watershed development schedules, 37
Weighted Average Realized Price (WARP), 88, 89, 90
Wheat,
  acreage elasticity estimates, 90-91
  annual average wholesale price, 79-80
  cost of cultivation, 74-76, 77-78
  economy, 52
  farm-harvest price, 78-80
  open market wholesale price, 74-75, 84-85
  price policy, 72-90, 95-97

procurement price determination, 74-90, 95-97
production response, 90-92
supply response regressions, 93
zonning policy, 84-87
zoning restriction, 96
Whitcombe, E., 49
Wholesale Price Index, 26, 321, 328
World Bank, 221, 223, 225, 228, 232, 241
World Economic Order, 298
World export trade, 30, 298-99
future strategy, 315-16
growth of, 314-15
World Market Prices, 382-53
World Trade Organisation (WTO), 24, 30, 292, 335

Yield elasticity, 93-95
Yield and incidence, 148-53

Zamindari systems, 49-50
Zonal restriction, 86